Robert Hill Institute

University of Sheffield

Energy, Plants and Man

by David Walker

with illustrations by Richard Walker

Second Edition

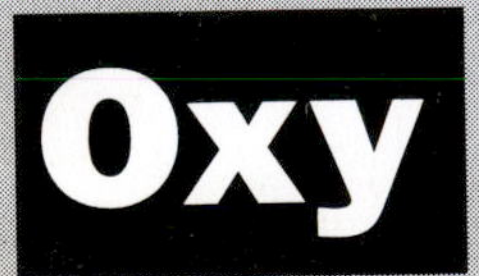

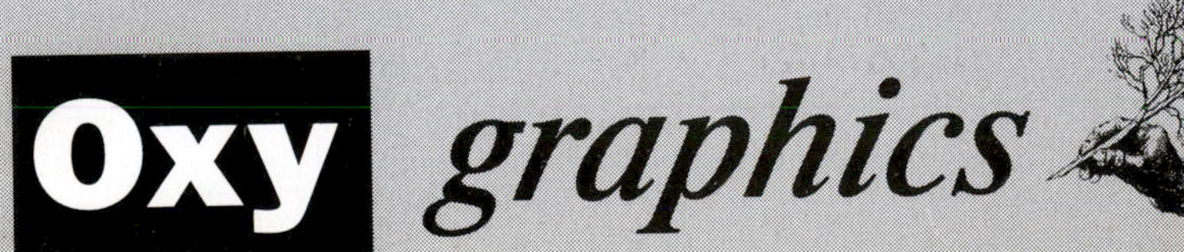

First published in 1979 (paperback only) by Packard Publishing Limited.
This revised edition published in 1992 by Oxygraphics Limited,
22, Montague Place, Brighton, East Sussex, BN2 1JE
Telephone/Fax: (0273) 695054

Second impression with amendments 1993

ISBN 1 870232 05 4

A CIP record for this book is available from the British Library

Distribution to the Book and Library Trade is by Packard Publishing Limited,
Forum House, Stirling Road, Chichester, West Sussex, PO19 2EN
Telephone/Fax: (0243) 537977

Exclusive distribution in North America by University Science Books,
20 Edgehill Road, Mill Valley, California 94941.
Telephone: (415) 383-1430 Fax: (415) 383-3167

Typeset at Robert Hill Institute, Sheffield
Printed and bound by the University of Sheffield Print Unit

"What is the use of a book", thought Alice, *"without pictures or conversations"*.

CONTENTS

FIGURES

FIGURES CONTINUED

FIGURES CONTINUED

TABLES

TABLES CONTINUED

Chapter 6

Chapter 7

Chapter 8

Chapter 9

PREFACE

The first edition of "Energy, Plants and Man" was published in 1979. At that time, the title seemed unexceptionable but, lest some might find it offensive now, may I emphasise that "Man" is used simply as the name of the species (*Homo sapiens*) and, like "ape", "sheep", "holly" or "ivy" carries no connotations of sex.

The 1979 edition was a mere 36 pages long and was then intended simply as an introduction to some aspects of environmental biology. In compiling a second edition it became impossible to ignore contemporary interest in the environment, pollution, the Greenhouse Effect, the possibility of global warming and the consequences of ever-increasing populations; hence the greatly increased length. I also found it impossible to be dispassionate or purely objective about these matters which have as much to do with economics and politics as with plants, biochemistry, or physics. I wish to make it clear therefore that, while the underlying "facts" are as sound and as objective as I could make them, the responsibility for the accompanying opinions are mine alone and would not necessarily be shared by others in the Robert Hill Institute, the University of Sheffield, or by any of the grant awarding agencies that have supported my work in the past.

David Walker, June 1992

ACKNOWLEDGEMENTS

Helping, in various ways, to transcribe thoughts and illustrations into printed text has involved much hard and tedious labour by Jane Evans, Karen Bailey, Joyce Harnden and, most of all, by Anne Dawson whose calming influence has been as important as her immense professional contribution. At the eleventh hour, switching from out-moded to more modern technology would have spelt disaster without the technical skills and generously given help of Martin Aberdeen, Mike Cheshire, Paul Leman and George Seaton, the unfailing backing of Terry Croft and long and unstinting sessions by Andrew Askew at the computer. There was much burning of midnight oil by Richard Walker who has been entirely responsible for the cartoons, the technical illustrations, the cover and the lay-out. Nothing would have been possible without Shirley Walker.

I am most grateful to the University of Sheffield, to Arthur Willis, David Lewis and Terry Thomas for seemingly endless forbearance and understanding (and for allowing me continued access to university facilities during a period of partial retirement); to the Agricultural and Food Research Council which was most responsible for providing me with a scientific base; to Peter Horton for taking over the chairmanship of what is now the Robert Hill Institute; to the Royal Society which gave me, with my late and greatly missed colleague Tom Delieu, the wherewithal, to design and construct the apparatus which has had such a major influence on my scientific thinking; to Hansatech Instruments for many years of fruitful co-operation. Most recently, I have been tremendously encouraged by a Humboldt Prize Award and a Leverhulme Emeritus Fellowship both of which, in different ways, have done much to further completion of this book.

All scientific books or papers are distillations of ideas which spring from experiment and experience, from colleagues past and present, and from the relevant scientific literature (see page 254 *et seq*). Long, long ago Charles Constable encouraged me to believe that one day, I might be capable of writing if I kept on trying. Meirion Thomas warned me about dispossessed gerunds and he and Robert Hill were my principal scientific guides and mentors. A year with Harry Beevers at 'Old Purdue' was an immensely valuable part of my scientific apprenticeship. Charles Whittingham gave me my first job. To mention everyone else by name that I am indebted to would be a bit like compiling a list of wedding guests - so easy to dismay or neglect by omission. I feel sure, therefore that all of those who have worked in my laboratory and all of those in whose laboratories I have worked, particularly in Australia, Germany and the United States, will understand if I simply thank them collectively; their work and mine is inextricably entwined but the responsibility for all errors, omissions and misconceptions remains mine.

David Walker, June 1992

Illustrations for the front cover and title pages include materials from a variety of sources including Leonardo da Vinci and a number of 'copyright free' woodcuts from Victorian magazines. The images of skeletons on page 1 are based on original work by W. E. Le Gros Clarke.

Richard Walker

Introduction

Energy & the Future of Man

ENERGY AND THE FUTURE OF MAN

"Any man's death diminishes me, because I am involved in Mankind;
And therefore never send to find for whom the bell tolls; it tolls for thee"
.... John Donne

Human predictions about the human condition have a way of turning out wrong. Only a short time ago, or so it seems, the pundits predicted standing room only in the western world by the year 2001. Now, although populations in many countries continue to increase at alarming rates, relative stability has been achieved in others in which contraception and social and economic pressures have combined to limit the size of families. In the United States, during the early days of the space age, it was predicted that, if Higher Education continued to expand at the going rate, the day would not be far removed when every man, woman, child and dog would become a Doctor of Philosophy. The fallacy of such dire prediction is self-evident and the underlying tendencies often self-correcting. In the United Kingdom, for example, the current destruction of the university system has progressed so far and so fast that PhD's may soon become an endangered species. Even so, those who may have some knowledge of the environment, or feeling for the environment, should not shrink from prediction or at least from providing a basis of fundamental knowledge which might help those who would like to curb the worst excesses of environmental exploitation. What follows is an excursion into that unsafe ground.

Green plants undertake photosynthesis. As we shall see, this is a process from which energy in light drives the conversion of atmospheric carbon dioxide into organic materials. As such, it provides us with most of our energy and most of our raw materials. It has also provided us with the air that we breathe because it produces oxygen as a by-product. Primeval atmospheres contained little oxygen and possibly rather more carbon dioxide than now. Photosynthesis has changed all this and now we have a great deal of oxygen (about 21%) and very little carbon dioxide (about 0.035%). Precisely how much carbon dioxide there is in the atmosphere has, until relatively recently, been the concern of only a few scientists bent on one of those esoteric pursuits which invite denigration by politicians and impatience by some taxpayers who labour under the misapprehension that they support science rather than science supporting taxpayers. Such scientists have measured the carbon dioxide content of the unadulterated air of Mauna Loa in Hawaii, with such precision that they have been able to see it fluctuating (but always moving upward) in accord with world oil prices. Until recently their labours have gone largely unremarked. Now, of course, the world at large has suddenly come to realise that the carbon dioxide content of the atmosphere is not only important but crucial to life on earth as we know it. The carbon dioxide which was extracted from the atmosphere by millennia of photosynthesis in the carboniferous age and locked into oil and coal reserves, is being returned, in a dramatically short time, by Man's profligate use of fossil fuels. Unless this process can be checked, it will change our planet out of all recognition. There are also many other areas for concern which, although they may be less dramatic, will also require fundamental solutions. In the long term, the return of carbon dioxide to the atmosphere will be eased because there will be no oil or gas or coal to burn. The energy crisis of the seventies has receded, for the time being, largely because the Organisation of Petroleum Exporting Countries (OPEC) has failed to limit production, because other countries, like the U.K., have disdained such regulation entirely and because the United States (which returns more carbon dioxide to the atmosphere than any other nation) has found it politically expedient to sell oil and gasoline to its citizens at prices which do nothing to deter use. Nevertheless, limitation of production will inevitably return as it becomes overtaken by

diminishing supply. At present, energy usage and carbon dioxide emissions are closely related. No doubt there are still large reserves to be discovered but oil and coal are finite. Accurate estimates are virtually impossible because there are so many imponderables but, if we were to continue as we are at present, oil might be good for another 50-100 years and coal for another 300-500. Long before that the nature of current usage will be vastly changed by rising prices. Even in the U.K. (a country notorious for the way in which it has squandered its coal and its oil) insulated lofts and double glazing are now more of a norm than an oddity. Energy-efficient light bulbs are not yet given away by public utilities, as they are in the United States, but at least they have started to appear in British shops. In the next century the burning of oil and coal for heating may come to be regarded in much the same way as burning paper money.

There are, of course, any number of possible alternative energy sources - some plausible, some not. For instance, many otherwise rational and intelligent people quite sincerely believe that nuclear fission is the only *practical* alternative to fossil fuels and naturally they have been encouraged, in this view, by the nuclear industry. Perhaps the most remarkable sophistry of all is the latest suggestion (by unlikely converts to "green" policies) that fission is preferable to fossil fuels because it liberates no carbon dioxide and would, therefore, spare the "greenhouse" effect. Quite apart from the dubious arithmetic involved (only about a third of fossil fuel is consumed in the generation of electricity), this is about as sensible as sharing your bedroom with a man-eating tiger as a protection against burglary - certainly unsafe, impossibly expensive and unlikely to be well received by your neighbours.

"Certainly unsafe......"

The main trouble with nuclear fission is that it produces radioactive products, terribly dangerous to life, which do not decay and cannot be destroyed. All that *is* possible is to attempt to contain them and, unless we are indifferent to the problems of our descendants, this means containing them for "ever". At the time of writing, the city of New York is reported to have started to create a pyramid of rubbish which, if ever completed, would be higher than existing skyscrapers. This is a strategy born of desperation while measures are sought to recycle 50% of new rubbish and to find more acceptable means of disposal. No one of sound mind (apart from the nuclear electrical generating industry and politicians more concerned with nuclear status than human life on earth) could even contemplate creating, quite unnecessarily, nuclear rubbish dumps infinitely less attractive than mere mountains of garbage. In the United Kingdom plutonium has already been detected in houses in the vicinity of the nuclear reprocessing plant at Sellafield. This single installation has also made the Irish Sea the most polluted in the world. Sheep farmers in Wales and reindeer herders in Lapland, are still unable to sell their animals for human consumption as a consequence of the radioactive fallout from Chernobyl. (There is also evidence that the scale and extent of Chernobyl fall-out was not recognised quickly enough in the U.K. to prevent children from drinking dangerously contaminated milk.) In what was the Soviet Union itself, where this single disaster is said to have cost $3.5 billion, it is no longer believed that nuclear generation of electricity can be regarded as a safe process and some such installations are being decommissioned both there and in the United States. Conversely, Electricity Generators in the U.K. have sought, by sustained advertisement, to persuade public opinion that there is no viable alternative, prompting the conclusion that if they really had such a good product to sell they might not need to protest so much. Those who believe in nuclear fission quite properly point to the appalling past record of human death and misery in coal-mining and comparable industries but there is no longer any doubt that nuclear disaster can occur. When it does occur, radioactive fall-out can reach out and damage human life and welfare well beyond national boundaries and on a scale which boggles imagination. It is also patently facile to suggest, as some have done, that Chernobyl was a product of inadequate technology or that it could never happen again. Nothing devised by man, or operated by man, has ever proved entirely safe in the long term. Today the problems are also compounded by international terrorism. Recent wars in the Middle East and the gratuitous murder of innocent travellers have underlined the fact that no nation will shrink from unleashing horrendous techniques if desperate and convinced that it has right on its side. There are also many more mundane considerations. Uranium reserves, like fossil fuels, are finite and fast breeder reactions bring new problems in their wake. Real costs are rarely taken into account. Quite apart from the need to secure nuclear plant from terrorist attack and the insoluble problem of safe disposal of radioactive waste, there is the price of insurance against accident and of decommissioning once reactors have become obsolete or unreliable. As it continued to be de-commissioned, the installation at Three-Mile-Island, in the United States, still consumed dollars much more efficiently than it ever generated electricity. It has been claimed that safe long-term storage of nuclear waste in the United Kingdom could cost as much as the Channel Tunnel. The British Government, intent on the "privatisation" of electricity generation finally had to concede, in the face of simple arithmetic produced by would-be investors, what the scientific community and the independent press had recognised or suspected all along, that the cost of decommissioning nuclear plants would be so immense that no one would want to buy it.

Recent demonstrations of cold fusion have created considerable interest but seem unlikely, at least at present, to be of vast significance. In principle, nuclear fusion could give us limitless energy but, although it occurs ceaselessly in the sun and is the ultimate driving force behind photosynthesis (a principal subject of this book) there seems to be no immediate prospect of it being a practical reality on earth. Even if it were to become practicable, 90 million miles from human habitation might seem the most desirable safe site for its operation. Fusion generates neutrons which must be contained. Containment gives rise to radioactive materials which, like products of fission, would need to be safely disposed.

Solar energy and the forces of wind, wave and tide seem much better bets. Although frequently derided and undervalued by the pro-nuclear energy lobby, there is no real doubt that these will come to make increasingly important contributions over the next two or three decades. Equally, there is little doubt that they would already be playing a much more significant role had not politicians the world over been seduced into diverting vast resources into nuclear fission (£17 billion on nuclear research alone in the U.K.) or determined to represent factories producing plutonium for bombs as electricity generating stations. It would be optimistic to suppose that the world will not experience considerable problems in meeting its energy requirements during the rest of this century but equally pessimistic to dismiss the possibility of new solutions. For example, solar photovoltaic (amorphous silica) generation of electricity, is already practicable, fast becoming cheaper and, at about 30% efficiency, would not call for inconceivably large areas of global surface (according to George Porter about 50 million hectares, or roughly twice the area of the United Kingdom, in order to provide existing world populations of 3-4 billion with 5 kW per head). Electrolysis of water would, in turn, produce a fuel (hydrogen) which, if ways can be found to safely contain its formidable reactivity with oxygen, could power surface vehicles, aircraft and the like. In the U.K., one hydro-electric facility, utilising the tidal energy of the Severn Bore could generate more electricity, *ad infinitum,* than the combined nuclear stations in Britain (about 10-15% of total requirement).

What *is* now manifestly clear is that the world should make much better use of the energy that it already has at its disposal. At least one authority in the field of energy conservation believes that present use of fossil fuels and electricity could be cut by 75% if rapidly developing techniques in energy saving were fully and properly utilised and that this could be achieved not only without detriment to world economies but with actual benefit. Mankind lived in equilibrium with the environment when populations were a fraction of what they are now and expansion only became possible as fossil fuels were used on an increasingly large scale. Advocating a downward adjustment of population size may seem unethical or simply unlikely to succeed but there is no doubt that such a decrease would ease the current problems and there is every likelihood that a decrease will occur by famine, pestilence or major global disaster if it can not be achieved by consent.

Even if the energy problem is solved, there will still be an increasing shortage of cheap organic carbon during a period in which oil and coal reserves decline and prices increase correspondingly. At present we use oil and coal not only as fuels but also for almost every conceivable *material* purpose. We live in a world of metal and plastic which is fashioned out of, or with the help of, the organic carbon of which these fossils are comprised. Now they are plentiful but, once they become scarce, we shall be obliged to fall back on contemporary photosynthesis to provide us with our needs. Past photosynthesis locked atmospheric carbon dioxide in oil and coal and, as shortages make these increasingly expensive, we shall come to rely more and more on products of contemporary photosynthesis such as wood. (Already the more far-seeing are turning to the fermentation of plant materials as a source of alcoholic fuels although it should be noted that while alcohol is a feasible alternative to gasoline in some places (as in Brazil) it is unlikely to provide a global solution because of the implications for land use and the high energy costs of distillation.) At first glance, the possibility of a real shortage of organic carbon seems ludicrous in the extreme. Annual photosynthetic carbon assimilation (the conversion of carbon dioxide into organic carbon by plants in the light) is of the order of 10,000 million tons of carbon dioxide fixed - much greater than all the oil, coal and minerals used every year. On the other hand a large proportion of this carbon enters products which are presently inaccessible or unused by man. Even the desirable commodities, such

as food, are in short supply in many places (although starvation is frequently a consequence of economics rather than real famine). The under-privileged go hungry, not because the world cannot grow enough food, but because they cannot afford it or because of political considerations. The European Community grows much more food than it knows what to do with. This is a reason, in some minds, for discontinuing research into means of increasing plant productivity while simultaneously turning a blind eye to the co-existence of food mountains in Europe and starvation in the Third World.

Famine in Ethiopia and elsewhere should remind us all that total food reserves are only about 40% of annual world consumption and that even this slender margin only exists because of the achievement of agricultural science. More people probably owe their lives to the introduction of new varieties of crop species and to the effect of selective herbicides on grain production than to the availability of antibiotics. However, even where famine on a major scale has become a thing of the past (as on the Indian sub-continent), it has only been prevented by major energy inputs in the form of chemical fertilisers and is still only too readily brought back by sustained drought.

The utilisation of energy, and of organic carbon is inextricably linked to economics and the pattern of human society. Food is expensive because energy is expensive. The most "advanced" forms of Western agriculture are often little more than inefficient ways of turning fossil fuels into edible products, of turning oil into potatoes, depending as they do on the manufacture of synthetic fertilisers, herbicides etc., and on the construction and operation of farm machinery. Conversely, primitive agriculture, which uses little energy other than that of the sun, the farmer, his family and his animals, is often equally sparing in yields and frequently fails to maintain large populations at acceptable levels of nutrition.

Linked to all of these considerations is the "Greenhouse Effect". Carbon dioxide and gases like methane (generated by mining, by cattle and by decomposition of vegetation and garbage) intercept infra-red radiation and therefore diminish heat losses to space. Deserts, very hot by day, are frequently very cold by night because water vapour also intercepts infra-red and nights are mostly colder when there is no cloud cover to act as a heat- retaining blanket. (Hence the "greenhouse effect", by analogy, because light entering a greenhouse, or glasshouse, warms the contents and its glass walls diminish the rate at which heat is lost to the exterior.) If there were no greenhouse gases, like water and carbon dioxide, the earth would be a frozen planet with an average surface temperature of about -15°C (rather than about +18°C as it is now). Until relatively recently, the natural greenhouse gases (such as carbon dioxide, water vapour, methane, nitrous oxide) trapped about 40% of the infra red radiation from the earth's surface and, from about 1850 to 1960, increasing carbon dioxide was the major contributor to Greenhouse warming. Since 1960, rising methane, nitrous oxide, and "refrigerator gases" (chlorinated fluorocarbons or CFCs) have become as great a problem as carbon dioxide because, like Heineken beer, they reach the parts that carbon dioxide doesn't reach (i.e. they absorb in parts of the spectrum in which carbon dioxide has not already exhausted any further capacity for infra-red absorption). For this reason, methane is 25 times more damaging than carbon dioxide and CFCs 10,000 times more damaging on a molecule for molecule basis.

A greenhouse would become even hotter if made of glass which transmitted ultra-violet radiation. This is why the destruction of the ozone layer by aerosol and refrigerator gases (CFCs) aggravates the greenhouse effect by allowing more ultra-violet radiation from the sun to reach the earth. (Ozone screens out ultra-violet like compounds used to prevent sun-burn). This is why, in letters to the editors of British newspapers, sighting the first sun-burnt skin has become more newsworthy than hearing the first cuckoo of spring.

Carbon dioxide in the atmosphere was diminished to something below its present concentration of about 350 parts per million by past photosynthesis and much of what was locked into coal and oil during the Carboniferous Era is now being restored to the air by human activity. In 1868, judging from bubbles trapped in polar ice and other evidence, atmospheric carbon dioxide was as low as 0.027%. Since 1957, carbon dioxide has increased from 0.0315% to about 0.0350% and is continuing to accumulate in an ever increasing fashion, threatening to double in concentration in the next 80 to 130 years. Contemporary photosynthesis cannot keep pace with this rate of increase even though, as we shall see, it is favoured by high carbon dioxide. The United Kingdom, for example, is currently producing about 10 tonnes of carbon dioxide per head of its population per annum and with a total area of about 24 million hectares consumes less than one third of this amount in photosynthesis. Moreover, much of this carbon dioxide does not remain "fixed". Carbon dioxide incorporated into materials like wood remain locked away for a long time but only mature forests constitute large repositories for fixed carbon dioxide and these are in balance with the atmosphere returning about as much carbon dioxide and methane via respiration and organic degradation as they fix in photosynthesis. Destruction of tropical rain forests, together with their untold genetic treasures, is a world tragedy which has contributed significantly to atmospheric carbon dioxide but, while they exist, these forests do not offset the unrelenting increases which result from burning fossil fuels.

Extrapolation of current increases in carbon dioxide suggests that many of those alive today will live to see a world in which the atmosphere contains twice as much carbon dioxide as it does now. Evaluation of the consequences of such doubling is much less certain. There is every prospect that world temperatures will increase by 2 or 3 degrees. It is much more difficult to predict what will follow because of the uncertainties relating to ocean currents and such notoriously fragile aspects of world ecology as the Arctic tundra. At worst, such temperature rises could change the world out of all recognition, as melting ice caps raise the level of the oceans and inundate the great cities built at sea level. At best it could unite world opinion behind measures to protect the environment. At its most facile, it could become yet another excuse to promote nuclear fission as a source of power.

What follows is an attempt to *introduce* some of the physical and biological realities which are involved in man's utilisation of energy. It discusses the source of biological energy, the manner in which solar energy is converted into chemical energy in photosynthesis and the implication of these relationships for plant productivity and for man. These matters are considered in the context of global warming, the greenhouse effect and what, if anything might be done to postpone or limit the consequences of increasing carbon dioxide. This book presumes only the most modest knowledge of biology, chemistry and physics and craves the indulgence of those who are much better versed in these subjects than the author. It does not seek to avoid all political bias and it trusts that the reader has some sympathy for Alice's sentiments about books.

Chapter 1

A Few Fundamentals

Atoms
molecules
bonds
oxidation
reduction
splitting water
hydrogen and oxygen, staying in the game
releasing energy
being one up...

MATTER AND ENERGY

1.1 Atoms, Molecules and Bonds

"The theory of the chemical bond, as represented in this book, is far from perfect" ....Linus Pauling

If we are to understand the energy requirements of Man, and of the plants on which he depends, we need to delve a little way into the nature of matter itself.

Before Queen Victoria came to the British Throne in the last century, the people of London used to bathe naked in Hyde Park because it seemed obviously sensible to undress before going into water. More recently, this has been regarded as improper, and (for bathing in public) people have worn correspondingly more or less clothing, according to the perceived morality of the day and the place. At one time there was thought to be an irreducible minimum called the "bikini". This was derived from the atom "test" in which a nuclear bomb was exploded on the Bikini Atoll in the Pacific. The connexion between the event and the garment is that the atom was also once regarded as the irreducible minimum. Dalton's atomic theory of 1803 proposed that all matter is composed of small indestructible particles, or atoms, which cannot be further divided by ordinary chemical means.

[Elements, the ninety odd chemical "entities" from which all other substances are formed, contain atoms which are characteristic of that particular element and are alike in all respects. Compounds, formed from the combination of two or more elements, are composed of molecules. These, in turn, are comprised of atoms joined together by various sorts of chemical bonds.]

As the world now knows to its cost, the atom is not indivisible but is itself composed of elementary, or fundamental, particles such as electrons, protons and neutrons. Hydrogen is the smallest, lightest and simplest atom. It can be regarded, for our present purpose, as a nucleus carrying a positive electrical charge (a proton or hydrogen ion, H^+) with a negatively charged particle, or electron, circling around it. In general, the atomic nucleus is much larger than the electrons associated with it and comprises more than 99.95% of the mass of an element. In the case of hydrogen, the proton and electron, taken together, are a bit like the earth with a single, orbiting, satellite or an Olympic athlete swinging "the hammer" at the end of a chain. Energy in the form of rocket fuel, is needed to lift a satellite into orbit and, the more distant the orbit, the more energy is required. Similarly, the electron may be considered to circle the nucleus at a specific energy level. More energy would be required to lift it further from the nucleus and energy would be released if it dropped to a lower level. In certain circumstances, electrons can also move between atoms and between molecules.

[In physical terms, such movement of electrons is described as an electrical current; in biologically analogous situations it is described as electron transport. It may also be noted that, by convention, an electric current is said to flow from positive to negative, whereas electrons actually flow in the opposite direction, from negative to positive. This is not as anomalous as it might at first seem because, as a negative charge moves in one direction, a positive charge must move in the opposite direction.]

The space satellite circling the earth is prevented from leaving its orbit by gravitational force, the athlete's hammer by its chain.

The electrons circling the atomic nucleus are similarly maintained in orbit by the electrical attraction between positive and negative charges. In a metal conductor, electrons can separate from their parent atoms and move freely through the metal, while the atoms which have temporarily lost their electrons, now positively charged, remain in a fixed position. If such a conductor, (e.g. a copper wire) is used to connect a "source" of electrons to a "sink" for electrons (as in joining the positive and negative terminals of a battery) electrons will flow from (negative) source to (positive) sink. The conductor will offer resistance to this flow of electrons as they collide with stationary atoms in the metal. In such a collision, the electron will transfer the kinetic energy (which it had acquired from the electrical field established by the potential difference between source and sink) to the vibrational energy of the atom. For this reason, conductors tend to become hot as electrons flow through them as in "electric fires" or heating elements. The thinner the conductor or wire, the greater the frequency of collision and the greater the internal generation of heat. All such objects can lose heat by radiation of electromagnetic energy (which moves through space at the speed of light). If the radiation is at wavelengths (see Section 2.2) which can be perceived by the human eye it is referred to as "visible light". We can "create" light in this way if we pass electrons from a source (the negative terminal of a battery), through the very thin wire (filament) in an electric light and back to a sink (the positive terminal of the battery). Chemical reactions in the battery maintain the electron flow. This is an example of energy transduction, the conversion of chemical energy into light and heat. Plants, as we shall see, can do the converse and convert light-energy into both electrical energy and chemical energy. In molecules, the orbiting electrons are shared by two or more atomic nuclei. As a result of this sharing process, the molecular orbitals are smaller than the combined atomic orbitals. Accordingly, molecular orbitals are at a *lower* energy level than corresponding atomic orbitals, just as it would require a smaller expenditure of energy to put a satellite into a close orbit, about the earth, rather than a more distant one. For this reason, the formation of chemical bonds between atoms is *always* accompanied by an energy loss. Similarly, an energy input is *always* required if a chemical bond is to be broken.

"energy input is always required if a chemical bond is to be broken"

Most human energy requirements are met by forming bonds between hydrogen and oxygen (see page 129). This occurs when fuels are burned. The fuels may be foods which are burned (or respired) in the body, or materials like oil and coal which are consumed in engines or furnaces. Conversely, photosynthesis is the process which produces, or has produced, such fuels. It involves splitting hydrogen-oxygen (H-O) bonds in water. To do this it utilises energy which is produced by nuclear *fusion* of hydrogen in the sun.

[Four protons (i.e. four H^+ or hydrogen ions) fuse to produce one molecule of helium. The fusion of one gram of hydrogen ions produces about as much energy as burning 16 million grams of oil.]

The energy produced by nuclear fusion fuels the furnaces of the sun and a fraction is transmitted to the earth as light (Chapter 2). It is intercepted by leaves and converted, by them, first into electrical energy and then into chemical energy. Apart from nuclear *fission* and ocean tides this is the source of virtually all energy on earth.

1.2 Oxidation and Reduction

Burning a fuel is an oxidative process. The formation of a fuel is a reductive process. Since this is central to the whole of biological energetics it is important that the concepts of oxidation and reduction (and the associated transport of electrons) are fully understood. It is better, so it is said, to give than to receive. This is true, in a sense, of oxidation and reduction which are also linked processes. There is always a donor and an acceptor. As one becomes oxidised the other becomes reduced and *vice versa*. The common feature is electron transport, or transfer. Electrons are donated in oxidation and accepted in reduction. Reduction implies diminution or becoming less and, if we remember that electrons are *negatively* charged, the terminology starts to make some sort of sense. Obviously accepting something negative can be regarded as a decrease or reduction.

Electron Transfer

If we look at a few examples, things should become clearer. Perversely, the most familiar example is perhaps the most difficult to understand in terms of electron transfer. If we take carbon (in a form such as charcoal) and burn it in a barbecue it becomes oxidised. Carbon dioxide is formed and energy is given off as heat and light

$$C + O_2 \rightarrow CO_2 + \text{energy} \qquad \text{Eqn. 1.1}$$

Man has been rejoicing in the energy released as heat and light in this and similar reactions since the beginning of time. It is not difficult to believe that those of us who live in cold climates prefer "open" fires to "central" heating because the warmth and flames and sound and smell of burning wood evokes some distant race memory. But, you might well ask, where in this particular process (i.e. burning carbon, as in Equation 1.1), can we see evidence of electron transport? Before we attempt to answer this question, let us first examine some other examples of oxidation and reduction.

Respiration, whether carried out by people or plants, is essentially the same process of "burning", or oxidising, organic compounds. The carbon containing compounds which are burned in respiration are most likely to be sugars. The combustion is more controlled but energy is produced and carbon dioxide liberated just as it would be if sugar were burned on a fire. Conversely, in photosynthesis, plants use the sun's energy to "fix" carbon dioxide into organic compounds. One of the major end-products of photosynthesis is, in fact, "ordinary" sugar (or sucrose, which contains 12 atoms of carbon) but the first formed products include more simple sugars containing 3, 4, 5, 6 and 7 carbon compounds. (Section 5.2)

Some biological oxidations also involve direct addition of oxygen but, more often, oxidation of a metabolite (a compound undergoing metabolism, such as a sugar) occurs mainly by removal of hydrogens. If we represent such a metabolite by MH_2 we can write an equation

$$MH_2 + A \rightarrow AH_2 + M \qquad \text{Eqn. 1.2}$$

In this equation MH_2 donates hydrogens to an acceptor (A). The donor, MH_2, becomes oxidised to give M. The acceptor, A, becomes reduced to AH_2. Because hydrogen is composed of a proton (i.e. a hydrogen ion, H^+ and an electron, e^-) transfer of hydrogens also always involves transfer of electrons.

$$\underset{\text{(atom)}}{H} \Leftrightarrow \underset{\text{(proton)}}{H^+} + \underset{\text{(electron)}}{e^-} \qquad \text{Eqn. 1.3}$$

Accordingly, electrons are transferred, or transported *from* the substance which is being oxidised *to* the substance which is being reduced.

Hydrogen Transfer

In metabolism, there are often long sequences, or chains, of hydrogen (or electron) acceptors. It is partly for this reason that "burning" a sugar in respiration, although it eventually involves the addition of oxygen, does not liberate energy quite so dramatically as it does if sugar is burned on a fire. Instead, energy is released in a trickle rather than a flood. In addition, much of the energy is not released at all, immediately, but is conserved, as chemical energy, in "high-energy" substances, such as ATP, which will be discussed later (Sections 4.6 & 4.7).

As we have now seen, hydrogen transfer always involves electron transfer (because a hydrogen atom contains an electron) but electrons can also be transferred by themselves i.e. without an accompanying hydrogen ion, or proton (H^+). Both in respiration and in photosynthesis, cytochromes of different sorts are important electron carriers (constituents of the electron transport chain. Cytochromes are a type of protein which contain iron (Fe) in the heart of their molecular structure. This iron can be represented as Fe^{++} (in its reduced or ferrous state) and by Fe^{+++} (in its oxidised or *ferric* state). The increase in the number of positive signs (in this case from two to three) indicates the loss of one electron just as it does in H^+ (a hydrogen atom which has lost an electron and therefore a negative charge). When a reduced cytochrome (Cyt Fe^{++}) is oxidised to Cyt Fe^{+++}, only electrons are transferred and the reduced electron acceptor can be written A^- to indicate that it is now the proud possessor of an electron carrying a negative charge.

$$\text{Cyt Fe}^{++} + \text{A} \rightarrow \text{Cyt Fe}^{+++} + \text{A}^- \qquad \text{Eqn. 1.4}$$

As it happens, electron acceptors are particular about what they will, or will not, accept. Some, like cytochromes, will only accept and donate electrons. Others, like quinones (which are also important electron carriers) like to have both an

Hydrogens Offered, Electrons Accepted

Electrons Offered, Hydrogens Accepted

electron and a proton; i.e. they are electron carriers only in the sense that they are hydrogen carriers and because accepting a hydrogen necessarily involves the acceptance of an electron. All of this may seem somewhat complicated but we need to understand it if we are to understand the basis of conserving energy in a chemical form. If a cytochrome is offered a hydrogen ($H \Leftrightarrow H^+ + e^-$) by a quinone (or a "quinol" as a "quinone" is more properly called when it is in the reduced state) it can only accept the electron (e^-) which is a constituent of that hydrogen atom. The proton (H^+) is left behind in solution and the solution, therefore, becomes more acid because of the consequent increase in its hydrogen ion concentration (see Section 1.7). Conversely, if the cytochrome seeks, in turn, to offer its newly acquired electron to yet another acceptor which is only prepared to accept hydrogens, a proton must be taken up from solution along with each electron accepted. When such acceptors are contained within a membrane this can result in protons being "moved" from one side of the membrane to the other during electron transport. As we shall see later (Section 4.7), this is the essence of Mitchell's Chemiosmotic Hypothesis which offers an explanation for the way in which proton gradients are established across membranes and how these gradients are then used in the synthesis of ATP (a so-called "high energy" compound).

What should always be remembered is that oxidation always involves electron transport. Electrons are transported *from* the atoms or molecules which are being oxidised to corresponding acceptors which therefore become reduced (and, in the sense that they have acquired electrons, more negative). If we return, for a moment, to the burning of coal or charcoal it is, perhaps, less obvious why the addition of oxygen to carbon also involves electron transport. In fact, in this and similar cases, the electrons move within the molecule. In this particular example, they move *away* from carbon *towards* oxygen. This is because oxygen is intrinsically more attractive to electrons (more oxidising) than carbon. The extent to which substances attract or repel electrons (their oxidation/reduction potential) is also very important in biological energetics and helps to determine the manner in which electron carriers are organised within membranes (see e.g. Fig. 3.5).

1.3 Electron Transport in Photosynthesis and Respiration

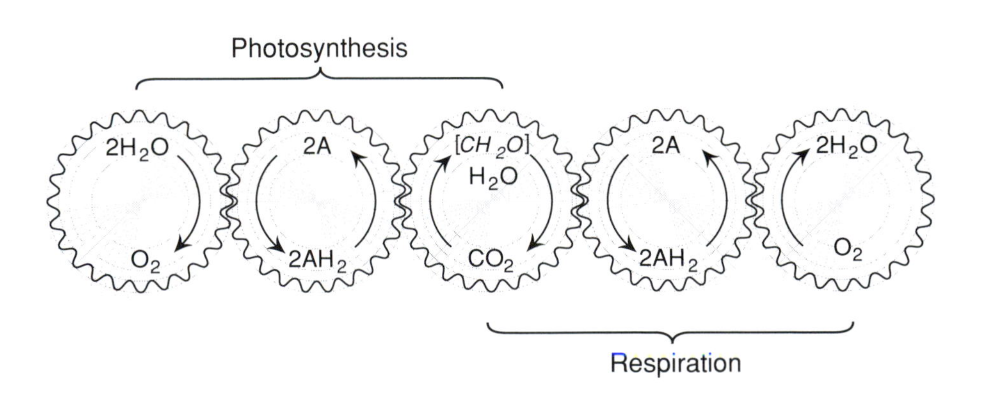

Fig 1.1 Relationship between photosynthesis and respiration. In photosynthesis H-O bonds are broken, O_2 is evolved, and CO_2 is reduced to CH_2O. In respiration CH_2O is oxidised. CO_2 is released and H-O bonds are re-established.

ELECTRON TRANSPORT IN PHOTOSYNTHESIS

In photosynthesis, light energy is used to break hydrogen-oxygen (H-O) bonds. The oxygen is liberated and the hydrogens (at times as hydrogen but at other times split into their constituent electrons and protons) pass through a series of carriers or acceptors (designated "A" in Fig. 1.1) to carbon dioxide. When carbon dioxide becomes reduced in this way we call it CH_2O because this is the simplest way to represent a carbohydrate (a compound containing the elements of carbon and water, i.e. "hydrated" carbon or carbo*hydrate*)

> [Some texts prefer to use $C_6H_{12}O_6$ rather than CH_2O to represent a carbohydrate but, while compounds with this formula actually exist, and CH_2O is just a symbol, there is little point in thinking about the fate of six carbons when one will suffice. Moreover, using $C_6H_{12}O_6$ could imply that glucose (which is one of several compounds with this empirical formula) was the immediate end-product of photosynthesis whereas, in reality, *it is not* (Section 5.2)].

In respiration CH_2O is oxidised. The hydrogens (or electrons) are passed, through a series of acceptors or carriers, to O_2 which becomes reduced to water (H_2O). Carbon is released as carbon dioxide (CO_2) and the energy which is derived from the re-formation of H-O bonds is made available for metabolic and other life processes. Incidentally, oxygen does not normally exist as a free atom (O) but as an oxygen molecule (O_2) so that, in order to derive the simplest possible equation, oxygen is sometimes represented as half a molecule ($\frac{1}{2}O_2$) rather than a single atom.

$$2H^+ + 2e^- + \tfrac{1}{2}O_2 \Leftrightarrow H_2O \qquad \text{Eqn. 1.5}$$

Man also sometimes burns CH_2O in his fires, furnaces and engines to meet his other energy requirements but, more usually, burns fossilised products of photosynthesis such as coal or hydrocarbons (C-H compounds) which are more reduced than carbohydrates.

All living organisms respire but only green plants can utilise light energy to split water and reduce carbon dioxide. Green plants make their own CH_2O; animals obtain it by eating plants or other animals. Compounds other than CH_2O are made in photosynthesis and consumed in respiration but the underlying principle is the same i.e. light energy is used to split H-O bonds and the restitution of these bonds releases energy for other processes.

THE SOURCE OF ORGANIC CARBON

1.4 The Products of Photosynthesis

When we look about us, much of what we see, other than stone and water, is either a photosynthetic product or something derived from photosynthetic products. This is obviously true of all living organisms, such as wood and natural fibres. In addition, coal, oil and natural gas (fossil fuels) were formed by photosynthesis in the past and metal, brick, glass and ceramics all require the expenditure of large quantities of fossil fuels to produce them. In processes such as smelting, fossil fuels may also serve as reducing agents (i.e. compounds which are ready to donate electrons or, in this instance, to accept oxygens).

$$\text{metal oxide} + C \rightarrow \text{metal} + CO_2 \qquad \text{Eqn. 1.6}$$

Even soil owes its present form and structure to the action of living organisms and one-fifth of the atmosphere is composed of oxygen released from water during photosynthesis (Section 8.3). The atmosphere is also the immediate source of organic carbon which is all derived from carbon dioxide. Although these facts are commonplace, many believe that the green plant obtains its *nutriment* from the soil. This belief seems entirely reasonable at first glance, and, in everyday life, we accept it without too much thought because we are used to gardeners who talk about "feeding" plants and to the extensive use of fertilisers. However, the fallacy of this supposition was demonstrated as long ago as the seventeenth century by Van Helmont who grew a willow in a tub of weighed soil. After watering for five years, the willow weighed 164 lbs more and the soil only 2 ounces less. Being an honest man, Van Helmont attributed the loss of 2 ounces to experimental error and very sensibly concluded that the tree had grown solely at the expense of the rain-water used for watering.

"Van Helmont Waters the Tree"

"But that Experiment of his which I was mentioning to You, he fayes, was this. He took 200 pound of Earth dry'd in an Oven, and having put it into an Earthen Veffel and moiften'd it with Raine water he planted in it the Trunk of a VVillow tree of five pound VVeight; this he VVatered, as need required, with Rain or with Difftill'd VVater; and to keep the Neighbouring Earth from getting into the Veffel, he employed a plate of Iron tinn'd over and perforated with many holes. Five years being efflux'd, he took out the Tree and weighed' it, and (with computing the leaves that fell during four Autumnes) he found it to weigh 169 pounds and about three Ounces. And Having again Dry'd the Earth it grew in, he found it want of its Former VVeight of 200 Pound, about a couple of Ounces; fo that of 164 pound of the Roots, VVood, and Bark, which Confituted the Tree, feem to have Sprung from the VVater".

- a contemporary account of Van Helmont's experiment (from Robert Boyle, "The Sceptical Chemist")

Despite what he said, Van Helmont was probably a better experimentalist than he imagined, and today's interpretation would be that the willow derived most of its nutriment from the air, as carbon dioxide, and some essential minerals from the soil The actual mineral requirement varies with the plant or crop but it is often less than 5% of the dry weight (Table 1.1). This is not to say that fertilisers are unimportant On the contrary, they are essential but they are not "nutriment", in the fullest sense of the word, any more than salt would be regarded as food in human diet. They become parts of a plant but are not a source of energy. Indeed large amounts of energy (in the form of fossil fuels) are used in modern agriculture to produce fertilisers (Table 6.2).

A large number of elements (including so-called trace elements which are needed in only small amounts) are required for sustained growth. The most important, in regard to the amount which is required, is nitrogen (N).

1.5 Chemical Composition of Leaves

All plants and animals contain protein. All proteins are made of amino acids and all amino acids contain some nitrogen, together with carbon and other elements (such as oxygen and hydrogen and, sometimes, sulphur).

Table 1.1

CHEMICAL COMPOSITION OF LEAVES	
	Fresh Weight
H_2O	Usually greater than 90%
	Dry Weight
C	About 45%
O	About 45%
H	About 5%
N	Often 1-3%
P,S,K,Na,Ca,Mg and Trace Elements (Together with N)	Usually 1-5%

Green plants are normally unable to utilise atmospheric nitrogen unaided although some (including beans and other legumes, and a few trees such as alder) can utilise nitrogen with the aid of certain micro-organisms. Most crops therefore depend on nitrate and ammonium compounds in the soil and, in principle, continued removal of grain and other products containing appreciable quantities of nitrogen should lead to progressively poor yields. In practice this does not always happen. For example, a plot at Rothamsted Experimental Station, in the UK, has produced more or less unchanging yields since 1848, perhaps because N_2 fixation by algae and other organisms has maintained the *status quo*, and presumably because the loss of other essential elements has been too slow to seriously deplete the soil's reserves. Nevertheless, although plants, like people, require only very small quantities of essential elements, these substances are essential and yields are improved by regular application of nitrogen (N), potassium (K), phosphorus (P) and other elements. Moreover, some modern crops have been bred to give maximum yields only in soils which have been copiously supplied with fertiliser. Much freely soluble fertiliser may be lost in drainage waters and create environmental problems such as massive algal growth in lakes in which the natural level of recycling minerals would otherwise be too low to support extensive growth. Fixed nitrogen is also lost to the atmosphere by bacterial action. Genetic engineering may conceivably extend nitrogen-fixing ability to species which lack it at present. This in the absence of unforeseen hazards, could constitute a major agricultural advance but it would involve expenditure of metabolic energy by the plant.

1.6 Ashes to Ashes

Carbon dioxide comprises only a little more than 0.03% (300 parts per million) of the normal atmosphere. Said quickly, this sounds small but, on sober reflection, it is *remarkably* small. There are more people with red hair per head of the average population than molecules of carbon dioxide in the normal atmosphere. This has many implications, including the fact that much of the available carbon dioxide is constantly recycled and much of it must have completed many cycles already. Plants fix carbon dioxide in photosynthesis. Plants, animals, microbes and people return it to the atmosphere through respiration or (in the fullness of time) by more permanent means.

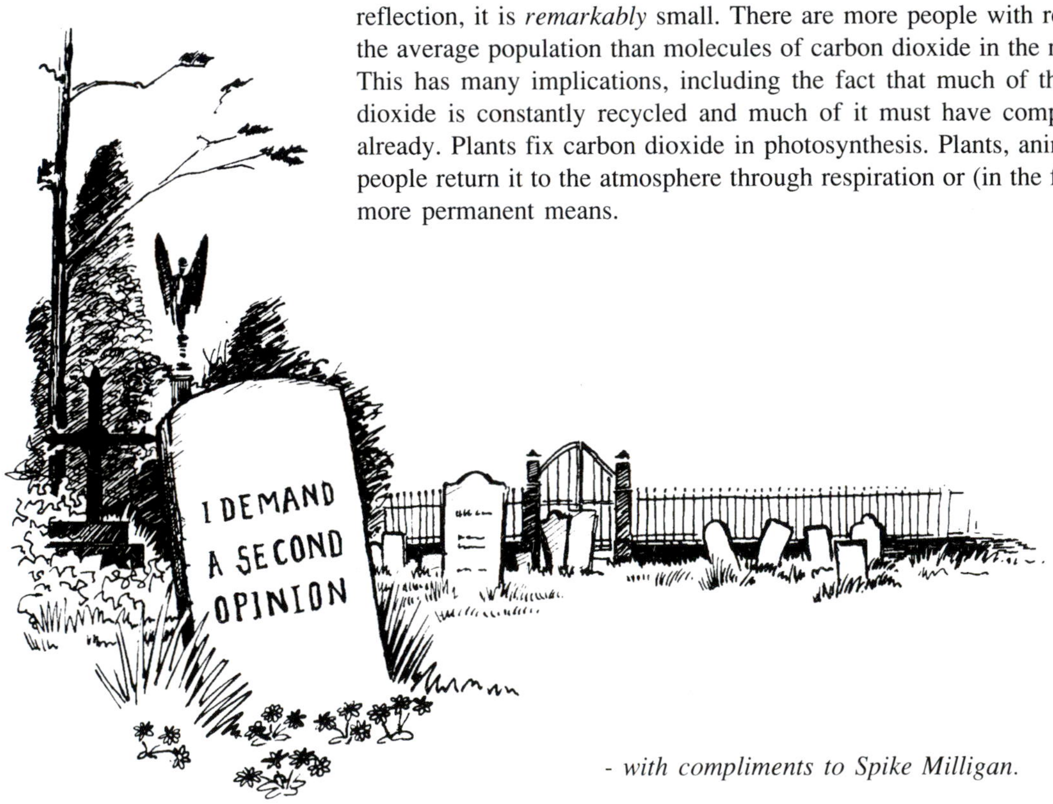

- with compliments to Spike Milligan.

In medieval times uptake and release of carbon dioxide was in some sort of balance just as large areas of untouched forest are now. In the last 200 years the huge explosion in the size of human population has only been made possible by a corresponding acceleration in the utilisation of fossil fuels (Chapter 8).

ENERGY TRANSDUCTION IN PHOTOSYNTHESIS

In one regard, photosynthesis is the source of organic carbon and material derived from organic carbon. In another, it is the major mechanisim of converting light-energy into the energy which Man uses every day for metabolic and other purposes.

1.7 An Analogy

Many of us learned, at mother's knee, that if two platinum electrodes are immersed in acidified water and joined by wires to an electric torch battery, hydrogen is evolved at one electrode and oxygen at the other. This "electrolysis" (electrical-splitting) of water is complicated in detail but relatively simple in principle. Water dissociates spontaneously to give very small quantities of hydrogen ions (H^+) and hydroxyl ions (OH^-)

$$H\text{-}O\text{-}H \Leftrightarrow H^+ + OH^- \qquad \text{Eqn. 1.7}$$

The H-O bonds in water are ordinary "*covalent*" bonds in which H and O each share a pair of electrons. On dissociation, both electrons stay with the hydroxyl ion (OH^- In consequence this gains a negative charge (becoming OH^- rather than OH) whereas the hydrogen atom (H) loses its electron and is left as a proton or hydrogen ion (H^+). Although this process occurs spontaneously it proceeds only to a very small extent before reaching equilibrium (when only one out of 10 million molecules has dissociated).

[This incidentally, is related to the fact that pure water has a pH of 7.0. The symbol "p" means "negative logarithm of" so that, if the hydrogen ion concentration is 1 in 10 million, i.e. if the logarithm of the hydrogen ion concentration is -7, then the negative logarithm will be +7. In theory the pH of water should therefore be 7. In practice it is frequently more acid because of dissolved carbon dioxide and other factors.. It should also be noted that both the nature of the ions which are formed and the mechanism of electrolysis are both more complicated than the simplified version given here but the basic principles are essentially the same.]

In a chemical reaction an unfavourable equilibrium can be displaced if it is possible to provide "sinks" for the products (i.e. if the products can be removed from the site of the reaction). In electrolysis, the electrodes are sinks.

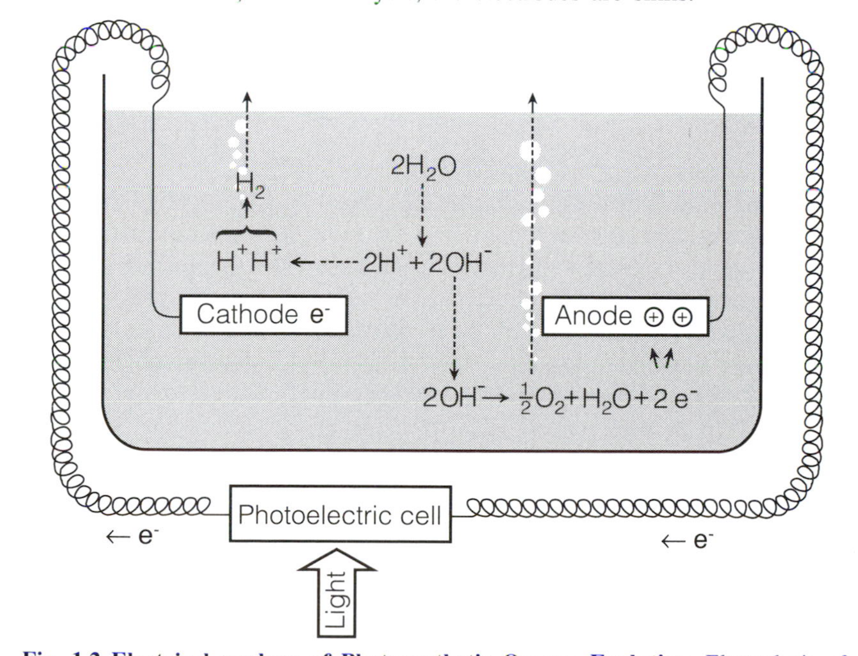

Fig. 1.2 Electrical analogy of Photosynthetic Oxygen Evolution. Electrolysis of water driven by an illuminated photoelectric cell rather than a chemical battery (See text).

The passage of the electric current (a stream of electrons) through the external circuit (Fig. 1.2) leads to an accumulation of electrons at one electrode (the cathode or "source" of electrons) and a dearth of electrons (visualised as positively charged holes from which electrons have been vacated) at the other (the anode or "sink" for electrons). Accordingly, protons accept electrons at the cathodes and are then evolved as hydrogen. At the anode, hydroxyl ions donate electrons in reacting together to give oxygen and water.

In the system illustrated in Fig. 1.2 the conventional battery has been replaced by a photoelectric cell. On illumination, the current for electrolysis is generated by the solar cell in an excitation process which is basically similar to the light excitation of chlorophyll (Section 3.1). Thus light energy is converted into electrical energy and this in turn is used to break H-O bonds. If the gases which are evolved from the electrodes are collected in a single vessel and a spark is passed, the H-O bonds are re-established with explosive violence, as the energy which has been trickled in from the photoelectric cell is instantly released.

"H-O bonds are re-established"

The electrolytic process is analogous to photosynthesis in the sense that light energy is used to break H-O bonds in a reaction, involving chlorophyll (Section 3.1) which is fundamentally similar to that which occurs in a photoelectric cell. As in Fig. 1.2, the photosynthetic process in green plants also involves the liberation of oxygen and the generation of an electric current but, of course, photosynthetic electron transport involves the transfer of electrons from carrier to carrier in a membrane rather than movement of electrons through a wire. As we have already seen, biological electron transport is an oxidation/reduction process. At one end of the chain of electron carriers, water (H_2O) is oxidised to oxygen (O_2) by virtue of hydrogen removal. In electrolysis this hydrogen is released as a free gas and (with the help of a bacterial enzyme) the green plant reaction can be made to do the same. In the intact green plant, however, most of the hydrogen is eventually transferred to carbon dioxide (CO_2) which is thereby reduced to carbohydrate (CH_2O). The process of transfer itself is basically one of electron transfer (Eqn. 1.3). Electrons are transferred at every stage but only some carriers can also accept a proton (H^+) along with an electron (i.e. accept a hydrogen, because $H^+ + e^- \Leftrightarrow H$). At such points along the chain, protons are released and at other points they are taken up. Finally, at the end of the chain, electrons (once part of water) and protons are added to carbon dioxide to give CH_2O.

In respiration (Fig. 1.1), H-O bonds are re-established and therefore energy is made available once again. Respiration does not proceed with explosive violence because hydrogen is not reunited with the oxygen in one step. Instead the energy is released, piecemeal, as hydrogen is transported to oxygen via a long sequence of intermediate carriers.

Most human energy requirements are met by forming bonds between hydrogen and oxygen, whether we are thinking in terms of metabolism, agriculture, industry, transport, building or domestic needs such as heating, cooking, refrigeration or whatever.

THE PHILOSOPHER'S STONE

1.8 How to be one up

"A good many times I have been present at gatherings of people who, by the standards of the traditional culture, are thought highly educated and who have, with considerable gusto, been expressing their incredulity at the illiteracy of scientists. Once or twice I have been provoked and have asked the company how many of them could describe the Second Law of Thermodynamics. The response was cold: it was also negative. Yet, I was asking something which is about the scientific equivalent of : - "Have you read a work of Shakespeare's ?"' ....C. P. Snow.

In this chapter it is claimed that photosynthesis winds the biological mainspring, that energy originates in the sun and that photosynthesis is the means of making this energy available to living organisms. That, you might think, is quite enough to be going on with but no self-respecting book which seeks to concern itself with energy and Man, let alone with plants, can get by without at least a nod in the direction of the Laws of Thermodynamics.

1.9 The Laws of Thermodynamics.

The Laws of Thermodynamics.

First Law	Energy can be neither created nor destroyed. The energy of the universe is constant.
Second Law	The entropy of the universe always increases. There is a universal tendency to chaos and disorder.
Third Law	The entropy of a perfect crystal of any element or compound is zero at absolute zero temperature.

As stated above these "laws" are more than a little daunting and, to mean anything at all, would propel us immediately into even less user-friendly concepts like *"entropy"*.

[Entropy describes the state or conditions of matter in regard to the random motion of the atoms and molecules of which it is composed. For example, as the temperature of water increases, or as it changes state from ice to liquid or from liquid to gas, this motion becomes more vigorous and less ordered. Entropy is therefore a measure of disorder and the second law supposes a universal tendency towards disorder. Herein lies the notion of cosmic futility or entropic doom. Only within a crystal at absolute zero does order reign supreme in the absence of thermal movement. Fleetingly, living organisms can reverse the general trend towards the random state by becoming more organised at the expense of their surroundings.]

If we are determined not to be daunted by the laws of thermodynamics we need only remind ourselves that they are not so much "laws" as summaries of endless human experience which tells us that we rarely get anything for nothing and that, if left to their own devices, everything in the universe tends to run down. In these terms they have been restated as follows:-

First Law	You can't win.
Second Law	You can't break even.
Third Law	You can't stay out of the game.

As stated above, the laws of thermodynamics are easy to remember. So easy in fact that we are now immediately one up on C.P. Snow's literary acquaintances and possibly on our friends. These laws also have clear implications for our present theme. Most of our energy comes directly or indirectly (as fossil fuels) from the sun. In either case it is finite. Fossil fuels once burned are gone for ever. Their energy is not destroyed but it is dissipated. There is no way that Man can easily recapture heat that has gone up a chimney; energy that has been used to move a vehicle from place to place. Only living organisms have the ability to create order (low entropy) out of chaos (high entropy) and this apparent reversal of the "second law" is only ephemeral - it is all ashes to ashes and dust to dust in the end. To recognise that the grave beckons, is not necessarily cause for despair even amongst those who see no prospect of resurrection. It may not happen for some time and consider what liberties might be taken with entropy in the meantime. The Laws of Thermodynamics should, on the other hand, behove us not to dissipate energy unwisely or unnecessarily. As we shall see, living organisms in general are very good at conserving energy and society at large would do well to emulate them.

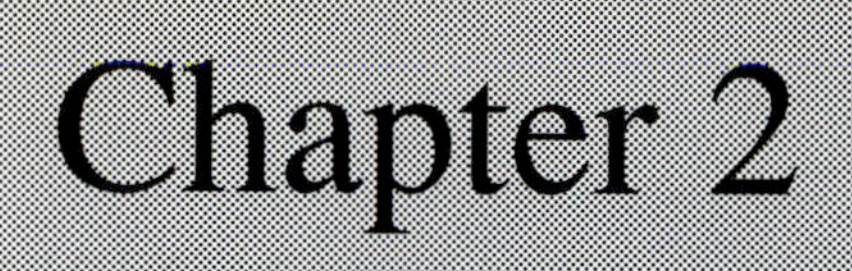

Chapter 2

Where It All Starts...

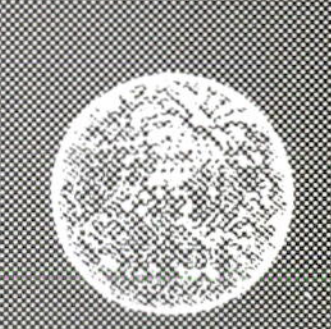

Sunlight
waves
particles
quantum leaps
getting excited
wealth and ambiguity

SOLAR ENERGY

2.1 The Sun

"Stand a little less between me and the sun" ... Diogenes

It has been claimed that many of the world's energy problems might be solved if physicists on earth could copy the nuclear fusion processes which occur in the sun (although there are those who would agree that a furnace of this type is best operated at a safe distance and that 90 million miles could be just about right). The incredibly high temperatures (many millions of degrees Celsius) within the sun, permit hydrogen ions to fuse. Helium is produced and vast quantities of energy (Section 9.4) are released as electromagnetic radiation. Some of this radiation travels through space, from sun to Earth. It is the source of most of Man's energy. That part which can be detected by the human eye is called "visible light" and it is light within this range of wavelengths which is utilised by green plants in photosynthesis.

2.2 The Nature of Light

"Nature and Nature's Laws lay hid in night.
God said, Let Newton be, and all was light" .....Alexander Pope

For some purposes, light can be regarded as an electromagnetic waveform, for other purposes, as a stream of particles. Newton suggested that light was composed of corpuscles travelling through space at constant speed and that there were different corpuscles for different colours. Much later, Thomas Young showed that light had the characteristics of a wave form and, subsequently, Lenard confirmed its particulate (or corpuscular) nature. Today, light remains an enigma but it can still be usefully described in these contradictory terms (i.e. particles and waveforms). Max Planck's conclusion that electromagnetic radiation (including x-rays, gamma rays, radio-waves) are absorbed or emitted only in discrete particles, or quanta, of energy remains central to modern quantum mechanics. In addition, Einstein, bringing together and interpreting earlier observations and ideas, formulated the concept of photochemical equivalence, i.e. that one quantum (or particle or corpuscle) of light (i.e., one *photon* if we are talking about *visible* light) brings about one photochemical event in any light-driven reaction.

[As we shall see later, 8 photons are needed to bring about the photosynthetic transport of hydrogen from water to NADP (Chapters 4 & 6) but this is because this process requires the movement of 4 electrons in 2 consecutive steps each requiring 1 photon per electron].

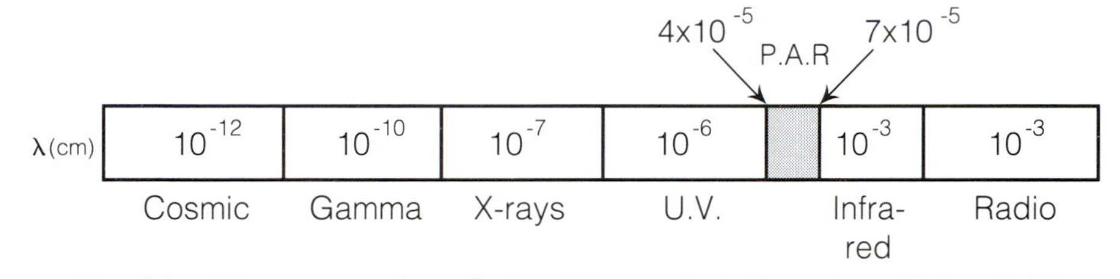

Fig. 2.1 Electromagnetic radiation. Photosynthetically active radiation (PAR) accounts for only a very narrow waveband. Note that values are expressed in metres (e.g. 4×10^{-5} m = 400 nm)

Described in the above terms, there is a relationship between colour, or quality of light and energy content. *"Visible light"* (i.e. light which can be detected by the average human retina) has wavelengths between 400 and 700 nanometres (1 nm = 1×10^{-9} metres = 1/1,000,000,000 metres). Below 400 nm (at the blue end of the spectrum) there is ultra-violet light, invisible to the human eye and, similarly, above 700 nm (at the red end of the spectrum), there is infra-red. *"Photosynthetically active radiation"* lies within the same wavelength limits as visible light (Fig. 2.1). White light is a mixture of all wavelengths between 400 and 700 nm and can be split into its primary colours by a prism or the corresponding action of raindrops which create a rainbow. Leaves appear green, to the human eye, when illuminated in white light, because their chlorophyll molecules absorb red and blue light and transmit and reflect green. The peak of human perception is in the green at about 555 nm (Fig. 2.2) not far removed from the intensity maximum of sunlight (about 575 nm).

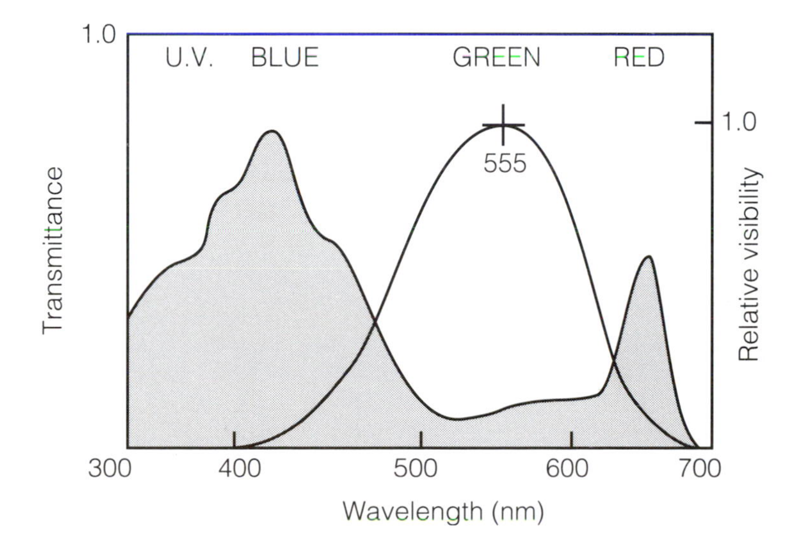

Fig. 2.2 What Man sees and what the green plant uses. By definition "visible" light is that perceived by the human eye (wavelengths between about 400 and 700 nm, with a perception maximum at about 555 nm). The green plant also uses light within this wavelength range but chlorophyll (shaded area) absorbs best in the blue and the red.

If we think of waves breaking along a shore, it is evident that they deliver energy as they pound a beach or breakwater. The amount of energy delivered depends on the height of the waves (their amplitude) and their frequency.

Waves of the same height will deliver more energy if they arrive more frequently. In the same way the energy content of photons is related to the frequency (and therefore the colour) of light at a given intensity. Blue light at 450 nm packs more energy per photon than red light at 650 nm. A quantum of ultra-violet radiation carries more energy than a quantum of infra-red.

The relationship between wavelength (λ) frequency (υ) and speed of light (c) may be expressed as follows:

$$\lambda \text{ (nm)} \times \upsilon \text{ (per second)} = \text{(nm per second)} \qquad \text{Eqn. 2.1}$$

$$\text{Alternatively } \upsilon = c / \lambda \qquad \text{Eqn. 2.2}$$

(frequency equals speed divided by wavelength)

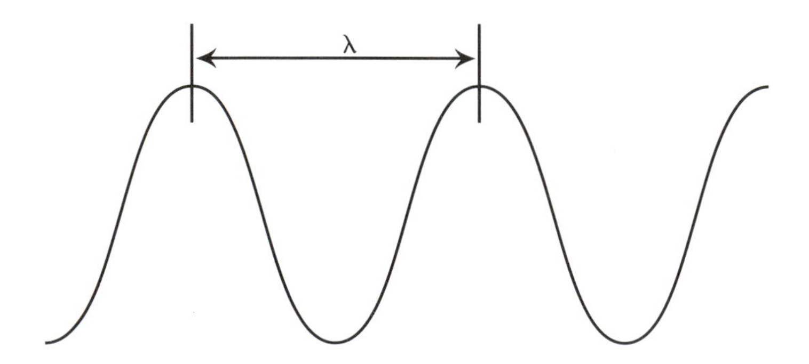

Fig. 2.3 Light as a waveform. The wavelength is the distance between peaks, the height of the waves (the amplitude) is a measure of intensity.

These days (see Section 2.4) the wavelength of light is usually measured in nanometers (nm) in accordance with the Système Internationale d'Unitès (S.I.). The

S.I. unit of length is the metre and this is sub-divided into:-

millimetres (1 mm = 1 m / 1000 or 10^{-3} m),

micrometres (1 μm = 10^{-6} m)

nanometres (1 nm = 10^{-9} m)

and so on. Before S.I. was established, a thousandth part of a millimetre was sometimes referred to as a micron and the terms mμ (millimicron) and nm (nanometre) are therefore synonymous. Physicists also used "Angstroms" (1/10 of a millimicron) so that:-

$$1 \text{ m}\mu = 1 \text{ nm} = 10 \text{ Å} \qquad \text{Eqn. 2.3}$$

To add to the multiplicity of units, the speed of light (c) was usually expressed in cm per second. In vacuum, $c = 3 \times 10^{10}$ cm/sec and, since 1 cm = 10^7 nm, this nice round number unfortunately becomes 3×10^{17} nm/sec.

2.3 The Energy Content of Light

The energy content (E) of one photon of light may be determined by multiplying the frequency (υ) by a value (h) called Planck's Constant. It is for this reason that light is sometimes depicted as $h\upsilon$ in equations such as 4.1 and 4.2). Note that the symbol traditionally used for frequency is the Greek letter Nu (υ), not v. Therefore

$$E = h\upsilon \qquad \text{Eqn. 2.4}$$

or, from equation 2.2,

$$E = hc / \lambda \qquad \text{Eqn. 2.5}$$

"Plank's Constant"

The value of Planck's Constant (h) is 6.6×10^{-27} erg seconds so that, for red light at 680 nm (which is most effective in photosynthesis),

$$E = 6.6 \times 10^{-27} \text{ erg.sec} \times 3 \times \frac{10^{17} \text{ nm}}{\text{sec}} \times \frac{1}{680 \text{ nm}}$$

$$= 2.9 \times 10^{-12} \text{ ergs} \qquad \text{Eqn. 2.6}$$

joules are a girl's best friend

There are 10^7 ergs to the joule and 4.184 joules to the calorie so that this becomes

$$\frac{2.9 \times 10^{-12}}{4.184 \times 10^{7}} \text{ calories} \qquad \text{Eqn. 2.7}$$

$$= 0.69 \times 10^{-19} \text{ calories}$$

or

$$\frac{2.9 \times 10^{-12}}{10^{7}} \text{ joules} \qquad \text{Eqn. 2.8}$$

$$= 2.9 \times 10^{-19} \text{ joules}$$

Clearly the energy content of a single photon is very small and it is usually put on a more convenient basis by the simple expedient of multiplying by Avogadro's number (6.023×10^{23}), which is similarly used to convert the weight of single atoms into grams, e.g. the mass of 6.023×10^{23} hydrogen atoms = 1 g). Thus the energy content of Avogadro's number (1 quantum mole) of photons (at 680 nm) is

$$0.69 \times 10^{-19} \text{ cal} \times 6.023 \times 10^{23} = 41.5 \text{ Kcal} \qquad \text{Eqn. 2.9}$$

or

$$41.5 \text{ Kcal} \times 4.184 = 174 \text{ Kjoules} \qquad \text{Eqn. 2.10}$$

For some purposes it is useful (e.g. Fig. 4.2) to convert ergs to electron volts (by dividing by a factor of 1.6×10^{-12}). Then we then find that 1 photon (at 680 nm) can accelerate 1 electron through a potential of 1.8 volts i.e.

$$\frac{2.9 \times 10^{-12}}{1.6 \times 10^{-12}} = \frac{2.9}{1.6} = 1.8 \text{ eV} \qquad \text{Eqn. 2.11}$$

The energy contents of monochromatic light at five visible wavelengths are expressed in each of these ways (Kjoules, Kcalories and electron volts) in Table 2.1.

TABLE 2.1

Energy content of monochromatic, visible, light

Wavelength	Colour	Kjoules /mole	Kcal/mole	Electron volts Photon
700 nm	Red	171	40.9	1.77
680 nm *	Red	174	41.5	1.81
600 nm	Yellow	199	47.7	2.07
500 nm	Blue	239	57.2	2.48
400 nm	Violet	299	71.5	3.10

* This is the wavelength of the red absorption maximum of chlorophyll

2.4 A Word About Units

"divers weights and divers measures, both of them alike are abomination to the Lord" ...Proverbs 20.10

In Section 2.3 we have arrived at values for the energy content of light expressed in calories and joules. We now have an immediate problem from which there is no escape and one which will come back to plague us again and again (see, for example, section 2.5). In the beginning, Man related his measurements to things that he had around him like hands and feet, the length of his stride (a yard), the distance between outstretched hands (a fathom), how much he could lift (a hundredweight) and so on. He also used other objects of fixed dimensions (such as the seed from the carob tree which almost invariably weighs 200 mg or one "carat" - as used by goldsmiths). As he became more sophisticated, of course, Man's units became correspondingly more sophisticated and, in many respects, a mess.

A contemporary attempt to create some order out of chaos was the introduction of Système Internationale d'Unitès (S.I.) based on metric units. However sensible this may seem it does not lend itself readily to all uses in all circumstances. Grams and millimetres have some obvious advantages over ounces and inches but the fact that Bavarian carpenters still use inches (generations after the introduction of the metric system into Germany), that pounds are not unknown in France, and that much of domestic building in the United States is still firmly based on 'Imperial' units, speaks for itself. Moreover, there is no getting away from the problem of translation or conversion. English is fast becoming a universal language. It is already the *lingua franca* of science and, if every future book were published in English, it would certainly make life easier for the English-speaking world. Nevertheless, without translation we would still be denied access to the foreign language literature. Similarly, even if we were to move universally into S.I. at this moment we would still be unable to understand older scientific measurements unless we were familiar with obsolete units. For this reason, in Section 2.3 and elsewhere, both calories and joules are used although the joule is now the preferred unit. Sadly, there is also an additional confusion created by dieticians (see Section 8.5). Thus, if and when we talk about calories in "every-day" usage (as we would if we were talking about the calorific content of food) we must bear in mind that, if properly

S.I. Prefixes

Prefix	Symbol	Value
tera-	T	10^{12}
giga-	G	10^{9}
mega-	M	10^{6}
kilo-	k	10^{3}
hecto-	h	10^{2}
milli-	m	10^{-3}
micro-	µ	10^{-6}
nano-	n	10^{-9}
pico-	p	10^{-12}
femto-	f	10^{-15}
atto-	a	10^{-18}

written with a capital 'C', the Calorie = 1 Kcalorie = 4.18 Kjoules. The calorie with a small 'c' is the amount of heat needed to raise the temperature of 1 ml of water 1°C (strictly speaking from 14.5°C to 15.5°C). So, for example, if we are told that the calorific content of 1 mole (180 g) of glucose is 672 *Calories* (= 672 K*cal*) this means that if we were able to ignite a pile of glucose weighing 180 g under a large beaker containing 10 litres of water (in such a way that we captured all of the heat released) we could expect a temperature rise of about 67°C (0.1°C per Kcal per 10 litres). This may seem like an excuse for introducing yet more tedious arithmetic but the intention is simply to give some feeling for the size and meaning of the units involved. We would all know that "a drop" of water is about 1millilitre (1ml) in volume and that if our immediate surroundings increased in temperature by 1°C we would scarcely notice the difference. This in turn gives us some idea of the size of a calorie as a measure of energy. The joule is also a familiar thing if we take it as the energy produced per second by the application of a power equal to 1 watt. For example, an electric light rated in terms of energy at 100 watts, will produce 100 joules of energy (in this case a mixture of light energy and heat energy) every second that it is switched on. As we have seen in Section 2.3, 1 joule = 10^7 ergs and this is a way of relating energy to mechanical work. The *International Joule* is the work expended per second by a current of one *International Ampere* flowing through a resistance of one International Ohm. These sorts of considerations allow us to start to relate the energy that Man needs for his metabolism and all other purposes to the sources of such energy (Chapter 6).

In the absence of units that we can all easily understand and relate to we find ourselves in immediate difficulties. You might suppose, for example, that a second is about the interval between the digits 1, 2 , 3, 4 .... and so on, counted in a measured way. For many purposes such a definition would be hopelessly inadequate but, if we came face to face with the most precise definition (below), we might find ourselves in difficulty.

"The second is the duration of 9,192,631,770 periods of the radiation corresponding to the transition between the two hyperfine levels of the fundamental state of the atom of caesium 133"

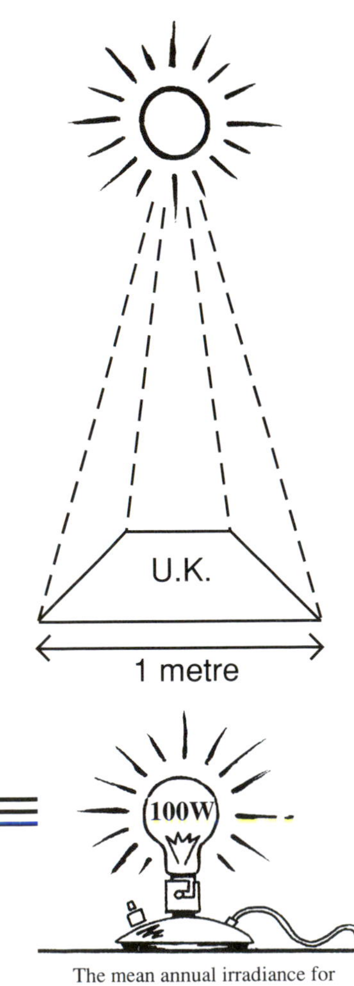

The mean annual irradiance for the U.K. is about 100W/m²

2.5 Light Intensity

In Section 2.3 we calculated the energy content (one Einstein) of one quantum mole (Avogadro's number of photons) of red light at 680 nm; a wavelength which is particularly effective in photosynthesis. Shorter wavelengths have a higher energy content. Sunlight does not contain all the colours in the same proportions and the corresponding calorific value for the intensity maximum (sometimes put at 575 nm) is about 50 Kcal (209 Kjoules). Similarly the energy delivered to the earth's surface depends upon the quantity of light (intensity) as well as its quality (wavelength).

For some purposes, *"light intensity"* is still measured in foot candles or some similar unit. As the name implies, this is the light given out by one standard candle and perceived at a distance of one foot. A standard candle emits 4π lumens (1 candle power) and if placed at the centre of a sphere of 1 foot radius produces a luminous flux, at its surface, equal to 1 lumen/sq.ft. If the radius is increased to 1 metre the same light is spread more thinly, the flux decreasing according to the square of the distance. Since there are 3.2808 feet to the metre and since 3.2808 squared = 10.8

it follows that one foot candle (1 lumen/sq.ft.) = 10.8 lux or metre candles (1 lumen/ sq. metre). The human eye is a very good device for comparing the intensity of weak light sources, but very poor for determining absolute intensity values. This is because the iris diaphragm closes in bright light, so protecting the retina. Accordingly we tend to over-estimate the intensity of light if we enter a well-lit room from a dark night. The light intensity in such a room is unlikely to be more than 200 ft candles, whereas in full sunlight it may be as much as 10,000 ft candles (100,000 lux). For accurate measurement the human eye (or instruments such as photographic "light meters" which were designed to mimic its performance) have been replaced by instruments which record the incident radiation in units such as watts per sq. metre per sec.

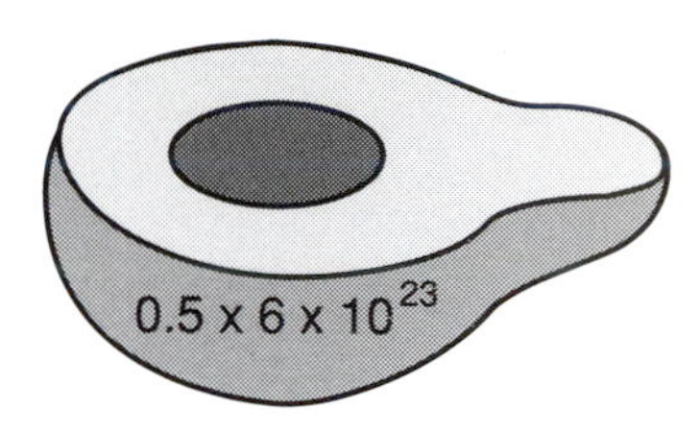

"Half an Avogadro Pair"

[It is not strictly correct to describe light energy arriving at a surface in terms of 'light intensity'. Intensity is a property of the radiating *source* e.g. a standard candle emits a flux of 1 candle power or 4π lumens. The corresponding measure of flux density or light received is 1 lumen/ft^2].

At the present time, relatively new instrumentation is fast displacing the need to cling to these old units, themselves a product of out-moded measuring procedures. Many contemporary light meters are, in fact, photon (Section 2.2) counters and give values of light *"intensity"* expressed as *"photon flux density"* (PFD), i.e. the number of photons which strike a unit area per unit time. For biological purposes, this area is often taken to be one metre and the time one second. It is also convenient (for reasons which will become clear in Chapter 6) to express the number of photons as µmole quanta or µmole photons.

[Avogadro's number of photons is called a *"quantum mole"*, or *"photon mole"*. One micromole of photons is 1 mole $\times$ 10^{-6}. The energy content of 1µmole of photons is called a "microeinstein" but this unit seems to be coming less favoured.]

Microeinsteins

The relationship between some of the units which are still used, or have been used, is given in Table 2.2 (overleaf) but it should be appreciated that it is often not possible to give precise equivalents because "white" light from the sun does not have

TABLE 2.2
Full sunlight (Global Irradiance) expressed in several ways.

Illuminance		Photosynthetically active component					
			per square metre per second				
lux	foot candles	Watts/m^2	ergs	joules	cal	photons	µmole quanta
100,000	10,000	500	5×10^9	500	120	1.3×10^{21}	2200

This table shows the approximate equivalence between full sunlight (expressed, at left, as illuminance) and the energy content of its photosynthetically active component, (right) assuming a mean wavelength of 575 nm.

precisely the same spectral composition as light from a traditional tungsten source (bulb, globe), or a fluorescent tube or a quartz-halogen lamp (typical of modern motor vehicles). Accordingly, these sources do not deliver the same energy at intensities which would be judged to be equal by some of the procedures which have been employed in the past. Precise evaluation is only possible for "monochromatic" light or when instruments are used (e.g. "spectroradiometers") which, in effect, count how many photons are delivered over the whole spectrum of wavelengths emitted by a particular source.

2.6 The Availability of Solar Energy

"Hail Smiling Morn, that tips the Hills with Gold"

More solar energy reaches the earth than might be imagined by those who live in a cold climate. At noon, on a clear day at the equator, each square metre of the sea's surface may be rated (according to the energy it receives) at 1 kW. Clearly this value will fall as the sun goes down or as the distance from the equator is increased. However, if averaged over 24 hours/day and 365 days/year it is surprisingly independent of latitude and cloud cover. Thus the mean annual irradiance in the Red Sea area is about 300 W/square metre, but the corresponding values for Australia and the USA are as high as 200 W/m^2 and 185 W/m^2 respectively and even, in the UK, the average is about 105.

2.7 Light as an Energy Carrier

In Section 1.2 we addressed the concept of charged particles. A proton (hydrogen ion, H^+) is an example of a particle (in this case an atom) which has acquired a positive charge as a consequence of "losing" a negatively charged electron (e^-). A *"dipole"* is a pair of oppositely charged particles separated by a very small distance. If the two charges can be made to oscillate along the line which separates them, such an *"oscillating dipole"* generates a moving, wave-like, force-field which moves away from its source with the speed of light (Fig. 2.4). The force field has an electric and a magnetic component; the two components in phase but perpendicular to one another. The amplitude of the waves gradually diminish as the force field moves away from its oscillating dipole source. As we have seen in Section 2.2, visible light (and other forms of electromagnetic radiation) can be partially described in these terms. Light intensity is determined by the amplitude of the waves (the bigger the waves the brighter the light).

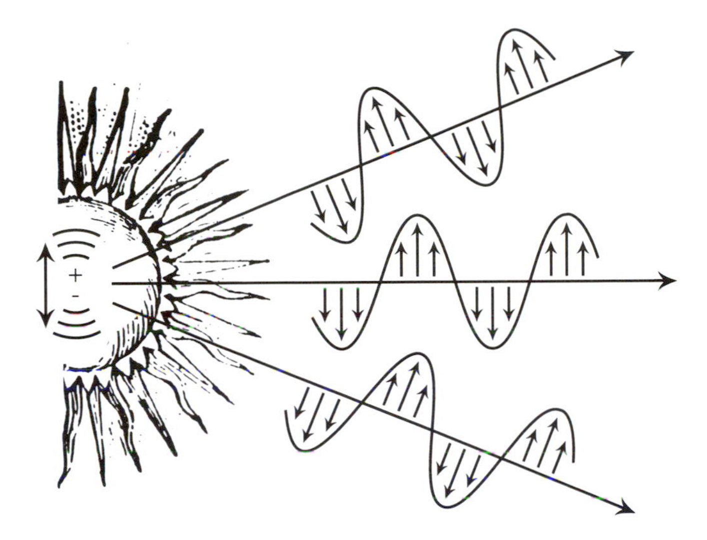

Fig. 2.4 Light as a force field. Electromagnetic waves radiated by an oscillating electric dipole. The electric component is shown by the short arrows. The magnetic component is not shown; it is perpendicular to, and in phase with, the electric component. The pattern moves away from the dipole with the speed of light, its amplitude decreasing with distance.

As we have seen in Section 2.2, this description of light is eminently useful but something less than the whole story. In the preceding paragraph we have talked about light intensity in terms of "amplitude" (the height of the waves) but elsewhere we talk about *"photon flux density"* and so far, in this section at least, we have had no mention of photons. We shall come back to this in a moment but let us first make a comparison.

"Just as an oscillating dipole generates an electromagnetic wave, the wave (a pattern of electric forces) can, in turn, set other dipoles into oscillation of the same frequency. The process is roughly like the resonant vibration of one tuning fork in the presence of another, energy being communicated from one fork to the other through the medium of compression waves in the intervening matter. The electromagnetic wave is thus a carrier of energy, as one can appreciate by standing

in the sunshine. Work must be done in maintaining the oscillation of a dipole, a corresponding amount of energy is radiated into the surrounding space, and some of this energy is absorbed by other dipoles when they are set in motion. These other dipoles in turn radiate energy as a result of their oscillation. We can thus visualise the interaction of light with matter as a process in which energy is exchanged between a collection of oscillating dipoles and a radiation field. The oscillation of the dipoles corresponds to the movements of electrons relative to positively charged nuclei in matter."

This analogy, drawn by Rod Clayton, is both useful and easy to understand in terms of a not uncommon human experience. To set a tuning fork "humming" we must strike it. A second tuning fork, set next to it, will then begin to hum. So long as we continue to strike the first, the second will continue to vibrate. In the sun we have dipoles induced to oscillate by nuclear fusion of protons. These oscillating dipoles radiate electromagnetic energy. Such energy (at wavelengths in the red and the blue which chlorophyll molecules can absorb) set dipoles in oscillation within the light-harvesting antennae of a plant's photosystems (Section 3.3). Energy transfer between sun and plant has been accomplished.

2.8 Getting Excited

So where do photons come in? The trouble was that the classical picture, described above, is inadequate. It simply could not account for all of the known observations. In an attempt to resolve these difficulties, Planck proposed (Section 2.2) that energy could only be exchanged in discrete parcels (quanta). This made it possible, amongst other things, to equate emission and absorption spectra (the wavelengths at which substances may emit or absorb light) with definite energy states of atoms. Absorption of a photon changes an atom, primarily by lifting an outermost, or valence, electron from a "ground state" orbital to a higher "excited state" orbital. The energy of the atom is thus increased by the energy ($h\upsilon$) of the absorbed photon.

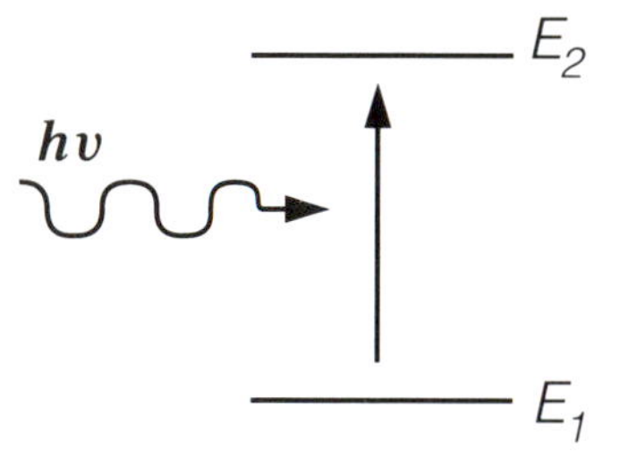

Absorption of light
$h\upsilon = E_2 - E_1$

[In terms of the classic theory the electromagnetic wave induces an oscillatory dipole in the atom, the tuning-fork begins to hum.]

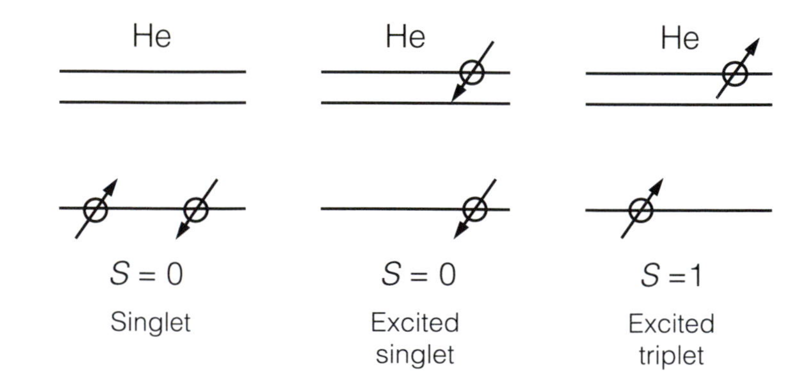

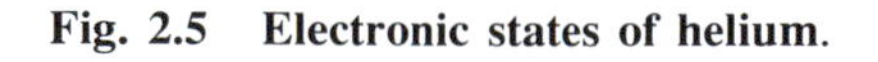

Fig. 2.5 Electronic states of helium.

If the electron drops directly back from the excited state to the ground state a photon is emitted (see Section 2.9). The transition from ground state to excited state (and *vice versa*) takes about 10^{-15} s. Electrons are like tiny spheres which spin on their axes. If a pair of electrons that are normally together, spinning in opposite directions, become separated by excitation they can continue to spin in an anti-parallel fashion or the direction of spin can change. The *"total spin quantum number, s"* can be 0, 1/2 or 1 depending on whether there are 2 electrons spinning in opposite directions (s = 0, or "singlet state") or two electrons spinning one way and a third in the opposite direction (s = 1/2, or "doublet" state) or two electrons spinning in the same direction (s = 1 or "triplet" state). An "excited triplet" is long lived and particularly reactive.

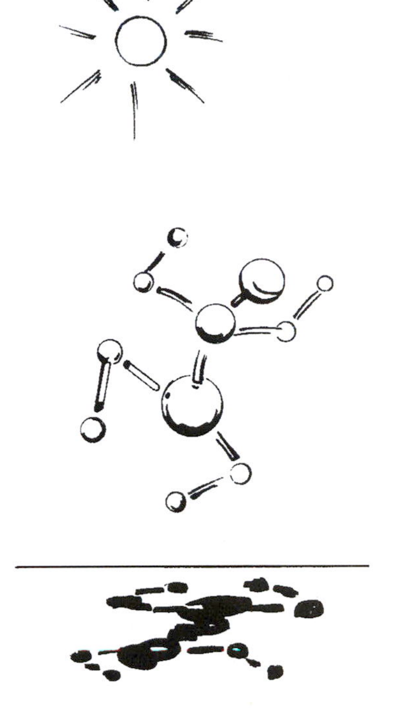

What applies to atoms also applies to molecules but there is the added degree of complexity brought about by the fact that there are now several nuclei involved as well as the vibratory and rotating movements of these nuclei in relation to the mass of the whole system which comprises the molecule. Accordingly, there are, in effect, a number of *"ground states"* and a corresponding number of *"excited states"* . The single sharp line of the atomic-absorption spectrum is replaced by a number of closely adjacent lines, each a consequence of a transition from a ground state to an excited state and each brought about by absorption of a photon with precisely the right energy content to effect such a transition. The height of the line is governed by the probability of the transition and the lines are usually too close to be resolved so that a characteristic absorption *band* is observed rather than the well-defined *line* of the atomic absorption spectrum which depends upon a single transition, consequent upon the absorption of one photon, rather than multiple transitions consequent upon absorption of a number of photons at different wavelengths.

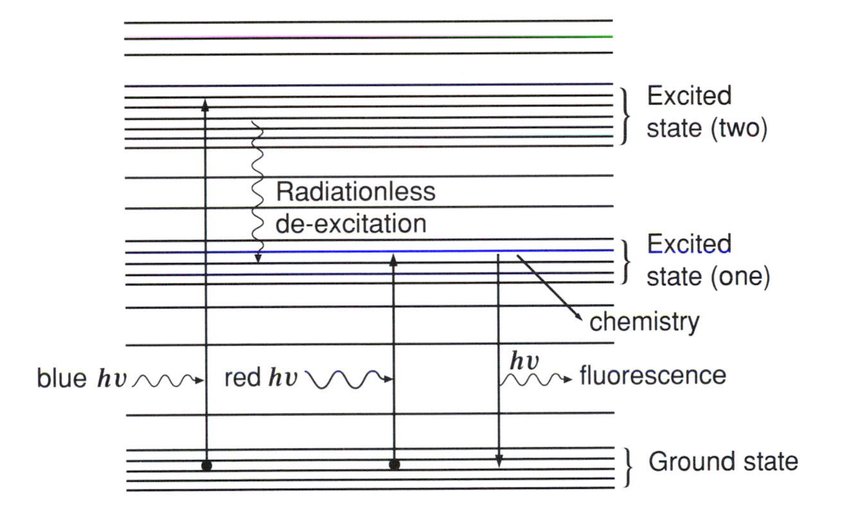

Fig. 2.6 Excitation of Chlorophyll by red and blue light. The parallel lines represent energy sub-state or electronic orbitals. Thus the energy delivered by the absorption of a blue photon (left) is sufficient to raise an electron to excited state 'two' from where it rapidly returns by a process of radiationless de-excitation, 'cascading' through sub-states, to excited state 'one'. A photon of red light (centre) only has enough energy to raise an electron to excited state 'one' but this excited state is sufficiently stable to permit useful chemical work and is, in effect, the starting point of all other events in photosynthesis. Particularly in organic solvents, excited state one can also dissipate energy by re-emitting light as (deep red) fluorescence.

Fluorescence, brought about by the emission of photons as electrons fall back from excited substates in molecules, occurs with a shift towards longer wavelengths i.e., "the Stokes Shift". This is because the excited states have an opportunity to lose energy as heat, in small steps between excited substates prior to the abrupt drop from an excited state to a ground state. Most molecules have two or more excited states and therefore two or more peaks in their absorption spectra. The tails of these peaks overlap, permitting easy thermal cascading (heat loss in dropping from one excited substate to another). This occurs so rapidly (in about 10^{-12} secs) that there is no opportunity for photochemistry or fluorescence.

[The excited states of molecules usually persist for for about 10^{-8} seconds, the triplet state may have a lifetime of about 10^{-4} to 10^{-1} seconds.]

It is for this reason that chlorophyll looks green yet emits red fluorescence when excited. Chlorophyll has two absorption maxima; (Figs. 2.2 and 3.1) one in the blue, at about 420nm, (corresponding to excited state 2) and one in the red, at about 680nm (corresponding to excited state 1). Because it absorbs red and blue light, and reflects and transmits green light, chlorophyll (and leaves) look green when illuminated in white light (which is, of course, a mixture of these dominant colours). If a solution of chlorophyll (e.g. in alcohol or acetone) is viewed from the side, however, (so that the retina of the observer is not flooded with green photons transmitted *through* the solution) a deep red fluorescence is seen. This only emanates from the red excited state because excited state 2 decays to excited state 1 by thermal cascading (dissipation of excitation energy as heat in going from one *sub*state to the next) too quickly (10^{-12} sec) to allow fluorescence emission from the blue excited state. Similarly, the red excited state is the starting point, the driving force, of the chemistry of photosynthesis. Photosynthesis derives no benefit from the extra energy of blue photons (Table 2.1) because of this rapid heat loss.

[It should be noted that chlorophyll fluorescence is emitted at a deeper (longer wavelength) red than the peak absorption in the red. This is because of the Stokes Shift (above) in which some thermal cascading through excited substates results in an energy loss and eventual emission of photons at longer (less energetic) wavelengths as electrons finally fall back to the ground states.]

FLUORESCENCE

2.9 A Rich and Ambiguous signal

Whole books have been written about chlorophyll fluorescence. Emitted from an illuminated chlorophyll solution, fluorescence is spectactular but otherwise not very exciting. Emitted from a leaf, it gives what Lavorel and Etienne once described as *"a rich and ambiguous signal"*. Chlorophyll fluorescence from leaves is a much smaller fraction of the excitation energy than that emitted from solution. So small a fraction indeed (about 3-5%) that it can only be detected by sensitive photodiodes, protected in instruments, by optical filters and other devices which protect against other wavelengths. Chlorophyll *a* fluoresces at about 685nm but, in leaves, the emission may be shifted further into the red, to a secondary maximum at about 740nm, by re-absorption of fluorescence by chlorophyll.

It has long been proposed that there must be an inverse relationship between chlorophyll *a* fluorescence emission from leaves and photosynthetic carbon assimilation. Fluorescence it was argued must be an indicator of "gainful employment". If photosynthesis is driven by the red excited state of chlorophyll the more excitation energy that is used for that purpose the less would be available for dissipation as heat or as light. This concept also gains support from so called "fluorescence induction" (the Kautsky effect). This effect is seen when leaves are

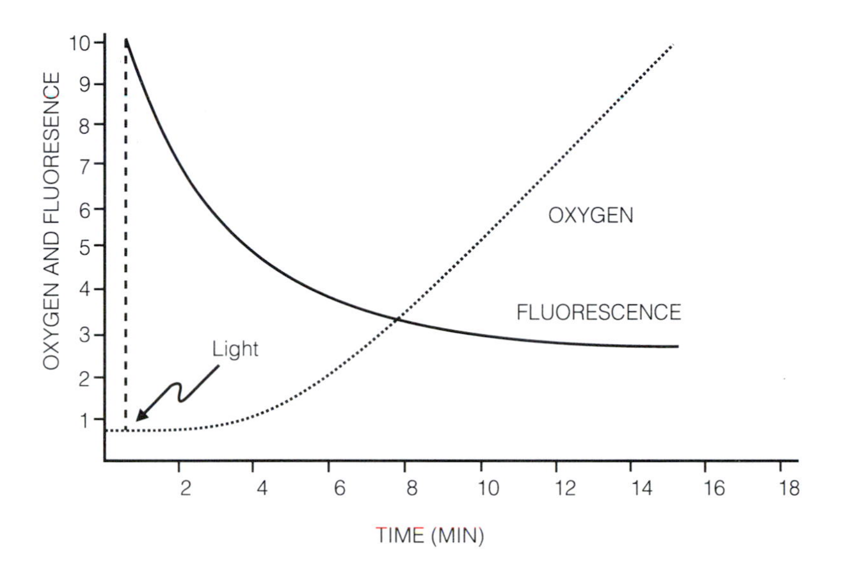

Fig. 2.7 The Kautsky Effect. Photosynthetic carbon assimilation starts slowly when a leaf in darkness is abruptly illuminated. During this induction period fluorescence rises rapidly at first and then falls, sometimes displaying secondary peaks, to a quasi steady-state level.

first illuminated after a period of darkness. Such leaves do not immediately commence to photosynthesise at maximal rates but only after a lag or induction period. Chlorophyll fluorescence on the other hand reaches a maximum within a very

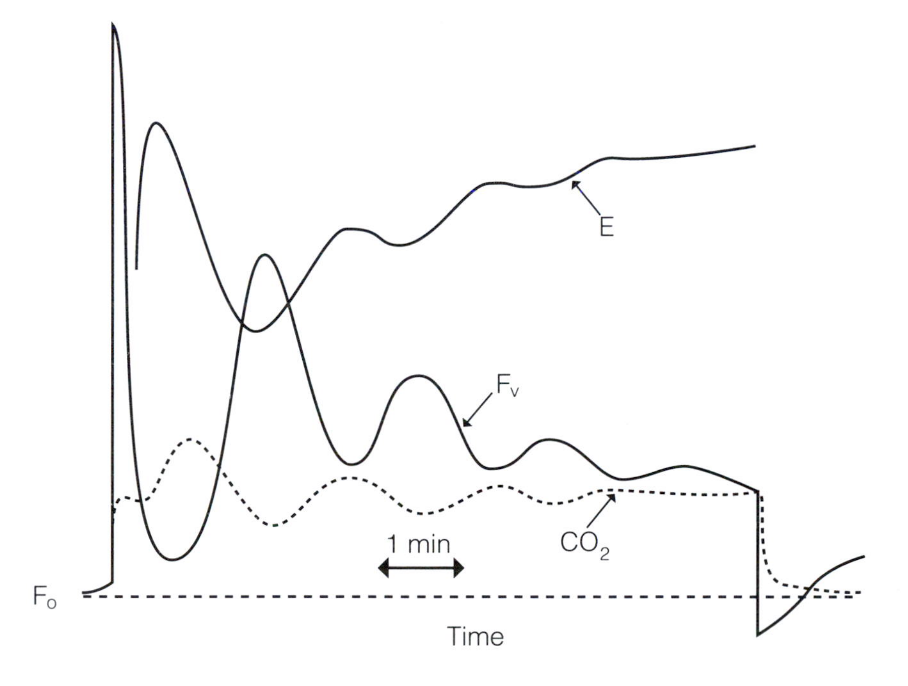

Fig. 2.8 Complex fluorescence kinetics. The broken line is CO_2 uptake and illustrates the broadly inverse relationship between fluorescence and carbon assimilation. "E" is defined by quenching-analysis (Section 2.10) and is the difference between maximal and variable fluorescence.

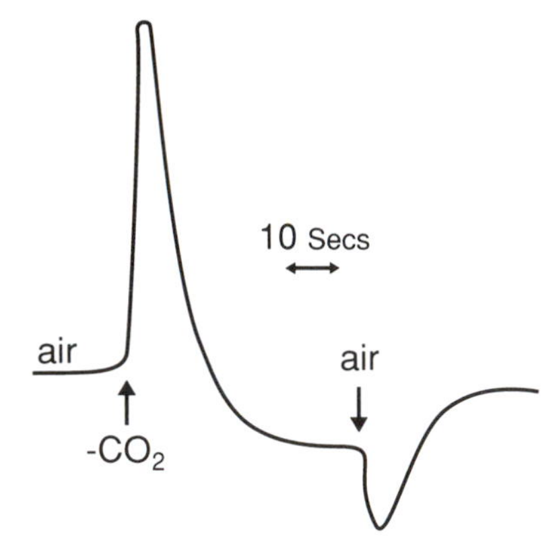

short time after illumination and thereafter falls, albeit sometimes in a complex way, through one or more secondary maxima, to a "steady-state" level much lower than the initial maximum. This is precisely what would be expected if it were supposed that the "dark biochemistry" took some time to get under way but that the photochemical events started immediately. Moreover, it is known that the photochemical events *do* start immediately (for example the Hill reaction, described in Section 4.1, in which chloroplasts reduce an artificial electron acceptor, starts at maximal rates without discernible lag). The problem was that, in the words of Lavorel and Etienne, fluorescence is such a *"rich and ambiguous"* signal. The "richness" of the signal lies both in the complexity of the initial rise and the secondary kinetics, often seen, in the subsequent fall. The "ambiguity" resides in the fact that, at one time, the complexity was such that it seemed to defy explanation and the fact that some changes were exactly the converse of those which might have been predicted. Perhaps the most important of these was the response (the "gas-transient", left) when carbon dioxide was abruptly removed from the atmosphere surrounding the leaf. This was bound to result in a similarly abrupt cessation of carbon assimilation. It would follow that there should be more excitation energy available for dissipation and reasonable to suppose, therefore, that fluorescence emission might increase. That, in fact, was observed but the abrupt rise in fluorescence emission which followed removal of carbon dioxide was soon overtaken by such a large decrease that, within a matter of seconds, the fluorescence signal had fallen below the level seen during unimpeded photosynthesis. Similarly, re-addition of carbon dioxide produced a downward excursion followed by a rise which took the signal back to the original level.

2.10 Quenching Analysis.

"Q-analysis"

Many of the ambiguities have now been explained in terms of two principal "quenching mechanisms". The term "quenching" is associated with work by Duysens and Sweers who proposed, many years ago, that there was an electron acceptor (associated with what is now described as photosystem II - Fig. 3.4) which would accept electrons when in its oxidised state. They designated this acceptor "Q" for "quencher" on the basis that, if excitation energy was used to lift electrons from water to Q this excitation energy would not be available for dissipation as fluorescence and, accordingly, fluorescence would be quenched. If, on the other hand, Q was, for whatever reason, in a fully reduced state, all of the excitation at that moment would be available for dissipation and that fraction dissipated as light would increase the fluorescence signal. Quenching (q_Q) associated with the oxidation reduction status of Q (which, incidentally, turned out to be a quinone) is now referred to as *"photochemical quenching"* (q_P). This is seen as the explanation for the initial rise in the fluorescence signal when CO_2 is suddenly removed from a leaf. Once reduced, Q (or more properly "Q_A", as the reaction centre in photosytem II is now called) normally passes electrons, via other electron carriers, to photosystem I and finally to carbon dioxide via NADP (Sections 4.1 and 4.2). In the absence of CO_2, $NADPH_2$ cannot be re-oxidised via this route (by CO_2), fluorescence quenching by oxidised Q_A *"relaxes"* and fluorescence emission rises. The second major quenching mechanism (formerly q_E for *"energy quenching"*), is now designated q_{NP}, for *"non-photochemical quenching"*). Precisely how this works is still not properly understood but it is associated with the establishment of a proton gradient (ΔpH) across the thylakoid membrane (Chapters 3 and 4). This proton gradient normally

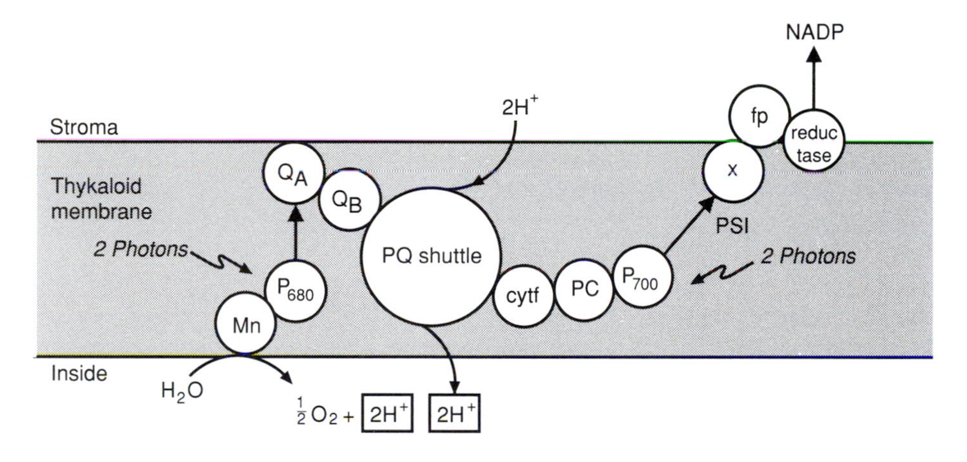

Fig. 2.9 Organisation of electron carriers in the thylakoid membrane. Note the inward movement of protons from the stroma into the lumen of the thylakoid. The proton gradient, discharging through the ATPase (see right margin) drives ATP formation from ADP and Pi. The nature of the carriers is considered in Section 3.9 & 3.10.

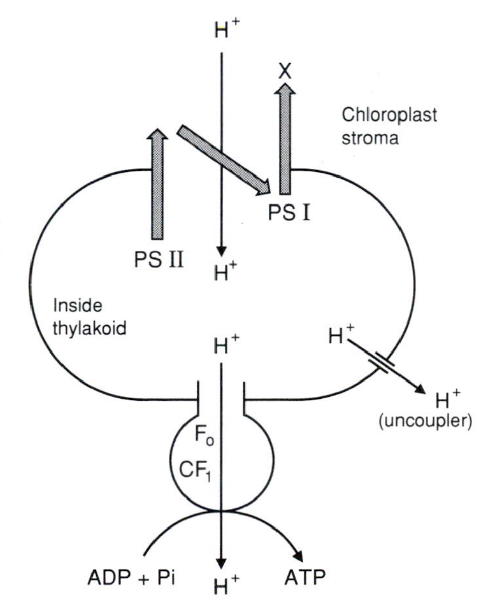

discharges through the ATPase (Section 4.7), generating ATP from ADP and Pi. In the absence of carbon dioxide, ATP is not consumed in photosynthetic carbon assimilation and ADP is therefore no longer available to discharge the proton gradient through the ATPase in this process. In these circumstances the un-discharged proton gradient appears to switch dissipation of excitation energy from fluorescence into thermal channels with the result that fluorescence is quenched. This, in part, is the source of the decline in fluorescence which pushes the fluorescence signal from the initial maximum (seen on the removal of CO_2) to a value even lower than before. When CO_2 is returned to the atmosphere surrounding the leaf the converse sequence of events occurs. Q_A^- can become re-oxidised in the presence of CO_2 so that its fluorescence quenching is re-imposed. ATP is consumed in carbon assimilation making ADP available to discharge the proton gradient. As the proton gradient falls, some dissipation of excitation switches back, from thermal channels, into fluorescence.

Major advances in the resolution of the fluorescence signal into its separate components and consequent understanding of the underlying trends has followed the introduction of "light-doubling" and the development of sophisticated apparatus by Schreiber and his colleagues. "Light-doubling" was introduced by Baker and developed by Horton and others. Saturating flashes of light were used in this

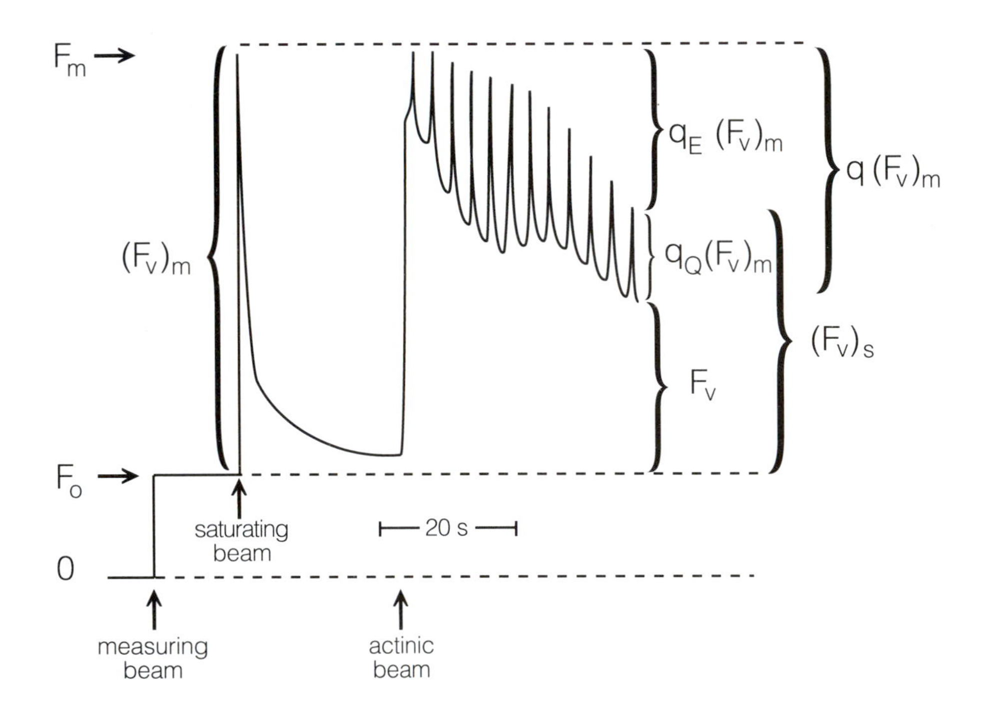

Fig. 2.10 Q-analysis Definition of quenching coefficients etc., according to Schreiber. F_O is the fluorescence displayed by a dark-adapted leaf in very weak modulated light. $(F_V)m$ is the maximal variable fluorescence first seen when a pulse of actinic saturating light is applied. F_V is the variable fluorescence seen, once continuous actinic light is applied and $(F_V)s$ are the peaks to which this fluorescence is then driven by pulses of saturating light. The derivation of q_Q and q_E, now called photochemical quenching (q_P) and non-photochemical quenching (q_{NP} or q_N), is explained in the text.

technique in order to drive Q_A fully reduced, thereby allowing the evaluation of maximum fluorescence in the absence of q_P quenching. A major contribution by Schreiber and his colleagues was to produce an instrument which combined optics with very sophisticated electronics. This allows fluorescence, initiated by a very weak measuring beam (the so-called F_o value), to remain undisturbed by the saturating flash, thereby allowing accurate measurement of the differences between F_o and F_{max}. The entire process, i.e. the light-doubling and the subsequent evaluation of the major quenching components by measurement and arithmetic has sometimes been referred to as *"quenching analysis"*.

In recent work, quenching analysis has been used to measure photosynthetic carbon assimilation. i.e. the rate of CO_2 fixation can be accurately determined by evaluating the light emitted from leaves as chlorophyll fluorescence. At Sheffield, for example, George Seaton has shown that the rates of CO_2 fixation in a number of disparate species can be determined solely by fluorescence, using parameters derived from a single reference species such as spinach.

Chapter 3

Harvesting The Sun

Putting out antennae to harvest light

Getting to the heart of it

Pictures in starch

Going downhill

3.1 The Photoreceptors

"The changing of bodies into light and light into bodies is very comfortable to the source of Nature, which seems delighted with the transmutation" ..... Isaac Newton.

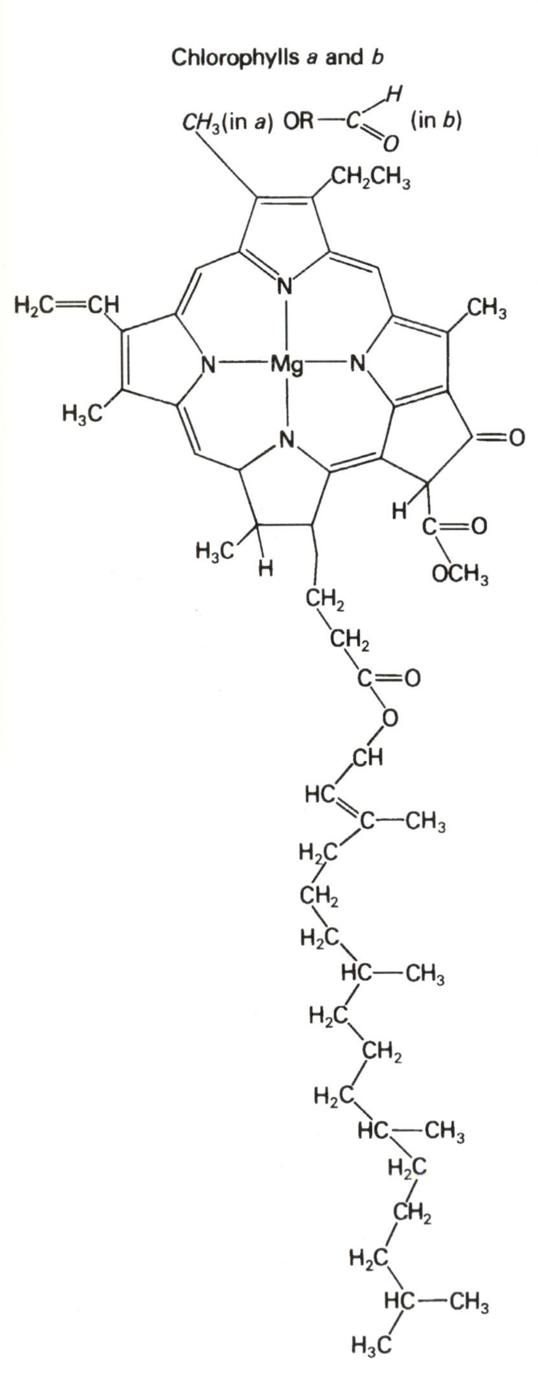

So far we have thought about a few fundamentals (Chapter 1) and examined the nature of light (Chapter 2). We have even taken a passing glance at the idea of light excitation. What we shall attempt to do now is to look at the way in which green plants harvest light energy. In following this, step by step, in the same sequence that it actually occurs, we should not lose sight of what we have already learned. What we are considering is essentially a process of energy transduction; the conversion of light energy to electrical energy in the excitation of chlorophyll. The transfer of electrons, from the photoreceptors, through the two photosystems which constitute the Z-scheme and the associated generation of assimilatory power (Chapter 4). Finally, this transient supply of energy (in the form of the high energy compounds ATP and $NADPH_2$) is consumed in the "dark biochemistry" of carbon assimilation (Chapter 5) in which atmospheric CO_2 is reduced (Section 1.3, 4.9) to the level of CH_2O and long-lived forms of chemical energy are therefore produced.

The conversion of light energy to chemical energy starts with "photoreceptors". In green plants the principal photoreceptors are the chlorophylls, *a* and *b*. These differ somewhat in chemical structure and physical properties but they have much in common, both with each other and with that other all important molecule (heme) which gives blood its colour. These are all substituted tetra pyrroles. In the chlorophylls, the nitrogen atoms of the four pyrrole units are linked ("co-ordinated") to a central magnesium atom and, in the hemes, to a central iron atom. In other respects the chlorophylls are very different from heme compounds. Most particularly they have an ester of a long-chain alcohol (phytol) as one of the side chains attached to the porphyrin ring. Phytol contains four isoprene units (isoprene is a constituent of rubber) and it is phytol which gives chlorophyll its distaste for water (makes it hydrophobic) and, conversely, its liking for association with fats (lipophilic). Accordingly it can fit happily into photosynthetic membranes or lamellae, which are complex structures based on lipids and proteins (approximately 50% protein, 40% lipid and 10% chlorophyll). For these reasons chlorophylls are largely insoluble in water but if a leaf is dipped briefly into boiling water to denature the proteins and then extracted with hot 80% alcohol (80 parts of alcohol, 20 parts of water) a bright green solution is readily obtained which shows a deep red fluorescence when illuminated (and viewed from the side). If we emulate Tswett, who was the first to do this, and run this solution through an icing sugar column we can separate chlorophyll *a* from *b*, which is slightly less blue (more yellow/green) in appearance. Using more modern techniques, such as thin-layer chromatography, we can get very good separation very easily and, following elution, look at the absorption spectrum of each pigment separately, using a spectrophotometer. Fig. 3.1 shows the results of such an experiment. There are major bands of absorption in the red and the blue parts of the spectrum which is why the chlorophylls look green in white light. It will be seen (Fig.3.1) that the absorption spectra of the two chlorophylls complement each other in the sense that they extend the range of light wavelengths which can be absorbed. We should also note that the absorption spectra are shifted a little, in aqueous ethanol, from the peaks that are displayed by an intact leaf. This is attributable in part to the fact that, in the leaf, chlorophylls are bound to proteins (Section 3.3).

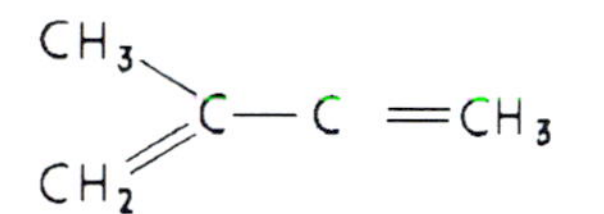

Isoprene

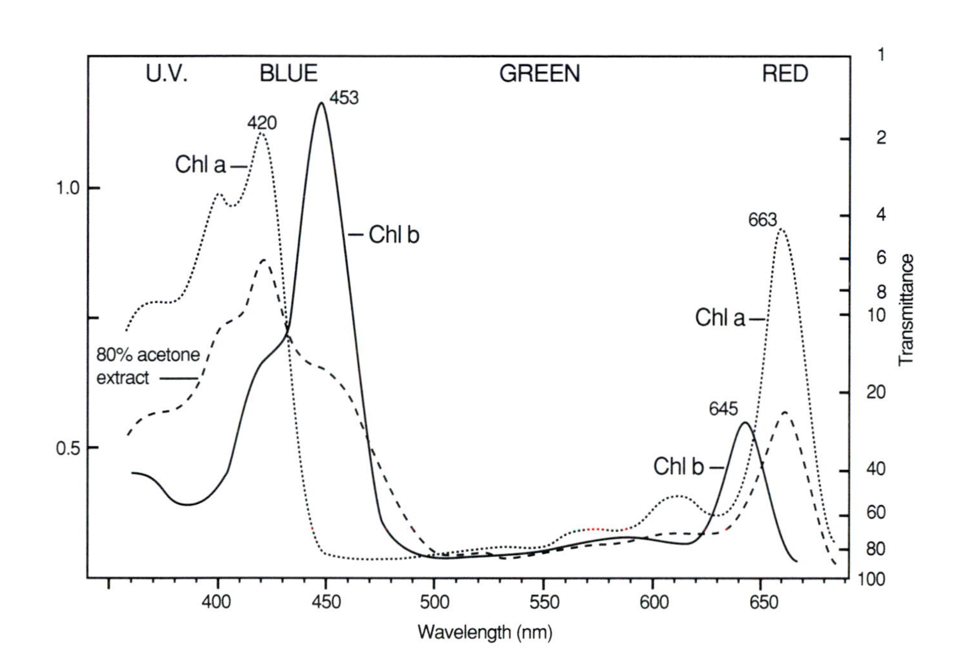

Fig. 3.1 Absorption spectra of chlorophylls in acetone. The absorption in the blue (left) and the red (right) leaves the "window" in the green which gives chlorophyll its characteristic colour when viewed in white light. Note that the peaks are shifted towards longer wavelengths when the pigments are associated with proteins in a leaf.

Chlorophylls are the first step in energy transduction. Excited by blue or red wavelengths of light, they initiate electron transport. The conversion of light energy to electrical energy is analogous to the reaction in man-made photoreceptors such as amorphous silica cells which generate electric currents when illuminated (c.f. Fig. 1.2). Each photon of red light which is absorbed by a chlorophyll molecule raises an electron from a ground state to an excited state and all of the available energy is transferred in this process. This excitation brings about an oxidation. Electron transport (Section 1.2) is initiated, as the electron is lifted into a higher orbit and a positively charged "hole" is created. Absorption of blue light causes even greater excitation (because of the higher energy content of the blue quantum) but the elevated electron then falls back into the "red orbit" too quickly to permit useful chemical work. Thus, whatever the quality of the light absorbed, the electron reaches the same energy level more or less immediately after excitation and all subsequent events derive from this common starting point (excited state I).

3.2 Photosynthetic Units

In a classic experiment which was to have major subsequent impacts, Emerson and Arnold subjected the alga *Chlorella* to brief flashes of high intensity light with a view to achieving saturation of the photochemical events while avoiding the limitations which would otherwise have been imposed by the "dark biochemistry" if it had been asked to contend with this situation on a continuous

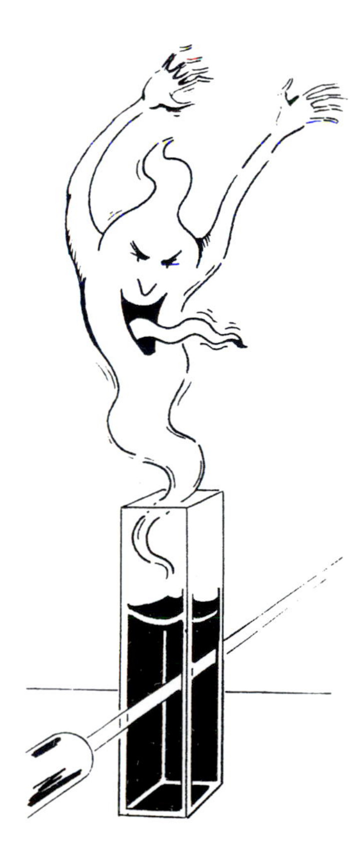

Absorption Spectre

rather than an intermittent basis. They also contrived to vary the chlorophyll content of the *Chlorella* by growing it under different light intensities. To their complete surprise, they found that saturation was effected when the number of photons supplied in a flash were sufficient to excite only about one out of every 2,500 molecules of chlorophyll in the sample being examined. They confessed that they were really unable to account for this result (which obviously implied that, at least in this situation, most of the chlorophyll was superfluous). It fell to Gaffron and Wohl to propose the concept of the "photosynthetic unit" as a way of explaining both Emerson and Arnold's findings and their own work (on varieties of plants, relatively deficient in chlorophyll, which were still able to achieve good rates of photosynthesis but only at high light intensities). The idea was that there might be photosynthetic units (assemblies of chlorophylls) in which most of the molecules were not *directly* concerned in the evolution of oxygen but in capturing and transferring energy to those few which were. Bearing in mind the quantum requirement of 8, there could, for example, be 8 such units each containing a reaction centre associated with about 300 molecules of "antenna" chlorophyll, concerned solely with light-harvesting.

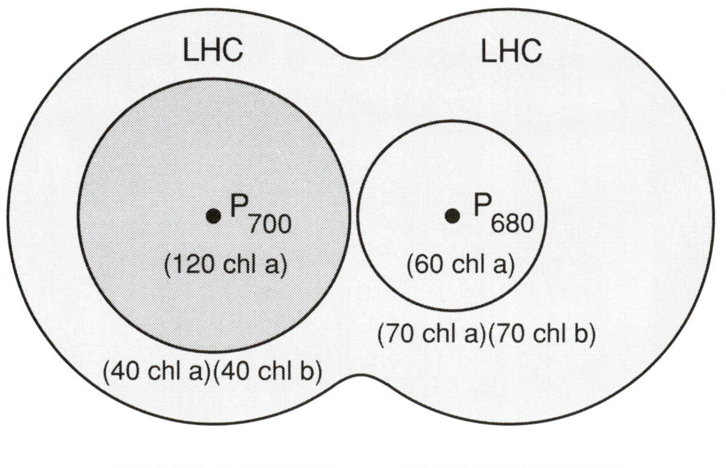

PHOTOSYSTEM I PHOTOSYSTEM II

Fig. 3.2 A photosynthetic unit. The "cores" of PSI (left) and PSII (right) are represented by the inner circles; light-harvesting complex, LHC. The areas correspond to relative proportions of each chlorophyll-protein complex (after Jan Anderson).

Some feeling for the underlying philosophy can be obtained by returning to the relationship between light intensity and rate of photosynthesis (Section 6.7). The first part of this relationship is linear. Above about 75 - 100 μmoles photons.$m^{-2}.s^{-1}$ it becomes curvilinear. At these low light intensities many plants seem to have a quantum yield of *about* 0.11 if they are not stressed. Indeed, the Z-scheme (Section 4.3), would dictate a minimum quantum yield of 0.125 (a quantum requirement, the reciprocal of quantum yield, of 8). Since we derive this from the rate of arrival of photons, in the linear part of the relationship, it follows that if, e.g., at a light intensity of 100,

$$\frac{x}{100} = 0.125 \quad (\text{or } \frac{100}{x} = 8)$$

then x must be 12.5 (μmoles of $O_2.m^{-2}.s^{-1}$). For a 100 sq.cm. leaf this would give us 0.125 (12.5 × 100/10,000) μmoles of O_2 per second. If 8 photons are needed to co-operate in order to produce 1 μmole of O_2, 0.125 μmoles of O_2 per second requires 1 μmole (0.125 × 8) photons per second. This fits our arithmetic. Our circular argument is sound. The leaf is receiving photons at the rate needed to evolve O_2 and, if there are about 5 times as many chlorophyll molecules (approximately 5 μmoles in a 100 sq.cm. leaf) as photons intercepted (5 μmoles of chlorophyll, 1 μmole of photons), then 4 out of 5 chlorophylls would not be involved. If we diminished the light by a factor of 10, the efficiency (the quantum yield) would be no less (1.25/10 instead of 12.5/100) but 49 out of 50 molecules of chlorophyll would not be excited at any one moment. Accordingly, it becomes more and more difficult to imagine how, in the absence of any organisation, 1 out of every 50 molecules of chlorophyll, randomly scattered throughout the leaf, could interact with precisely the same efficiency as if they were adjacent. If, however, these chlorophyll molecules were part of an "antenna" which "harvested" incoming photons; which funnelled excitation energy to a "centre" (in which the combined energy of the photons brought about the evolution of oxygen), the problem would not arise.

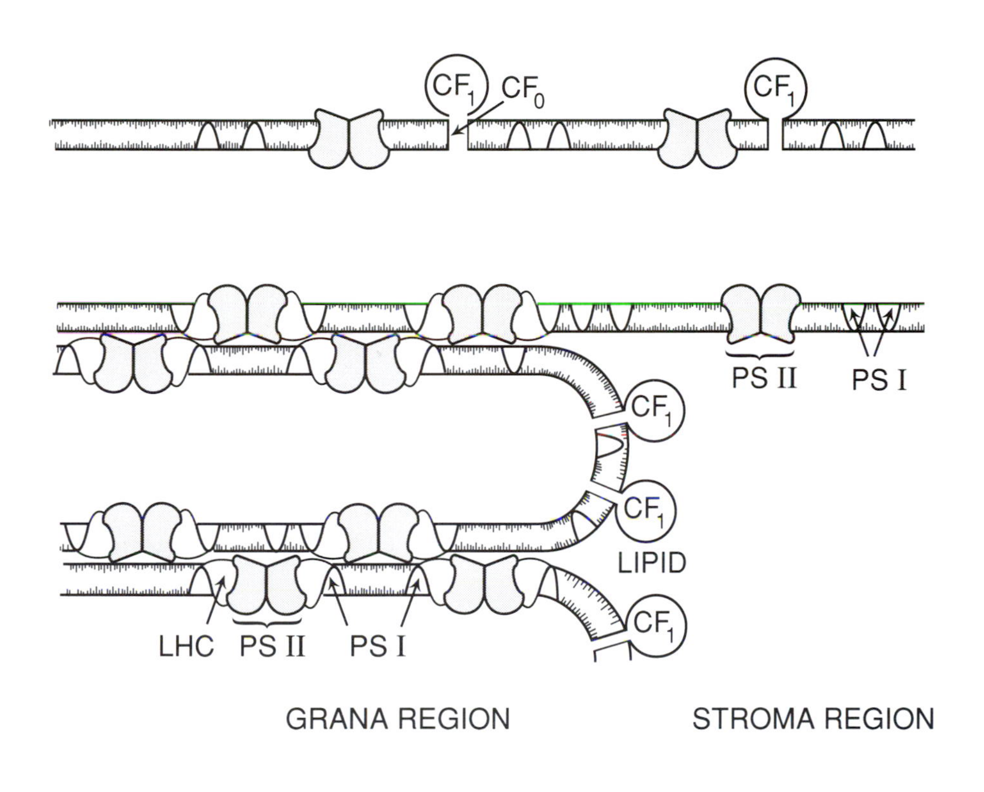

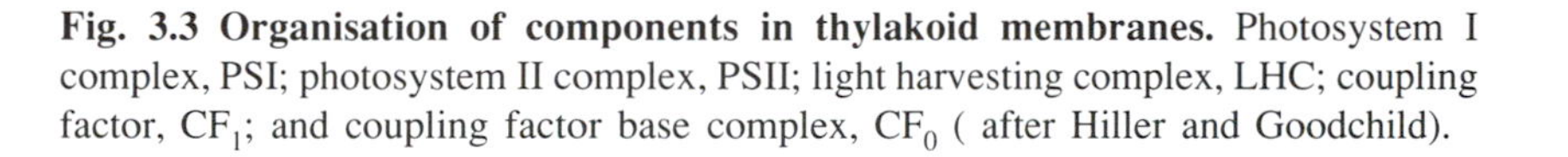

Fig. 3.3 Organisation of components in thylakoid membranes. Photosystem I complex, PSI; photosystem II complex, PSII; light harvesting complex, LHC; coupling factor, CF_1; and coupling factor base complex, CF_0 (after Hiller and Goodchild).

It must be borne in mind that this earlier work pre-dated the Z-scheme (Section 4.3) and work on chlorophyll-protein complexes (Section 3.3) and that much still remains to be done in assigning function to structures known to be present but not yet properly understood. For this reason the "photosynthetic unit" has meant, and will continue to mean, rather different things to different people. For the present we will take the simplest view that the reaction centres (Sections 3.7 and 3.8) of photosystems I and II are located in *one* such unit (Fig 3.2) comprising about 400 molecules of chlorophyll, most of which are incorporated into the antennae.

3.3 Light-harvesting Antennae

The light-harvesting chlorophyll-protein complexes (LHC) occur in all higher plants and green algae. These contain chlorophyll *b* (and often carotenoids) as light-harvesting accessory pigments and it is probable that all of the chlorophyll *b* is located, with approximately equal amounts of chlorophyll *a*, in the antennae. As the name of these complexes implies, their function is to act as antennae which harvest incoming photons and funnel them to the respective reaction centres of PSII and PSI where the photochemistry of oxygen evolution, water splitting, and electron transport to NADP actually occurs. This is why (Section 3.2) light saturation, with respect to chlorophyll excitation, occurs when less than 1% of the total chlorophyll is excited. Within the antennae (and the other components of the photosynthetic unit concerned with light-harvesting) excitation energy is transferred by "hopping" from one molecule of chlorophyll to another. This is sometimes described as transfer of an "exciton". The transfer of an exciton has much in common with the transfer of an electron. Both can be regarded as transfer of a particle and the process can be represented by:-

$$L^*R \rightarrow LR^* \qquad \text{Eqn. 3.1}$$

where "L" and "R" stand for "left" and "right", respectively, in order to avoid confusion between transfer from primary donors to primary acceptors (Section 1.2) and the asterisk indicates the particle which is transferred. In LHC, this particle is an exciton. Once the exciton has reached the reaction centre of PSII (Section 3.7) "L" represents the primary donor (in its first excited single state), "R" is the primary acceptor and the particle which is transferred is an *electron* starting on its journey through PSII, via PSI, to NADP and carbon dioxide.

3.4 Downhill All the Way?

Within the antenna, excitons (Section 3.3) migrate "downhill" from carotenoids

"Wir sind übern Berg" with acknowledgements to Weber

and chlorophyll *b* at the periphery, via chlorophyll *a*, to the reaction centre which constitutes a trap from which there is no escape. Chlorophyll *b* absorbs photons of higher energy (shorter wavelength) than chlorophyll *a* which, in turn, has absorption maxima at wavelengths lower than that of the reaction centre itself (Section 3.7). There are, however, a number of forms of chlorophyll *a* (which cannot, at present, be separated from the leaf but the presence of which can be inferred from computer analyses of absorption spectra) and if these were structurally arranged in such a way (Fig. 4.2) that the longer wavelength (lower energy) forms were arrayed about the reaction centre it would decrease the number of hops which an exciton would make by a factor of ten, compared with a purely "random walk" (or random hop). Whether or not such structural organisation exists, exciton transfer is very rapid and very efficient (the overall state of excitation of chlorophyll molecules within a photosynthetic unit lasts less than 1 nanosecond and the observed quantum efficiency at this primary level is close to 100%).

3.5 The Heart of the Matter

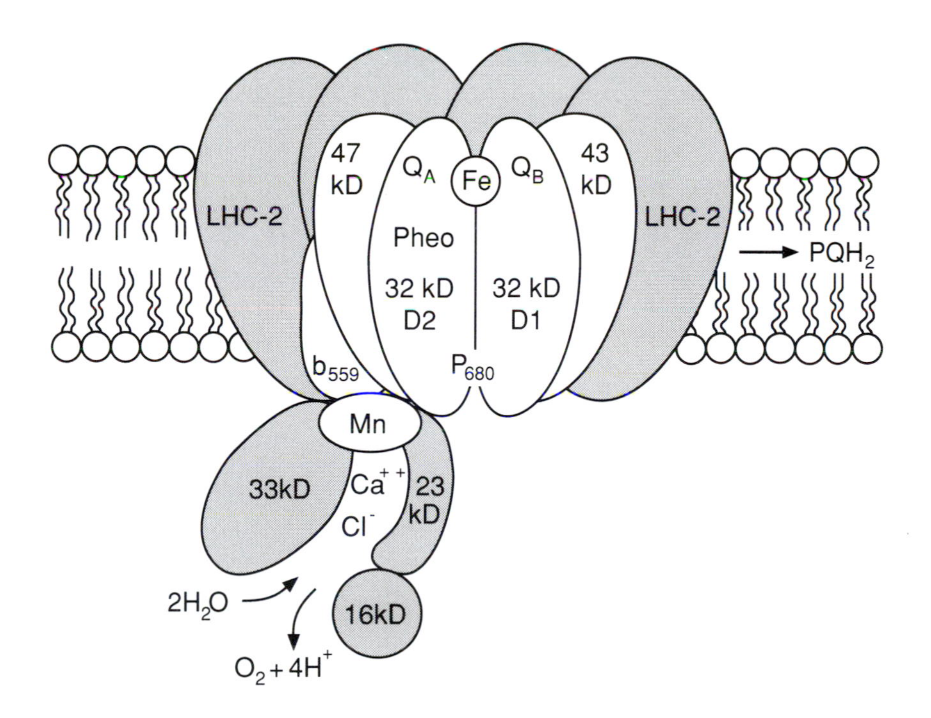

Fig. 3.4 Photosystem II. Diagrammatic representation of PSII and its associated components as a light-harvesting complex in the thylakoid membrane (Courtesy of Jim Barber). The water-splitting side gives rise to powerful oxidising species which are harboured in the D1 polypeptide which binds Q_B (Photoinhibition [Section 5.9] involves damage to this Q_B binding protein).

The light-harvesting complex or antenna (Section 3.3) funnels excitation energy into the reaction centres (Sections 3.7 and 3.8). No one knows, for sure, if there are separate photosynthetic units (Sections 3.7 and 3.8) for PSI or PSII with separate antennae or if the structure is closer to that postulated by Anderson in Fig. 3.2. This shows one unit with common antennae complexes, although there is more chlorophyll *b* close to the reaction centre of PSII than close to the reaction centre of PSI. In any case, there is general agreement that there are two reaction centres, one for each photosystem; that these have absorption maxima at 680 nm (PSII) and 700 nm (PSI); that they constitute traps for the excitons which migrate through the light-harvesting antennae (Section 3.3) and that they are the sites at which light-driven electron transport commences, in PSII, and is boosted (by PSI) as electrons travel onwards to NADP.

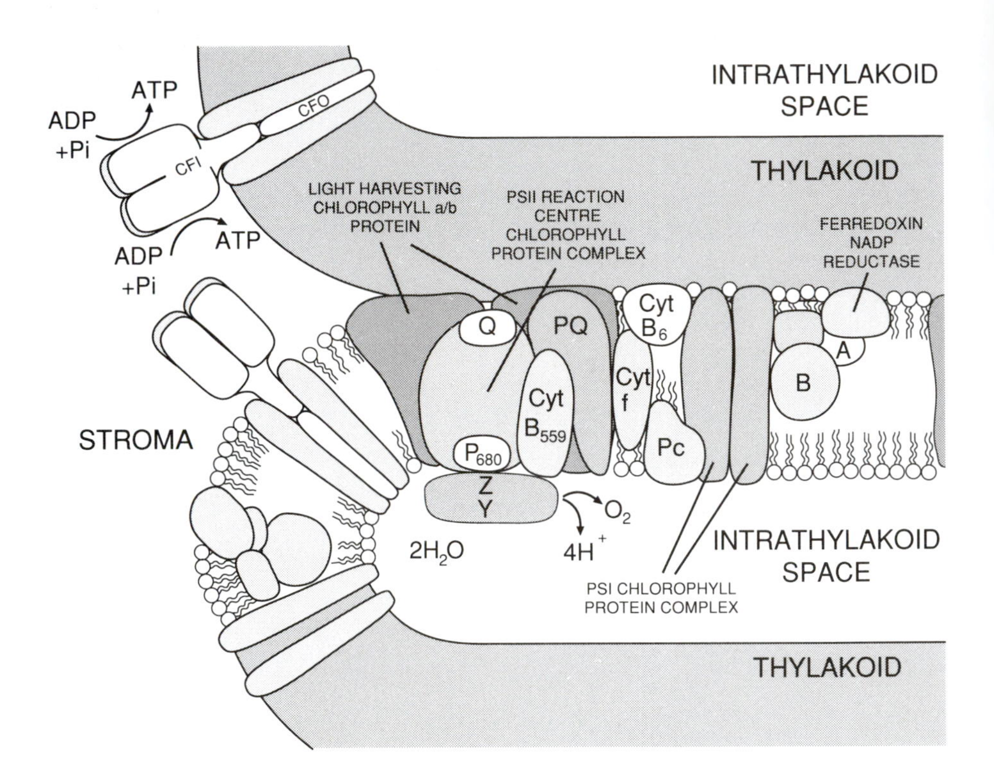

Figure 3.5. The thylakoid membrane. The light-harvesting chlorophyll protein complex funnels excitation energy to P_{680}, the reaction centre of PSII. From there, electrons are "raised", via P_{680}, to Q_A and then "downhill" through plastoquinone and cytochrome *f* to P_{700}, the reaction centre of PSI. Light excitation of PSI raises electrons a second time to an iron-sulphur protein, the primary electron acceptor of PSI and then again "downhill" through ferredoxin to NADP. Electron transport creates a proton (H^+) gradient across the thylakoid membrane which discharges through the ATPsynthase (CF_1, CF_0) generating ATP from ADP and inorganic phosphate (Pi).

3.6 Chlorophyll Protein Complexes

Chlorophyll does not exist within chloroplasts as a separate chemical entity but in association with proteins. Such chlorophyll protein complexes are more varied than was once thought and differences in function related to differences in structure (e.g. differences in the ratios of chlorophyll *a* to chlorophyll *b* etc., and the amino acid content of the respective proteins) are still largely unresolved. It is clear, however, that much of the chlorophyll is involved in "light-harvesting" and that "Light-Harvesting (Chlorophyll-Protein) Complexes" constitute antennae which funnel excitation energy to the reaction centres (Figures 3.4 & 3.5). The reaction centres themselves (see Section 3.7) are embedded in core chlorophyll protein complexes and contain β-carotene and electron carriers (such as cyt b_{559} which is closely associated with the reaction centre of PSII). Most of the thylakoid proteins are organised into four intrinsic protein complexes (PSII, PSI, cyt b/*f* and the ATP synthase) and these together with lipids account for the bulk of the thylakoid membrane. The thylakoid however, is not in any sense a homogeneous structure. There is both marked asymmetry (e.g. electron donors to PSII and PSI at the inner surface and acceptor sites at the outer surface) and "lateral heterogeneity" (i.e. the way in which the complexes are distributed along the length of the membrane. Both of the aspects of structure have profound implications for function.

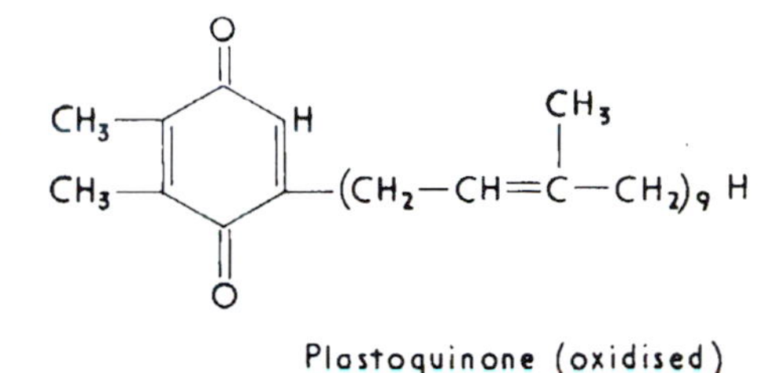

Plastoquinone (oxidised)

3.7 PSII Reaction Centre

The reaction centre in PSII accepts electrons from the water-splitting oxygen-evolving complex and uses light energy to raise them to a higher energy level from which they can run downhill though a series of electron acceptors to PSI which, with the help of longer wavelength photons, kicks them further upwards *en route* to ferredoxin and NADP (Section 4.2).

The primary electron donor in the PSII reaction centre is pigment 680 (P_{680}), a special form of chlorophyll *a* with an absorption maximum, as the name implies, at 680 nm. Upon excitation of P_{680}, an electron is passed uphill, via an intermediary acceptor, phaeophytin, to the primary acceptor Q_A (see e.g. Fig. 2.9). Q_A is a covalently bound form of plastoquinone. Electrons pass from Q_A to Q_B, a diffusible form of plastoquinone, and then out of PSII, down a thermochemical (energy) gradient to PSI and beyond.

3.8 PSI Reaction Centre

My own lasting and endearing recollection of an irreverent, ertswhile Dutch scientist, Bessel Kok, was of a head, wearing glasses and a serene expression, swimming gently into the distance in a New England lake. Back in the laboratory he had made many a notable observation including, in 1956, the fact that there was a small light-induced absorption change at about 700 nm which he ascribed to "pigment 700". The change in P_{700} was subsequently shown to result from photo-oxidation of specialised chlorophyll *a* by far-red light. This chlorophyll was designated the primary electron donor in the reaction centre of PSI which nestles within a chlorophyll protein complex (Section 3.3). When illuminated it transfers an electron to the primary acceptor in PSI which is believed to be a membrane-bound form of the iron-sulphur protein, ferredoxin (an electron carrier in the chain of carriers between PSI and NADP). When P_{700} is oxidised to P_{700}^+ by a photon entering PSI, the "hole" so created is filled by an electron which has been passed, from PSII via plastoquinone, plastocyanin and cytochrome *f* (Section 3.9). P_{700} is involved in both linear (non-cyclic) and cyclic electron transport.

3.9 Electron Carriers

In the preceding sections in this chapter, we have looked at light harvesting, at the capture of photons and the initiation of electron transport. We shall return to this topic again in Chapter 4 but we need, for the moment, to anticipate that discussion and remind ourselves, very briefly, that two photosystems are involved in the transfer of electrons from water to NADP and that associated with these photosystems there are a number of electron carriers (Section 1.2). Precisely where structure ends and concept begins need not concern us too much for the moment but it is convenient to think of the two photosystems joined by a bridge of electron

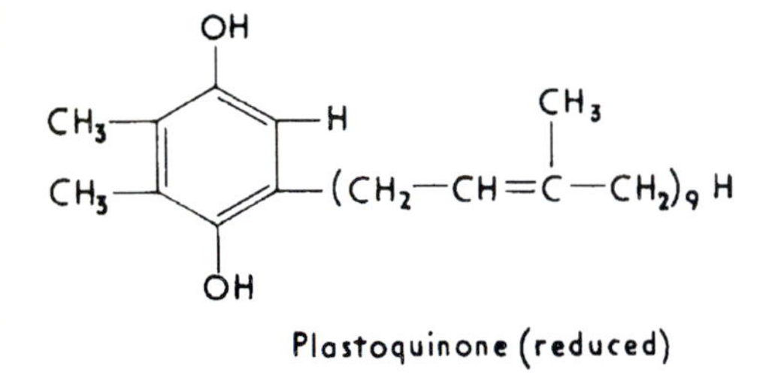

Plastoquinone (reduced)

carriers (Fig. 3.12). These include plastoquinone, plastocyanin (a blue copper protein) and a cytochrome (cyt. *f*) discovered by Robert Hill which only occurs in leaves (hence *f* for *frons*). As we shall see in section 4.7 the nature of these electron carriers is crucial. They have to have the right oxidation/reduction potential (willingness to become oxidised or reduced) but, in addition, some will only accept electrons whereas others will accept hydrogens (section 1.2). This results in the formation of a proton (H^+) gradient across the thylakoid membrane which brings about the conservation of energy, in the form of ATP, as these protons discharge through an ATPsynthase (Fig. 2.9).

3.10 Beyond PSI

"Tom, Tom the piper's son
Stole a pig and away he ran
But the only tune that he could play
Was over the hills and far away"

To some, "Beyond PSI" is not over the hill but almost beyond the pale. In a sense, they are right. It is possible, as Arnon and Whatley showed in their classic experiment (Section 4.8) to divide photosynthesis into a light part and a dark part. It is possible to get isolated chloroplasts to generate assimilatory power (ATP, $NADPH_2$), in the light and then encourage them to use this to "fix" a little CO_2 in the dark. In actuality, *in vivo*, the early photochemical events are closely linked to reactions which can be regarded as part of the "dark biochemistry". Some of these do not even occur in the chloroplast. Sucrose synthesis, for example, occurs in the cytosol but it utilises triose phosphate which is synthesised in the chloroplast and returns inorganic phosphate (via the phosphate translocator - Section 4.11) to the chloroplast in order that the synthesis of triose phosphate from CO_2 and inorganic phosphate might continue. If the inorganic phosphate content of the cytosol is manipulated experimentally, say by feeding phosphate through the cut petiole of a leaf, early photochemical events, such as fluorescence emission from chlorophyll *a* in PSII (Section 3.7) is affected within minutes. Similarly, some of the enzymes involved in the "dark biochemistry" are only "switched on" in the light. All of this reminds us that, while we can divide photosynthesis into photochemistry and "dark biochemistry", *the latter actually occurs in the light* and any separation of these two aspects is simply a matter of convenience which has little basis in reality. Having said that, if we seek to make such a division at all, PSI is where we would do it. All of the photosynthetic pigments and their associated proteins, the two photosystems, the electron carriers and the ATPsynthase (Fig. 2.9 and section 4.7), are located in the thylakoid membranes. Beyond PSI is the "dark biochemistry" of carbon assimilation. Of course, there is the problem of where the thylakoid ends and where the stroma begins but this is perhaps not of vast importance in the present context and we shall note it, as best we can, as we continue.

The job of the constituents of the thylakoid membranes is to absorb photons and to use this excitation energy to generate ATP and $NADPH_2$). So far, in this chapter, we have taken a brief look at the components involved in the transfer of electrons from PSI to the primary acceptor in PSI. Where do they go from there? The answer, of course, is to NADP but there are some bits and pieces in between. The primary acceptor in PSI passes electrons to iron-containing sulphur proteins which in turn transfer them to ferredoxin, a soluble iron-containing protein but one which seems to be "associated" with the thylakoids. A flavoprotein, also tightly associated with the stromal surface of the thylakoids, called "ferredoxin-NADP reductase", then passes these electrons to NADP. Also in some circumstances, electrons may be passed, from about this point, to the plastoquinone pool (thereby giving rise to cyclic electron transport - section 4.5) or to molecular oxygen (the Mehler reaction or pseudocyclic electron transport). If, for the sake of convenience, we lump most of these together with PSI it is clear that by the time that NADP has accepted electrons the photochemistry involving the thylakoid is over and the "dark biochemistry" (Chapter 5) of the stroma is about to begin.

3.11 Where it All Happens

So far we have considered the structure of "the photochemical apparatus", as it is sometimes called, but not where it is located. We know now that all these processes of energy transduction occur within the thylakoid membranes but where are they within the leaf? It has been known, since the last century and the work of Sachs in Würzburg, that starch in leaves is normally formed only in illuminated chloroplasts and, since starch was then regarded as the end-product of photosynthesis, chloroplasts were identified as the sites of photosynthesis. Hill's classic experiments (Section 4.1) raised doubts about this because his isolated chloroplasts evolved oxygen but did not "fix" carbon dioxide. Eventually, we learned how to be nice to isolated chloroplasts. So nice, in fact, that they would fix carbon dioxide at rates which were almost as good as the intact leaf. But they would not make sucrose, regarded (since the work of Benson and Calvin) as the major end-product of photosynthesis. Now all is resolved although there are endless details still to be added. Photosynthetic carbon assimilation occurs in the stroma of the chloroplast (Fig. 4.8) where it leads to the formation of triose phosphates which can be converted in the stroma of many (but not all) plants to starch. Alternatively it can be transported to the cytosol, where sucrose is synthesised. The photochemical apparatus (chlorophylls, reaction centres, electron carriers, etc) are located in the thylakoid membranes within the chloroplast. These are embedded in the stroma, a protein gel in which the carboxylating enzyme, RUBISCO (section 5.2) is a major component. The stroma is contained within a double envelope (section 3.13). The chloroplasts themselves (Fig. 3.8) are located in several layers of cells between the upper and lower epidermis of the leaf (which is perforated with stomatal apertures which permit exchange of O_2 and CO_2 between these tissues and the external atmosphere). In dicotyledenous leaves (i.e., in species like herbs and deciduous trees which produce two cotyledons or "seed leaves" when their seeds germinate), there is often a single or double palisade beneath the upper epidermis. These are cells which look like a fence (hence the name) when seen in cross-section and they comprise an effective layer of chloroplasts, maximising light interception. Deeper within the leaf is the spongy mesophyll which still contains chloroplasts but which has an open cellular structure to facilitate gaseous exchange. C4 species (Fig. 5.11) have a different ("Kranz" type) anatomy.

One thing worth remembering about chloroplasts is that there are an awful lot of them and they are very small. They are *about* the same size and shape as cyanobacteria (Section 8.3) and red blood corpuscles (but then it may not help to know that something is the size of a plover's egg unless you have one in your hand). Rabinowitch tells us that chloroplasts are flattened ellipsoides about 5 microns (5μ = 5 μm = 5000 nm) long but that might not help a lot either. Nor might it help, too much, to know that 5 μm (5000 nm) is about ten times the wavelength of green light or about as long as 50,000 molecules of water laid end to end. Perhaps, we can begin to get an idea of size and number if we remember that, in a leaf as big as your hand (which tells you immediately that this is a very approximate figure), there are about

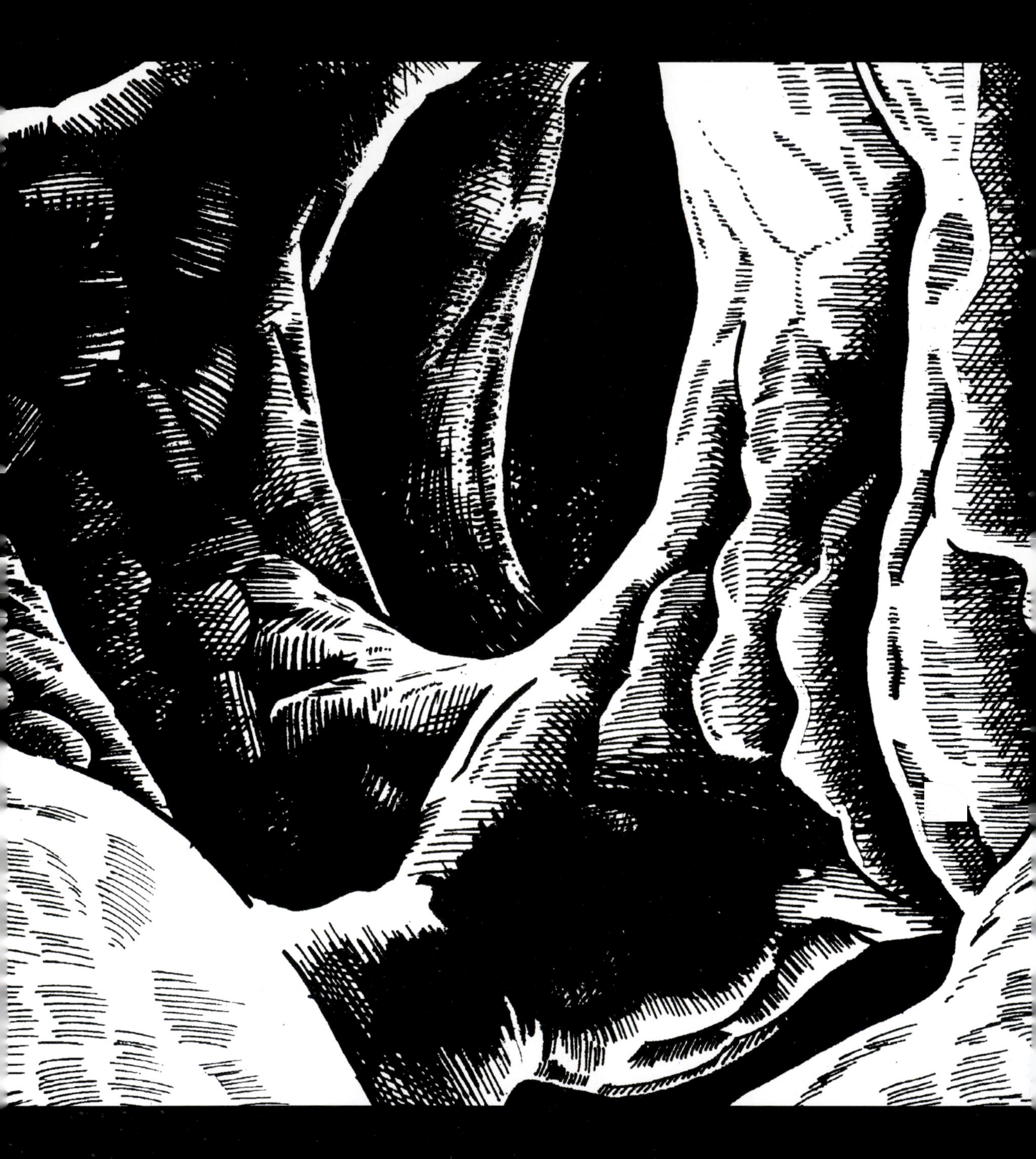

Fig. 3.6 Three dimensional view of the interior of the spongy mesophyll of a leaf (after Troughton & Donaldson). To illustrate the open nature of the structure of this tissue which facilitates gaseous exchange. Carbon dioxide, entering the leaf via the stomatal apertures diffuses freely through the cavernous inter-cellular spaces and dissolves readily in the thin film of water which always bathes these surfaces. The outline of the chloroplasts lying immediately beneath the cell walls (see also Fig 3.11 and 3.12) gives these surfaces a patterned appearance and the fact that the chloroplasts are distributed in this way within the cells maximises both light interception and transport of CO_2. The thin film of water is constantly replenished by the transpiration stream which conveys water and mineral ions from the roots, via the vascular tissues of the leaves. Transpiration itself, in which water evaporates within the leaf and diffuses as a vapour through the stomata, is regulated by the size of the stomatal aperture but there is an obvious trade-off between limitation of water loss and restriction of carbon dioxide entry

3 to 5 × 10^9 chloroplasts - i.e. *about as many chloroplasts as there are people on the earth.* This is why "starch pictures" (Figs. 3.8-3.10) are so well defined. These, if you are not familiar with them, are the things that old-fashioned spy stories were made of, like invisible ink. They are pictures created in starch (within chloroplasts) in a leaf rather than in particles of silver in a conventional black and white photographic negative. If you have access to one or two reagents and a slide projector you can make one for yourself.

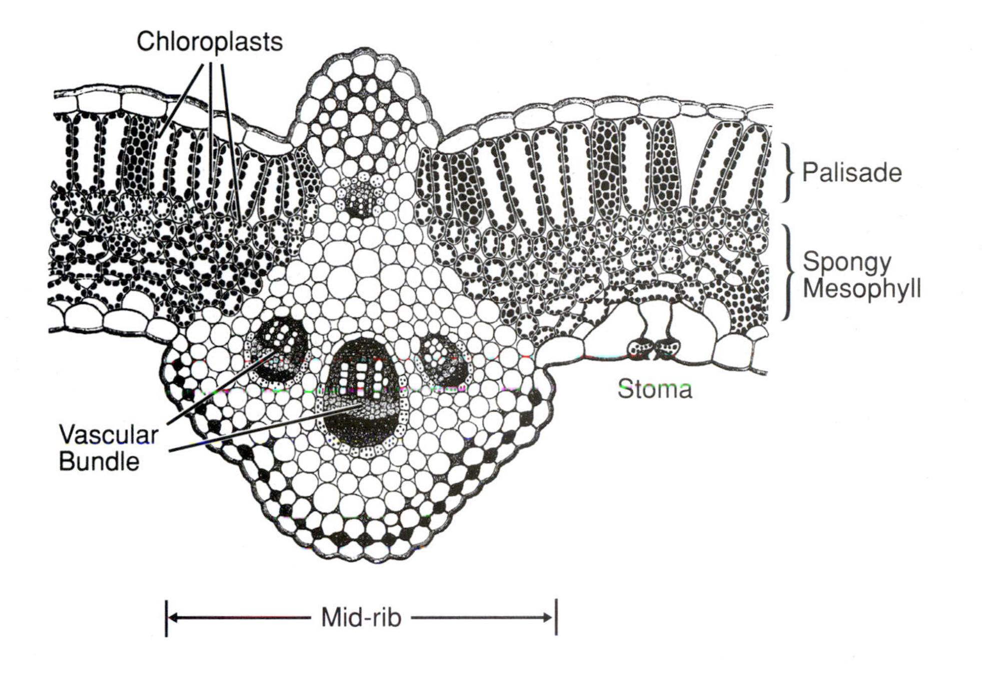

Figure 3.7. Cross section of a C3 leaf. (After Sachs) Shows location of chloroplasts (black bodies lining periphery of palisade and mesophyll cells - see also Fig. 3.12). This section is through a mid-rib in which the main vascular tissue (responsible for supply of water and mineral salts and removal of sugar etc.) is embedded. Some leaves have more than one layer of palisade cells; all have a spongy mesophyll layer with spaces between the cells which permit free circulation of air, so that CO_2 (entering via the stomata) can be fixed in carbon assimilation. In starch forming C3 leaves, all of the chloroplasts will come to contain starch grains after illumination.

3.12 Starch Pictures

What you do is this. First find a leaf. In principle any leaf will do but some do not make starch and others are not too easy to handle. Molisch, who started all this, used *Tropaeolum* (which gardeners, for some reason, usually call *Nasturtium*). *Geranium* (which gardeners ought to call *Pelargonium*) is better. The first trick is to get rid of all its starch. In the leaf, as you recall, starch (a polysaccharide made up of "glucose" units) is present only in the chloroplasts where it is presumed to represent photosynthetic product in excess of that which could be used by the rest of the plant in daylight hours. Certainly in a healthy leaf, in its prime, much of this starch is "mobilised" (i.e. converted to simpler molecules such as triose phosphates and glucose) by night and transported out of the chloroplast. So if we put a "geranium" in the dark for 48 hours (72 hours may do it grievous bodily harm) we

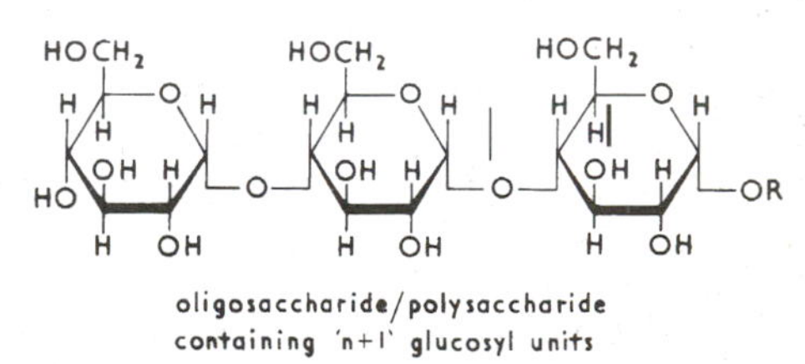

can expect to find it free of starch. Accordingly, if we remove the chlorophyll from the leaf (so that it now looks pale cream in colour) and stain for starch we will not find any. How to remove the chlorophyll? Chlorophyll, as you recall, can be dissolved in organic solvents but leaves are full of water and solvents such as petrol (petrol ether) do not mix with water and are messy and dangerous things to handle. Acetone is fine but it has an awful smell and, in the normal way of things, **is far too dangerous to use** (because of flash ignition of acetone vapour). Ethanol is charming but you may have problems with the law, customs and excise, or whatever. If in doubt settle for methanol (methylated spirit) containing 10 to 20% water. Kill the leaf by dipping briefly in near-boiling water. Transfer to warm methanol in a beaker and heat carefully (best over a hot plate and in a fume cupboard or hood to avoid fire risk or, failing this, support the beaker over a container of hot water in a well ventilated space). Once the chlorophyll has been leached from the leaf (it takes about 15 minutes and it may help to use more than one lot of methanol) rinse it well in water and add a few drops of iodine solution (0.2 g. iodine + 5 g. potassium iodide in 100 ml of water). Nothing will happen, except that, if you add too much iodine the leaf will lose its pleasant cream colour and become pale brown. But this, of course, is the equivalent of attempting to take a photograph with the lens cover in place. Before proceeding to extraction and staining you must first make your picture. The easiest, and least satisfactory, thing to do is to obscure part of the leaf with a piece of card and put the geranium, pot and all, into bright sunlight for half an hour or so. Subsequent killing, extraction and staining will produce a black image against a cream background - cream where the card has stopped the light getting to the leaf. "Real" pictures of good quality can be achieved by obscuring part of the leaf with a photographic negative and illuminating overnight with a 100 W lamp at a distance of a foot or so above a trough of water to filter out heat. If you want to be really fancy, project an image on to a de-starched leaf using a slide-projector and a lens (to keep the image small). In this case it is important to hold the leaf between glass to prevent it moving and, ideally, to place black cloth soaked in bicarbonate solution behind the leaf to provide it with plentiful CO_2 and diminish the possibility of light being reflected back into the leaf from the surface of the glass. These devices are all intended to shorten the illumination time to 15-20 minutes and keep the leaf perfectly still during this period. (Even the fathers of modern photography could not get good images if their subjects moved during exposure of the film).

Figures 3.8-3.9 are examples of what is possible. This is how to smuggle the plans out of the country as an image, created in starch within a leaf, now dried and slipped innocuously between the pages of a book, waiting to be mounted and stained with iodine. All very well, you might think, but where is its relevance to the subject of this book? I have included them for several reasons. The first is that I needed a schematic diagram of chloroplast structure and what nicer way than to persuade chloroplasts to make me one? The second to re-create the original observation by Sachs in the last century that starch was only found in illuminated chloroplasts and that these must therefore be the organelles responsible for photosynthesis. The third to give a glimpse of chloroplasts at work because they, after all, are what we depend on for our energy and our existence and here is visual evidence of transduction of light energy to chemical energy in the form of CH_2O. The fourth to show that even plant biochemistry can be beautiful.

The successive enlargements of the starch picture in Figs. 3.9 to 3.10 illustrate the degree of resolution which can be attained in this way given the immense number of chloroplasts in a leaf. In the highest magnification the outlines of stomatal apertures (through which CO_2 enters the leaf) can be seen because starch in stomata is not mobilised in darkness like that in the mesophyll and palisade tissues and is therefore stained by iodine regardless of prior illumination, or the lack of it.

STARCH PICTURES

Fig 3.8 (Top right) The figure of "Innocence" was originally drawn by Pierre-Paul Prudhon (1758-1823). Here it is executed in starch within a "geranium" leaf which has been illuminated through a photographic negative (see text). Every detail of the image is made up of many millions of starch grains. This starch is formed only in the chloroplasts which have been illuminated, hence the definition is as good as that which would be obtained with photographic paper and provides a striking visual demonstration of photosynthetic carbon assimilation.

Fig 3.9 (Below) Starch picture created in a "geranium" leaf by illuminating the starch-free leaf through a photographic negative. The picture contains a facimile of the title of the first work by Molisch on this subject in 1922 and below it, the outline diagram of a chloroplast.

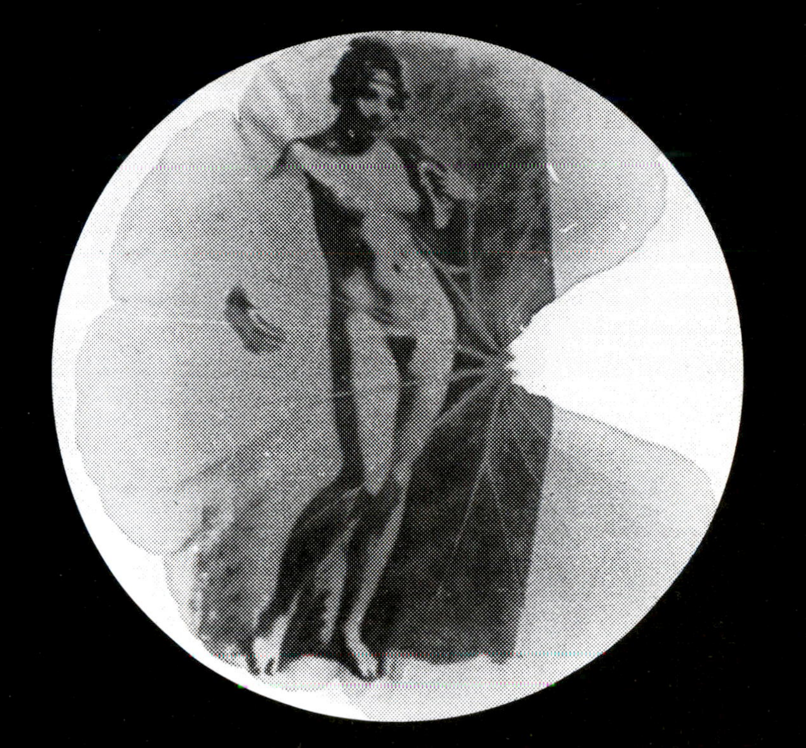

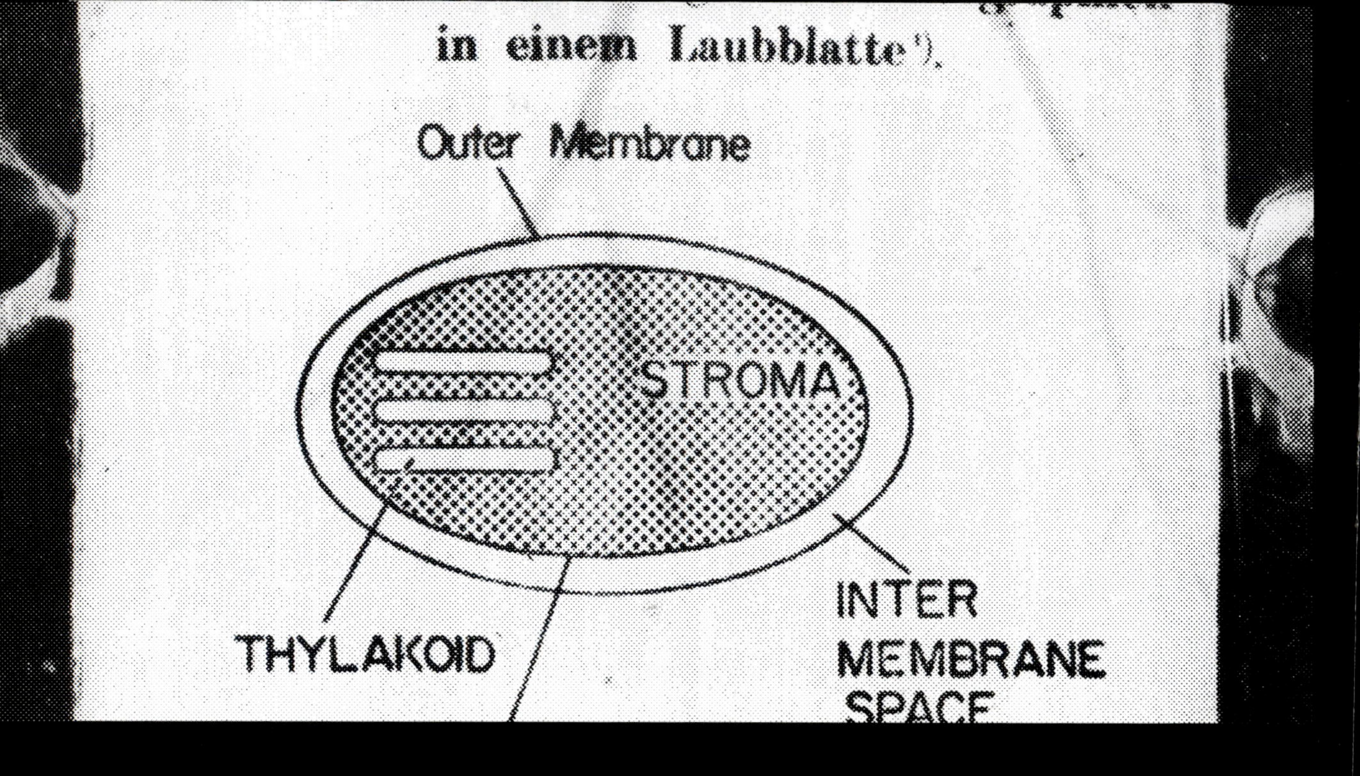

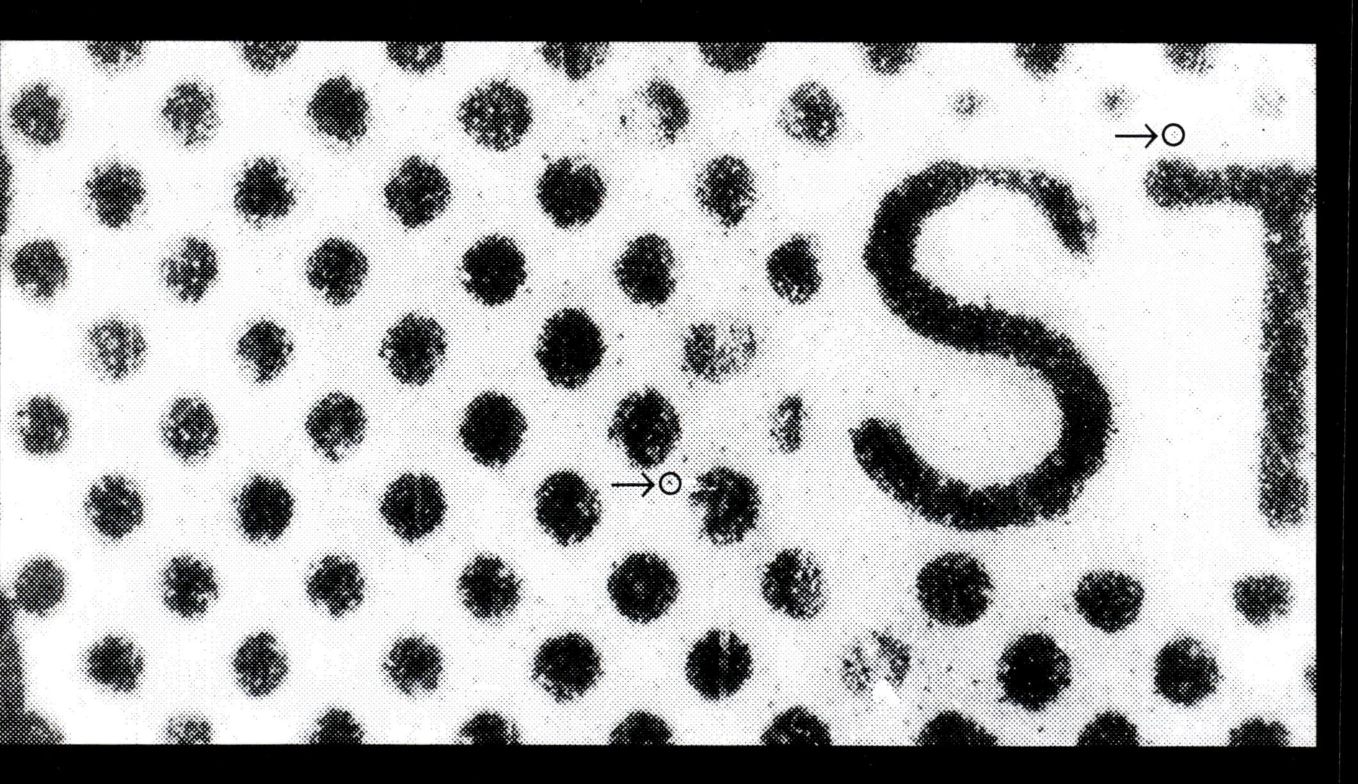

Fig. 3.10 Starch Pictures. Successive enlargements of Fig. 3.9. In the lower of these it is just possible to see the starch grains in the guard cells of the stomata. These cells do not lose their starch entirely in the dark and therefore appear as small dots (arrowed), providing a convenient scale by which the definition of the picture can be judged. It will be seen that the much larger black, circular, machine-made areas used by the draughtsman in the diagram of the chloroplast are reproduced with remarkable precision. There is no discernible formation of starch in the darkened areas, even in the immediate vicinity of the illuminated cells.

3.13 Structure of the Chloroplast

As the diagram within the leaf (Figs. 3.9 and 3.10) indicates, the basic structure of the chloroplast is very simple. The detailed structure is, on the other hand, extremely complex and still far from being fully elucidated or properly understood. That there is still so much to learn is not for want of effort or interest but it should be appreciated that, even in regard to the overall structure, we are talking about an object so small that we can comfortably accommodate as many chloroplasts as there are people in the world within the confines of a leaf of average size. When it comes down to detail we are essentially talking about molecular structure and very complex molecules at that. These, apart from those already touched upon in other sections of this chapter and elsewhere are largely beyond the scope of this book. For present purposes, Fig. 3.10 will suffice to define the major

compartments. First of all there are the outer envelopes. The first of these is freely permeable and, in the words of the song, can be regarded as *"the skin, that keeps the rest in"* although there is reason to believe that there are one or two bits and pieces of importance between the two envelopes. The inner envelope is certainly extremely important in function because it houses *"the phosphate translocator"*. (Section 4.11). This works like a door guarded by a demon which only allows certain molecules to pass through and then only in exchange for other molecules which pass out. This is the principal channel of export of photosynthetic products from the chloroplast to the cytosol (the compartment of the cell within which other organelles such as nucleus, mitochondria, etc., are located). Triose phosphate produced by operation of the Calvin Cycle (Chapter 5) is exported from the chloroplast via this route in exchange (Fig. 3.11) for inorganic phosphate (Pi). Much of the triose phosphate is used in the synthesis of sucrose in the cytosol, prior to movement of this sucrose to other parts of the plant. Sucrose synthesis from triose phosphate involves release of inorganic phosphate so there is a constant cycling of Pi between chloroplast and cell brought about in this way.

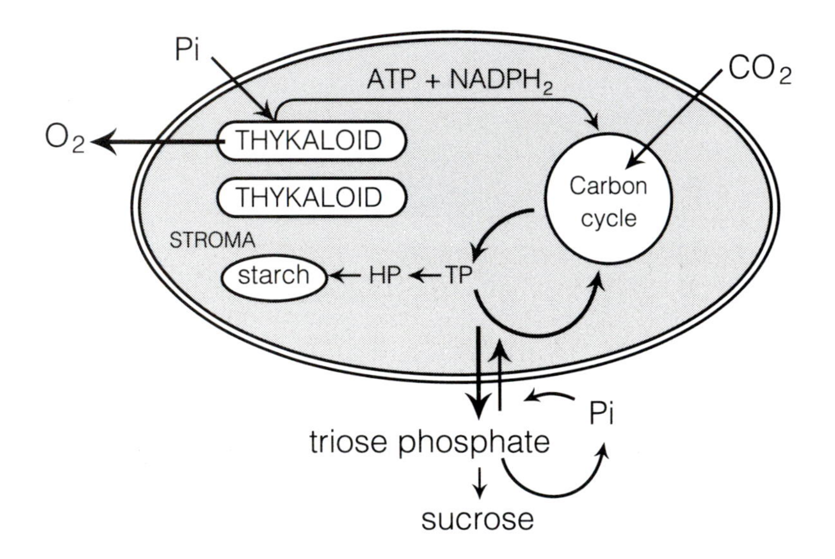

Fig. 3.11 Metabolic sites in the chloroplast including the Pi translocator.

Within the inner membrane there is the stroma. This can be thought of as a protein gel containing the enzymes of the Calvin Cycle; inorganic molecules like Pi and dissolved carbon dioxide and oxygen; ions like K^+, Na^+, Mg^{++} and Cl^- etc. The largest single constituent is RUBISCO, the carboxylating enzyme (section 5.2), which is present in astonishingly large quantities. Lying in the stroma is the thylakoid compartment. The name is from the Greek and derived from its perceived similarity to baggy trousers, traditionally worn in some areas of the Mediterranean. There is probably only one thylakoid compartment within each chloroplast but it is very difficult to be sure because this sac is often folded repeatedly upon itself (Fig 3.14) to form stacks of *"grana"* (so called because they look like "grains" when viewed alone in the light microscope). The thylakoid membranes are green, unlike the chloroplast envelopes which are yellow. They house the chlorophylls, some carotenoids, and the electron carriers (such as plastoquinone, cytochromes, plastocyanin and ferredoxin - section 3.9) all embedded in a protein-lipid matrix. The degree of organisation within the thylakoid membranes is immense including, as it does, the light-harvesting complexes, reaction centres, etc., and the ATPsynthase which is concerned in ATP synthesis (Section 4.7). During electron transport, protons are moved from the stromal space, which therefore becomes more alkaline, into the thylakoid space which becomes more acid. The discharge of protons through the ATPsynthase drives ATP formation so that the whole system is roughly analogous to a hydroelectric system in which electricity is "stored" by night by pumping water into a reservoir from which it is discharged the next day through an electricity generating turbine. In photosynthesis, light energy generates the electrical

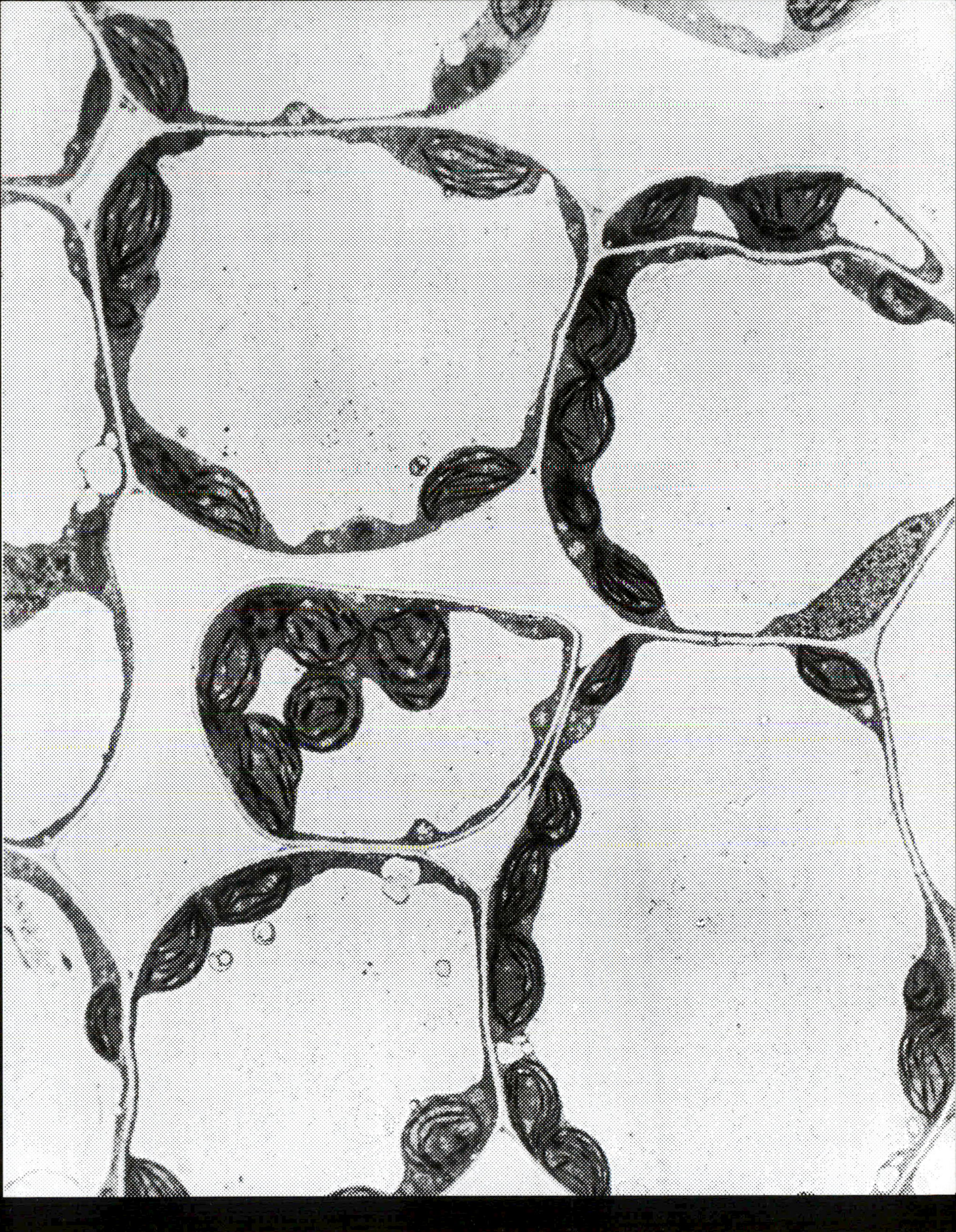

Fig. 3.12 Electron micrograph showing chloroplasts lying in a thin film of cytoplasm against the walls of spongy mesophyll cells. For the location of these cells see Fig. 3.7. The centre of each cell is occupied by a large central vacuole. The spaces between cells facilitate gaseous exchange.

- Courtesy of A. D. Greenwood

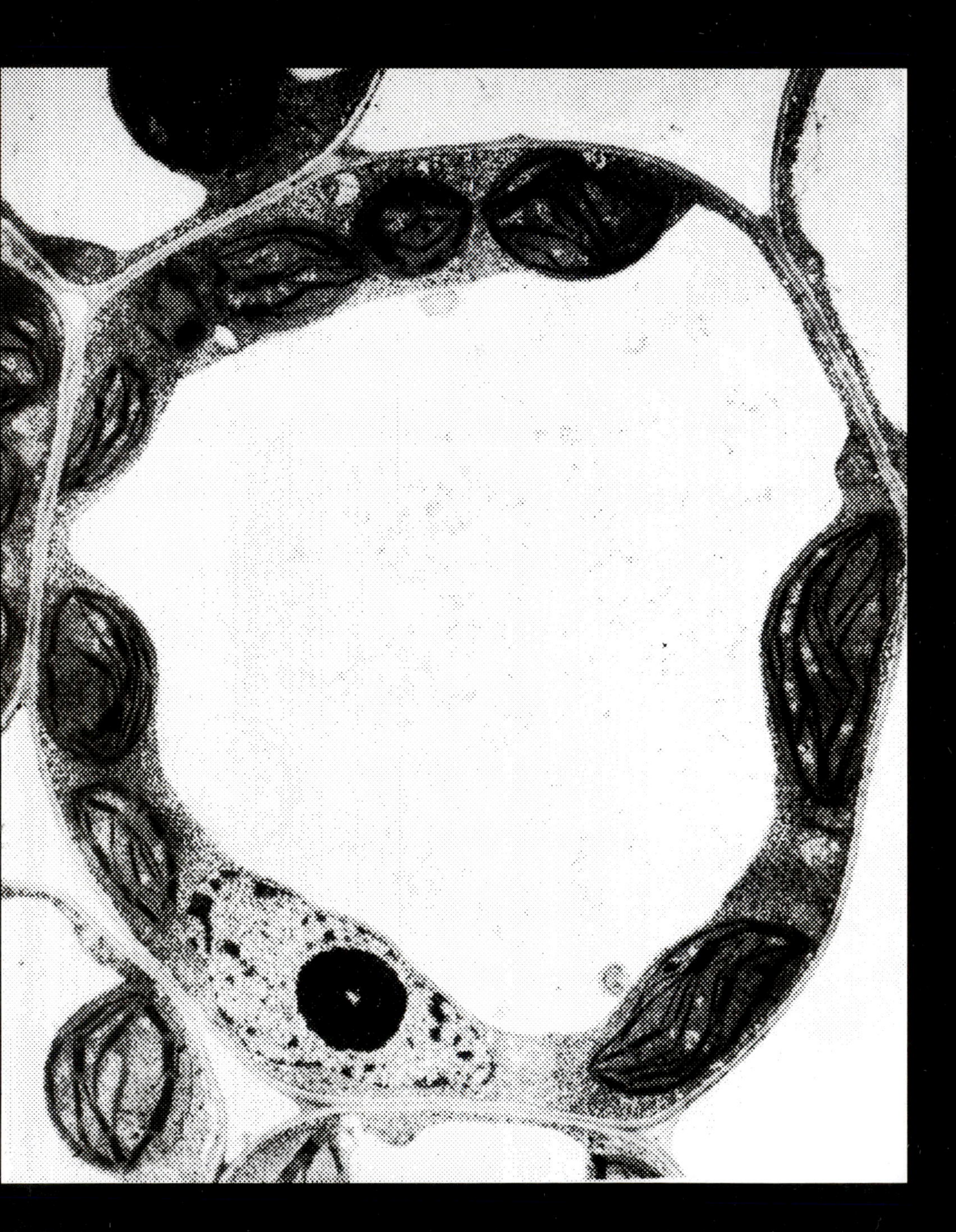

Fig. 3.13 Enlargement of Fig. 3.12. Showing thylakoid membranes within chloroplasts embedded in cytosol surrounding the large central vacuole and bounded, in turn, by the mesophyll cell wall.
- Courtesy of A. D. Greenwood

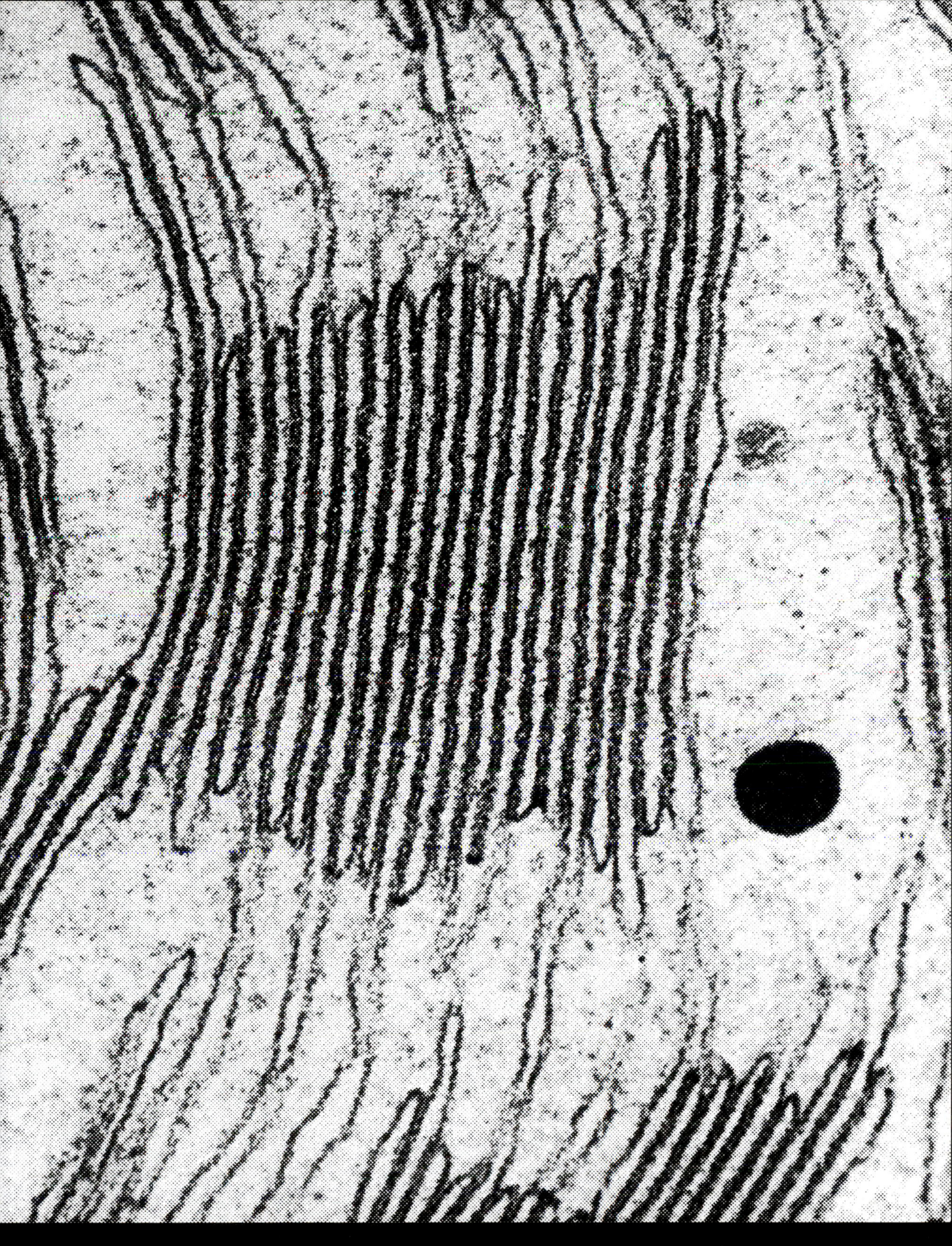

Fig. 3.14 Further enlargement of thylakoid membranes. Showing stacking (so called "grana", because of their granular appearance under the light microscope) of thylakoid membranes (c.f. figs. 3.3 and 3.5).

- Courtesy of A. D. Greenwood

energy which pumps the protons from the stroma to the thylakoid compartment. As the protons discharge through the ATPsynthase, energy is conserved in a chemical form as ADP is joined to Pi to give ATP (section 4.7).

3.14 Summary

In this chapter we have examined the physical basis of light-harvesting and the transfer of electrons from water to NADP. We have seen that chlorophyll *a*nd other photosynthetic pigments are located in thylakoid membranes lying in the stroma of the chloroplast. Chlorophyll is green because it absorbs red and blue photons whilst transmitting and reflecting green photons. Absorption of photons causes chlorophylls to become excited and "excitons" migrate down antennae (light-harvesting chlorophyll protein complexes) to the reaction centres of photosystems I and II. Excitation of P_{680} and P_{700}, in the reaction centres of PSII and PSI respectively, initiates electron transport. Electrons are raised to a higher energy level thereby creating the "holes" which accept electrons from water. From Q_A the primary electron acceptor in PSII, electrons run "downhill" through a series of electron carriers including plastoquinone (PQ), plastocyanin (PC) and the cytochrome b/*f* complex, to P_{700} in the reaction centre of PSI. Here further excitation by incoming photons boosts electrons on their way to the primary acceptor (A) of PSI, ferredoxin, NADP and eventually to carbon dioxide. Back at PSII, electrons released from water fill the "holes" created by excitation of P_{680} and oxygen is evolved. Four photons, acting consecutively in each of the two photosystems, combine to evolve one molecule of oxygen and reduce two molecules of NADP. The very last components of the electron transport chain are located in the stroma of the chloroplast, the protein gel of enzymes in which the thylakoid membranes are embedded. As such, NADP and ferredoxin lie at the boundary between the photochemical events and the so-called "dark biochemistry" of the Calvin Cycle (Chapter 5). The entire process, for which these components are responsible, is one of energy transduction in which light energy is converted first into electrical energy (electron transport) and then into chemical energy (energy rich compounds such as ATP and $NADPH_2$).

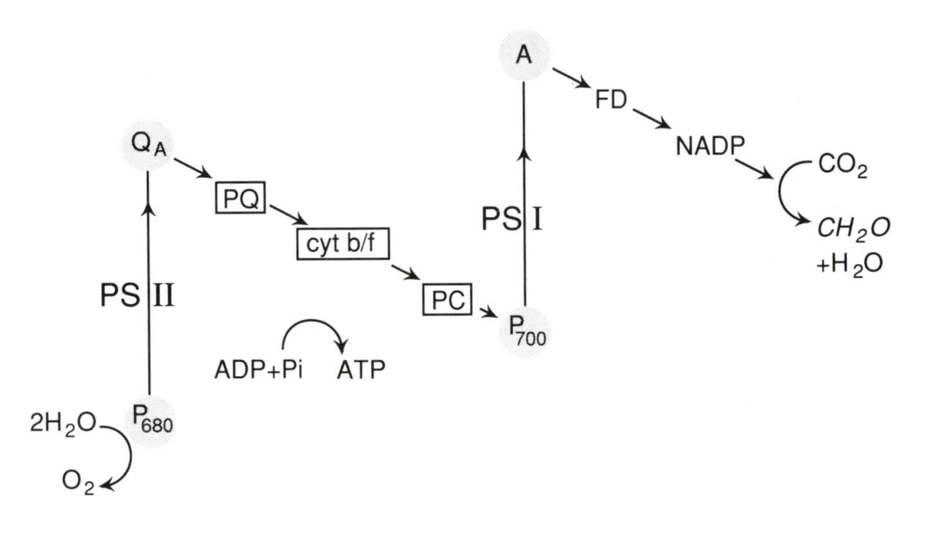

Fig. 3.15 Principal components of electron transport route from H_2O to CO_2. For details see below

Chapter 4

Assimilatory Power

New equations for old

Alphabetic schemes

Cycling and straight chains

Assimilatory power

The light before the dark

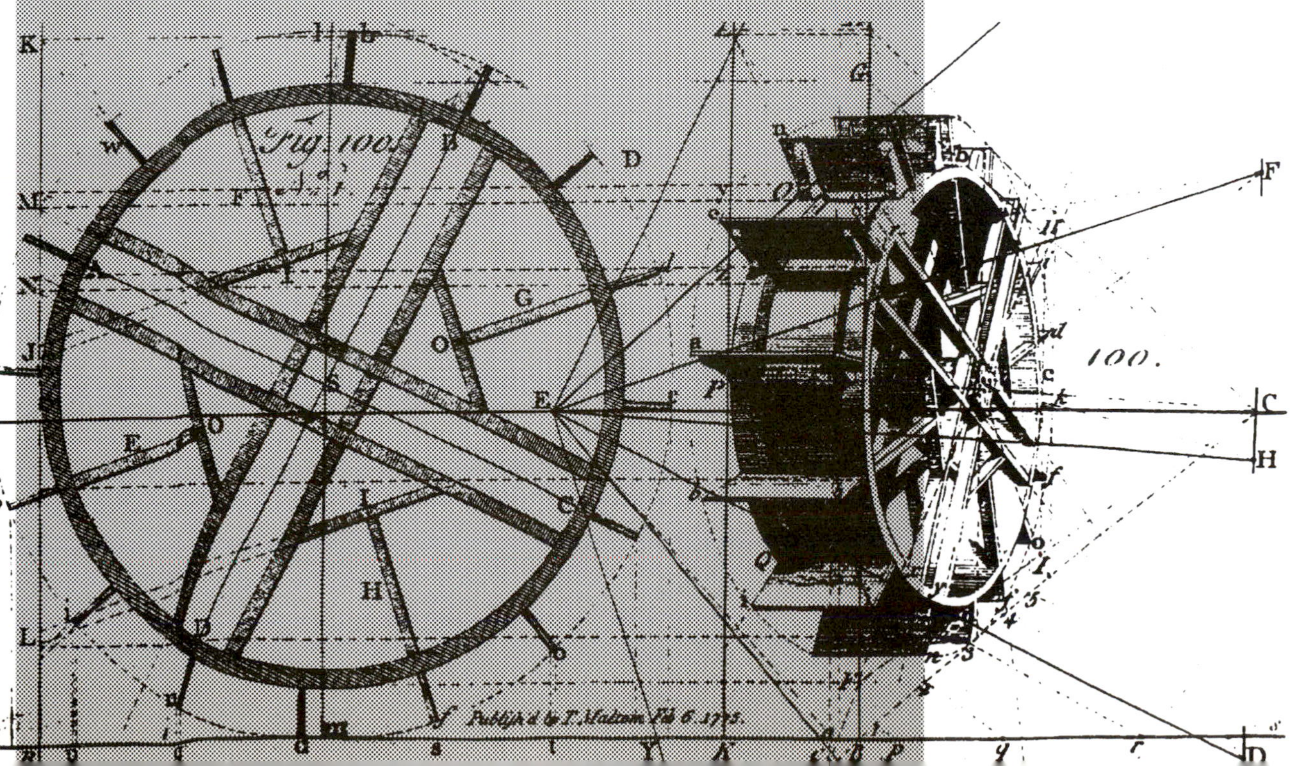

4.1 Equations Old and New

Traditionally, the overall equation for photosynthesis has often been written as :-

$$6CO_2 + 6H_2O \overset{h\nu}{\rightarrow} C_6H_{12}O_6 + 6O_2 \qquad \text{Eqn. 4.1}$$

There is nothing wrong with this equation to the extent that it succeeds in summarising most of the essential features of photosynthesis - the utilisation of light energy i.e. *hv* Section 2.3), the release of oxygen and the formation of a carbohydrate ($C_6H_{12}O_6$) from carbon dioxide. In other regards it is misleading. For example, the first real "product" of photosynthesis is not a free hexose sugar, as $C_6H_{12}O_6$ implies, but rather a triose phosphate (i.e. a sugar containing three atoms of carbon, rather than six, joined to a terminal phosphate group - Section 5.2).

Although sugars such as glucose and fructose (both of which have the same empirical formula, $C_6H_{12}O_6$) may eventually appear in plant leaves as a result of photosynthesis they are not formed directly from carbon dioxide and are not even particularly important compounds in leaf metabolism. Many, *but not all*, leaves accumulate polysaccharides, such as starch, which are multiples of $C_6H_{12}O_6$ units. These are mainly temporary storage materials rather than immediate products. For the moment, therefore, we can simplify equation 4.1 by substituting CH_2O for $C_6H_{12}O_6$ while remembering that CH_2O is merely a convenient short-hand way of describing a carbohydrate (all carbohydrates, i.e. hydrated carbon, contain CHOH groups). Our simplified equation therefore becomes:-

$$CO_2 + H_2O \xrightarrow{h\nu} CH_2O + O_2 \qquad \text{Eqn. 4.2}$$

But even this equation, like equation 4.1, disguises some important underlying facts. One is that, equation 4.3, as written, is energetically unfavourable. In order to proceed it must consume "assimilatory power" (Section 4.8) in the form of ATP and $NADPH_2$. Another is that, if H_2O is added to C to give CH_2O, it looks as though the oxygen evolved is derived from the carbon dioxide and, in such a case, (no matter how the equation is juggled) part at least of the evolved oxygen would have to be derived from that source. However, at Cambridge in the 1930's, Robert Hill succeeded in isolating active chloroplasts from green leaves and showed that these green organelles (Section 3.13) would evolve oxygen when illuminated in the presence of an artificial electron acceptor (Section 1.2). No carbon dioxide was involved in the Hill reaction, which can be written, as follows, if the acceptor (the Hill oxidant) is represented by "A":-

$$2H_2O + 2A \xrightarrow{h\nu} 2AH_2 + O_2 \qquad \text{Eqn. 4.3}$$

In short, it was clear that, at least in these circumstances, the oxygen resulted from the photolysis (light-splitting) of water.

[Hill used potassium ferricyanide to accept electrons but a great many compounds, such as the dye, 2,6-dichlorphenol indophenol (which loses its blue colour as it is reduced), can be used for this purpose. Moreover, provided the acceptor has an oxidation/reduction potential (see legend to Fig. 4.2) which allows it to interact with carriers in the Z-scheme it can be either an electron acceptor, as such, or a hydrogen acceptor (c.f. Section 1.2).]

The overall equations for photosynthesis, and for the Hill reaction, can be made compatible by using "AH_2" to represent the hydrogen donor required for the reduction of carbon dioxide:-

$$CO_2 + 2AH_2 \rightarrow CH_2O + 2A + H_2O \qquad \text{Eqn. 4.4}$$

If equations 4.3 and 4.4 are now added together, and the A's cancelled, we get:-

$$2H_2O + CO_2 \xrightarrow{h\nu} CH_2O + O_2 + H_2O \qquad \text{Eqn. 4.5}$$

If we go one step further, and cancel one molecule of water at either side of the equation, we would be back to equation 4.2. However, if we leave equation 4.5 unchanged, we have an equation which is essentially the same as equation 4.2 but which does not insist that some of the oxygen evolved must come from the carbon dioxide. Moreover, if we bear in mind the derivation of equation 4.5, it tells us that photosynthesis involves the reduction and reoxidation of some substance, A, an electron carrier. In the plant, the real compound "A" is a coenzyme (a non-protein molecule) which works in conjunction with an enzyme (i.e. a protein catalyst). The particular coenzyme concerned is called nicotinamide adenine dinucleotide phosphate (NADP). It is, in this sense, the natural Hill oxidant or hydrogen acceptor (c.f. Section 1.2 Eqn. 1.2). Like other compounds which function in oxidation/ reduction reactions it also functions as a hydrogen donor, in this instance passing hydrogens to carbon dioxide.

4.2 The Reduction of NADP

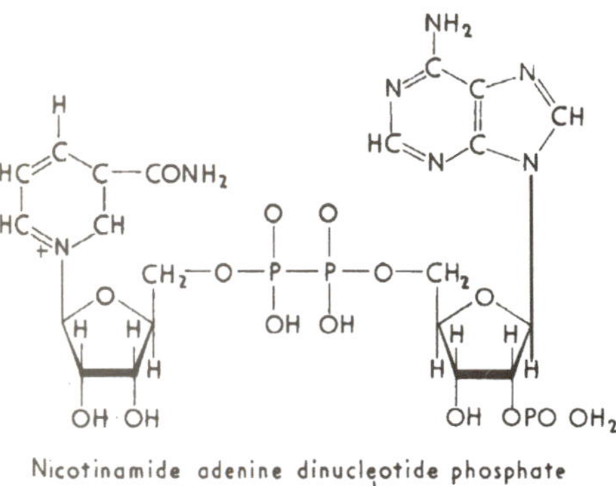

Nicotinamide adenine dinucleotide phosphate
(NADP or $NADP^+$)

It would be nice if we could simply write the "natural" Hill reaction as :-

$$NADP + H_2O \rightarrow NADPH_2 + \tfrac{1}{2}O_2 \qquad \text{Eqn. 4.6}$$

and, indeed, you may sometimes find it written just like that. Striving for too much simplicity can bedevil understanding, however, and (having read and digested Section 1.2) you should not feel too uncomfortable with the fact that hydrogen transfer always involves electron transfer (because $H \Leftrightarrow H^+ + e^-$), and that some oxidants (acceptors) can accept hydrogens whereas others can only accept electrons. Not to be outdone, NADP can accept *both*. Accordingly, if we wish to represent the reduction of NADP more accurately we should write :-

$$NADP + H_2O \rightarrow NADPH^- + H^+ + \tfrac{1}{2}O_2 \qquad \text{Eqn. 4.7}$$

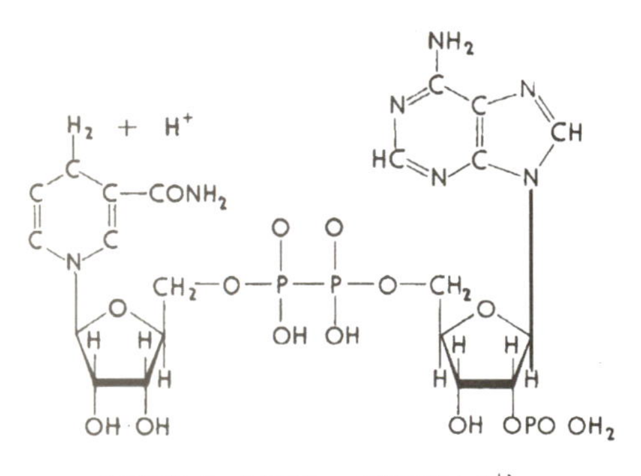

$NADPH_2$ (or NADPH or NADPH + H^+)

indicating that the NADP molecule has been able to accept *one* hydrogen and *one* electron rather than two hydrogens, as equation 4.6 would imply. When acceptors leave hydrogen ions (H^+) behind in solution, as NADP has done in equation 4.7, there are two additional consequences. One is that these acceptors become negatively charged (having acquired an e^-) and the other is that the surrounding solution becomes more acid because its hydrogen ion concentration has increased (i.e. its pH has *decreased*). As discussed in Section 1.7, water spontaneously dissociates, to a small extent, into hydrogen ions (H^+) and hydroxyl ions (OH^-).

$$H_2O \Leftrightarrow H^+ + OH^- \qquad \text{Eqn. 4.8}$$

[There are 10^{-7} gram ions of H^+ and OH^- in water which is neutral (i.e. neither acid nor alkaline) but it is more convenient to express the hydrogen ion (proton) concentration [H^+], as its negative logarithm (p), so that $[H^+]\times 10^{-7}$ becomes pH 7. Accordingly, we can arrive at a pH scale, or measure of acidity, which ranges from zero (most acid) to 14 (most alkaline) and reflects the extent to which acids give rise to H^+ ions. If hydrogen ions (protons) are formed as the result of a reaction such as that described by Eqn. 3.7, the pH will fall (say from 7 to 6); i.e. the solution will become more acid.]

All of this is mentioned again here for two reasons. One is to explain why, if we wish to represent the reduced form of NADP as accurately as possible, in an abbreviated way, we should write $NADPH^- + H^+$ rather than $NADPH_2$ (even though we shall continue to use $NADPH_2$, or even NADPH, in this text where it is most appropriate to do so). The other is that, although photosynthetic electron transport involves a number of electron carriers, only some, like NADP, will also accept protons. This, as we shall see (Section 4.4), generates a change (Δ) in pH. This is important, because the ΔpH so generated, drives the formation of ATP (adenosine triphosphate) from ADP (adenosine diphosphate) and Pi (inorganic phosphate). Together, ATP and $NADPH_2$ constitute *"assimilatory power"* (Section 4.8) i.e., so-called high energy compounds which are consumed in the assimilation of carbon dioxide. They are, in effect, the first stable chemical entities in the transduction of light energy, via electrical energy to chemical energy in the photosynthetic process. As such, they occur in chloroplasts in relatively small quantities and are utilised (Section 4.9) as rapidly as they are generated (Section 4.7). Sucrose, the major end-product of photosynthesis, is also a "high energy" compound in the sense that it yields a lot of heat if burned in a calorimeter and generates many molecules of ATP etc., if consumed in respiration but it serves a very different function (transport and storage of carbon and energy) and, as such, is accumulated in large quantities in leaves.

4.3 The Z-scheme

Following excitation of chlorophyll by light (Chapter 3), the first steps in the conversion of electrical to chemical energy involve the generation of "assimilatory power" (Section 4.8). This is accomplished by the operation of the Z scheme of Hill and Bendall (through which electrons follow a Z-shaped path, through two photosystems and a connecting bridge). This process may be compared to the unlikely sequence of events depicted in Fig. 4.1.

Fig. 4.1 The Z-scheme by analogy.

In this scheme, the man on the extreme left delivers energy in the form of a hammer-blow. This projects a ball upwards into the hands of the second man who drops it down a steep incline. As it runs downhill it drives a wheel (in the bridge between the two platforms). On the right, the process is repeated except that the final catcher then throws the ball into a net. In the Z-scheme proper (Fig. 4.2), the hammer-blows are photons delivering energy to chlorophyll molecules in photosystems II and I respectively (the physical nature of the two photosystems is discussed in Chapter 3). The ball is an electron. Each electron therefore requires the expenditure of two photons (two quanta of energy) to lift it, from ground level, into the net (NADP). The passage of electrons down the thermochemical (energy) gradient drives the wheel (the ATP-generator). Each NADP accepts two electrons. Four photons are therefore required to bring about the reduction of *one* molecule of NADP.

The positively charged "holes", created by light-excitation of chlorophyll molecules in PSII, are filled by electrons donated by water so that eight photons are required for the reduction of two molecules of NADP and the associated evolution of one molecule of oxygen (Eqn. 4.9). In equation 4.9 the formation of one molecule of ATP is also indicated (Section 4.7), but the ratio of ATP to $NADPH_2$ could be 1.3 to 1, or even higher. Alternatively, extra ATP could be formed by cyclic electron transport (Section 4.5) involving only one photosystem, or by the reduction of molecular oxygen instead of NADP. Either alternative would lead to more ATP and a lower quantum efficiency (greater quantum requirement - Fig. 4.2 Section 6.6).

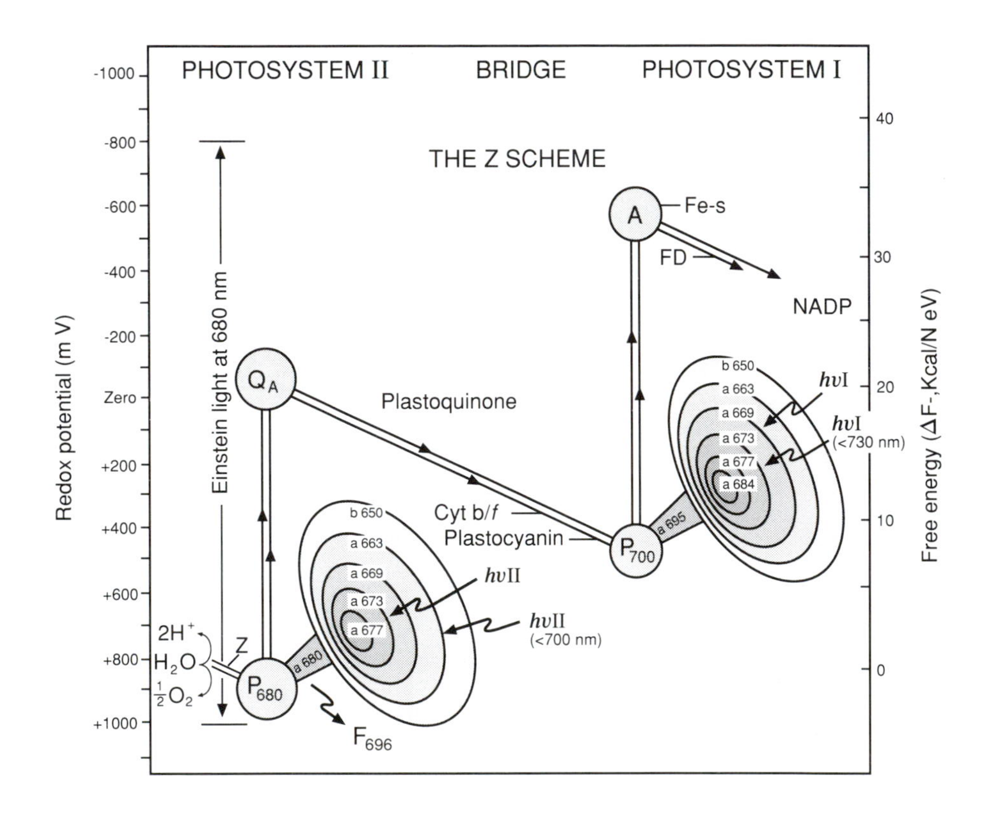

Fig. 4.2 The Z-scheme . This is both an energy diagram and a map of electron flow. Redox potentials (left) are like football league tables or tennis "seeds". Under given conditions, a team at the top of the league is likely to beat a team below it. Similarly, compounds with negative redox potentials are able to reduce compounds with more positive potentials. Thus, the transfer of electrons from water to NADP (involving the co-operation of two photosystems) leads to the creation of a hydrogen donor ($NADPH_2$) used to reduce CO_2 to CH_2O. The energy needed to drive this electron transfer (right) is seen to be about 26 Kcal (109 Kjoules) per electron (on a molar basis) and enters through the two photosystems. Two light-harvesting antennae (Section 3.3) of chlorophyll molecules funnel photons into P_{680} and P_{700} in PSII and PSI respectively (Sections 3.7, 3.8 and 6.6). Other letters (e.g. PQ, for plastoquinone) mark the probable position in the energy diagram of various carriers through which electrons are passed on their way from water to NADP. Each photon (quantum) entering each photosystem can lift one electron to a higher (more reducing) level so that, in order to transfer 2 electrons from H_2O to NADP, 4 photons are required. Since 4 electrons are transferred for each molecule of O_2 released ($2H_2O + 2NADP \rightarrow 2NADPH_2 + O_2$) the whole sequence has to occur twice and the quantum requirement thus becomes 8 per O_2. Some of the energy released as electrons flow "downhill" across the "bridge" between the two photosystems is conserved as chemical energy in the form of ATP (Section 4.7).

4.4 Non-cyclic or Linear Electron Transport

In vivo the donor is H_2O and the acceptor NADP but, in experiments with isolated chloroplast, both artificial donors and acceptors are sometimes used.

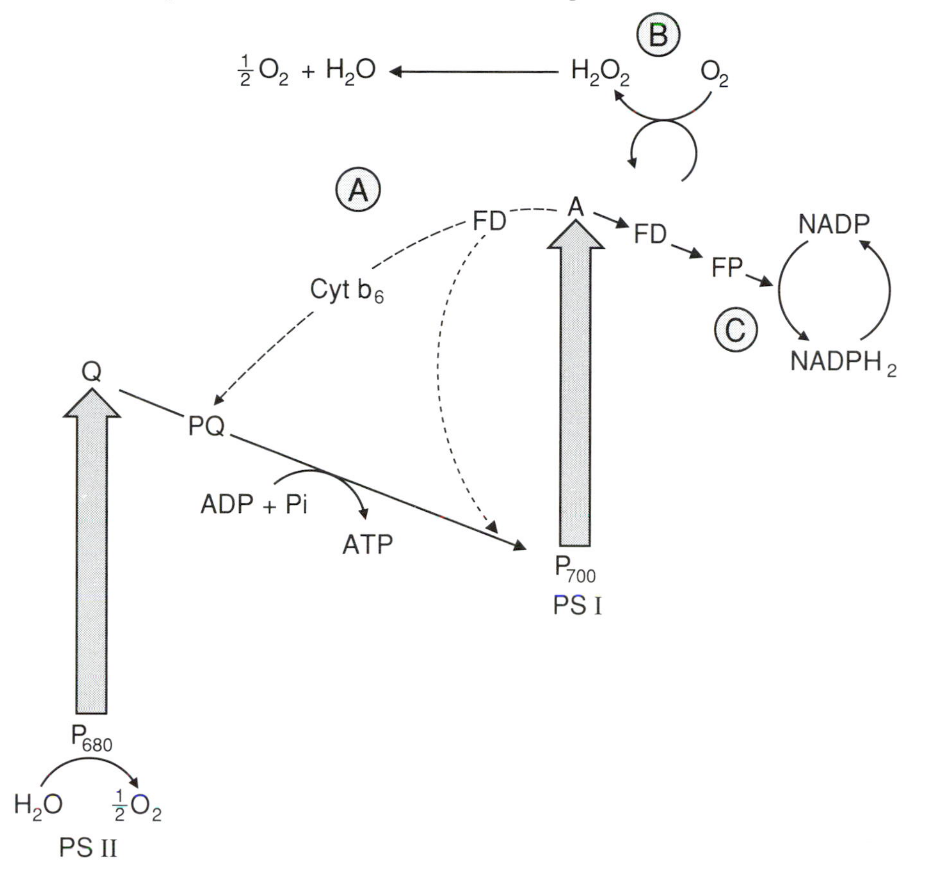

Fig. 4.3 Possible routes of electron transport. A, cyclic; B, pseudocyclic (Mehler reaction) and C, non-cyclic or linear. See Fig. 3.5 for details of structure. The hydrogen peroxide produced in the Mehler reaction can be broken down, by catalase (in the cytosol) or further reduced.

"Non-cyclic photophosphorylation" is a term sometimes used to describe the generation of ATP (Section 4.7) driven by linear electron transport. Non-cyclic electron transport in the Z-scheme (Fig. 4.2) involves the "linear" transfer of electrons from a donor to an acceptor. If the intermediate carriers are omitted, in order to arrive at a relatively simple equation, non-cyclic photophosphorylation looks like this :-

$$2NADP + 2ADP + 2Pi + 2H_2O \rightarrow 2NADPH + 2H^+ + 2ATP + H_2O + O_2 \qquad \text{Eqn. 4.9}$$

but there is continuing uncertainty about the precise stoichiometry of this reaction, i.e. whether or not, in all circumstances, there is only one molecule of ATP generated per molecule of NADP reduced. The Calvin cycle consumes 1.5 molecules of ATP for every one molecule of $NADPH_2$ and, even though it is possible that additional ATP is generated by cyclic electron transport (Section 4.5) to make up any shortfall, it seems likely that linear electron transport may be responsible for most of the ATP required in carbon assimilation.

4.5 Cyclic Electron Transport

As the name suggests, electrons in this process travel in a cycle, rather than in a straight line from A to B. Cyclic electron transport in the Z-scheme usually implies a return of electrons from a point at "the top" of PSI (possibly bound ferredoxin) to the plastoquinone pool. In this way, electrons would cycle endlessly round PSI without giving rise to oxygen evolution or NADP reduction. On the other hand, as the electrons passed back down "the bridge" from plastoquinone to PSI a ΔpH would be created, just as it is in linear electron transport, and this would drive ATP generation (Section 4.7). So called cyclic photophosphorylation, i.e. ATP generation driven by cyclic electron transport, may make up the short-fall (Section 4.4) in ATP needed for carbon assimilation. If this occurs, it will increase the quantum requirement above 8 (see legend to Fig. 4.2) because of the photons needed to "lift" additional electrons through PSI. In practice, quantum requirements, measured under ideal conditions, are closer to 9 than to 8, possibly indicating the occurrence of this process.

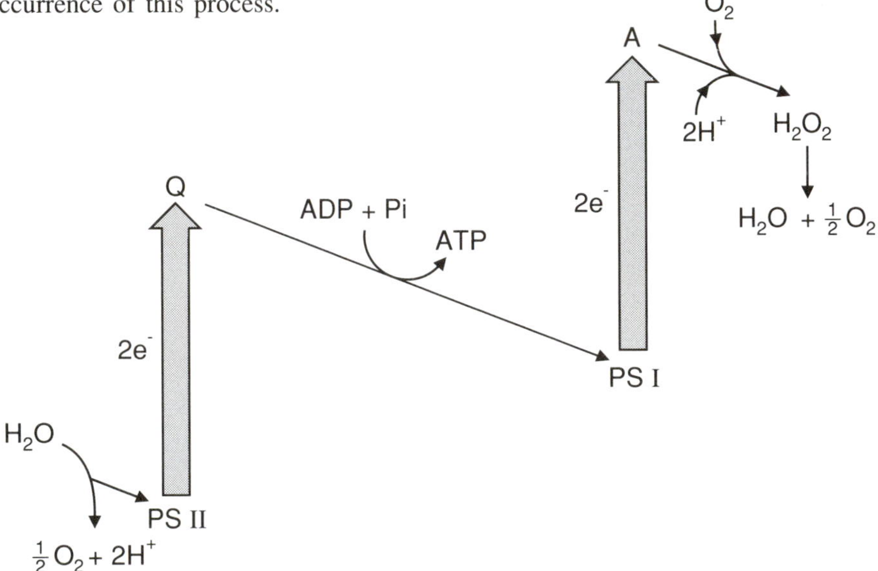

Fig. 4.4 Pseudocyclic electron transport (Mehler reaction). Although, in the presence of catalase, there is no oxygen evolution, this is not a true cycle (c.f. Fig. 4.3). Instead oxygen evolution is masked by equimolar oxygen consumption. Nevertheless, electrons traverse both photosystems and are finally used to reduce oxygen rather than NADP. The hydrogen peroxide may also undergo further reduction to water.

Pseudocyclic electron transport is a term sometimes applied to the Mehler reaction in which molecular oxygen serves as an electron acceptor in PSI rather than NADP:-

$$H_2O + O_2 \rightarrow H_2O_2 + \tfrac{1}{2}O_2 \qquad \text{Eqn. 4.10}$$

but in fact, this involves linear electron transport. "Psuedo" is a good description because the electron transport involved is actually linear, just as if NADP rather than O_2 was the acceptor.

The concept of a cycle in the Mehler reaction comes from the fact that if the hydrogen peroxide (H_2O_2) so generated is broken down as fast as it is formed by the enzyme catalase (Eqn. 4.11):-

$$H_2O_2 \rightarrow H_2O + \tfrac{1}{2}O_2 \qquad \text{Eqn. 4.11}$$

there is no *net* consumption of water or evolution of oxygen as the reaction runs its course In such circumstances (Eqn. 4.10 and Eqn. 4.11) the oxygen released by photolysis of water exactly balances that fraction of the oxygen reduced by electron transfer (Eqn. 4.10) and subsequently released by "catalatic dismutation" of the hydrogen peroxide so formed (Eqn. 4.11). Chloroplasts prepared for experiments by normal procedures are usually contaminated by catalase and the Mehler reaction can then only be followed, by measurement of O_2 uptake, if cyanide or azide is added in small amounts to inhibit the catalase. The Mehler reaction is readily demonstrated in experiments with isolated chloroplasts. *In vivo*, hydrogen peroxide may also be reduced to water.

4.6 Hydrolysis of Adenosine Triphosphate (ATP)

The hydrolysis of one mole of ATP to ADP plus inorganic phosphate (Pi) is accompanied by a release of energy (about 7 Kcal or 29 Kjoules) - see Eqn. 4.12. Thus if ATP is consumed in a metabolic reaction (and ADP and Pi are released) it can be regarded as having contributed about 7 Kcal of energy to that reaction. For this reason ATP is frequently spoken of as an "energy donating" molecule.

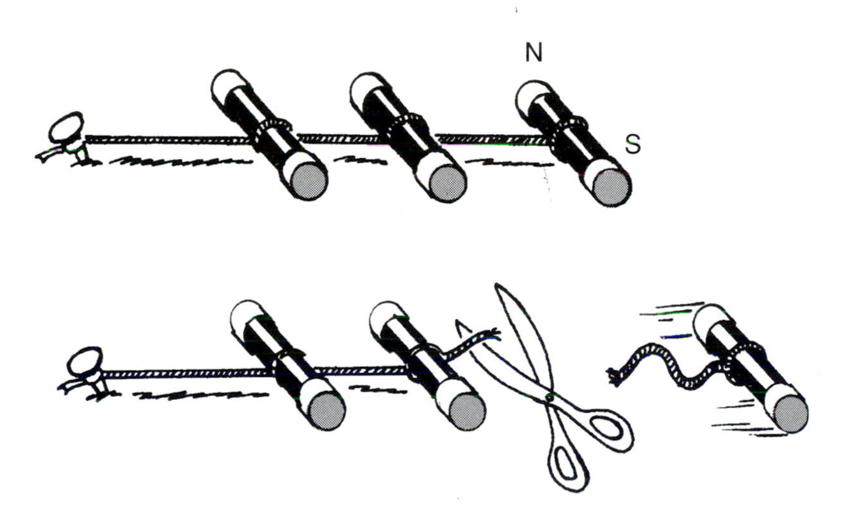

Fig. 4.5 ATP hydrolysis by analogy

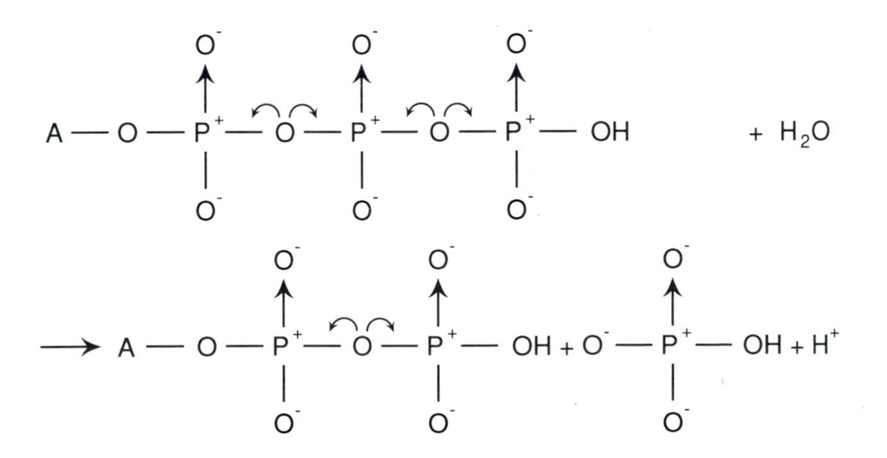

Fig. 4.6 Hydrolysis of ATP to ADP and Pi. (Where "A" is adenosine)

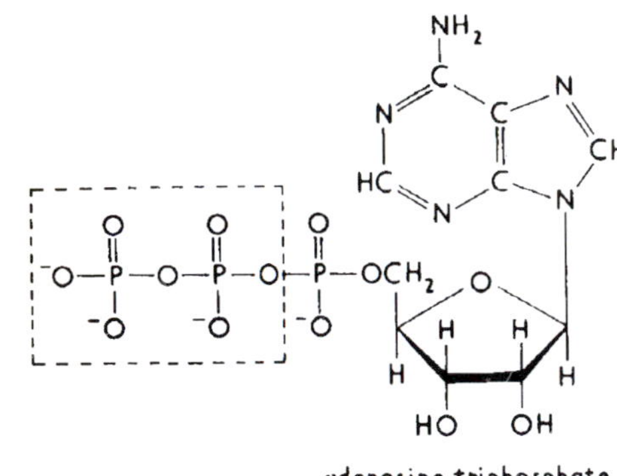

adenosine triphosphate (ATP)

Confusion sometimes stems from the implication that energy is released when a "high energy bond" (e.g. one joining ADP and Pi) is broken. This offends the principle that bond dissociation always requires an energy input (Section 1.1). No comfort can be immediately derived from a careful consideration of all the bonds which are broken and re-formed during the hydrolytic process because at first glance these are identical and no net gain or loss could be predicted. On this basis *"what is gained upon the roundabouts is lost upon the swings"*. In fact, the bond values depend not only on the atoms which they join, but also on the molecular environment in which they exist. ATP is subject to internal stress (because of electrical charges generated by ionisation and opposing resonance within the molecule) and may be compared to bar magnets (with opposing poles) joined by pieces of string. Energy would be needed to cut these bonds but the magnets would then spring apart, because of the magnetic repulsion, producing a net energy gain. Pi is more stable as a separate entity than as part of a molecule of ATP so that, even though equal numbers of O-H and O-P bonds are broken and re-formed during hydrolysis, the new bonds are stronger than the old and energy is released. Conversely an energy input is required when ADP and Pi are condensed to give ATP and H_2O.

4.7 The Generation of ATP

ATP is, like NADP, another co-enzyme but one which is involved in phosphate (Pi) transfer rather than hydrogen transfer. It can also be regarded as energy "currency", or a source of energy in metabolic reactions. If ATP is hydrolysed,

$$ATP + H_2O \Leftrightarrow ADP + Pi \qquad \text{Eqn. 4.12}$$

energy is released for the reasons outlined in Section 4.6. Conversely, if ADP is esterified (i.e. if Pi is joined to ADP in a reaction in which water is released,

$$ADP + Pi \Leftrightarrow ATP + H_2O \qquad \text{Eqn. 4.13}$$

energy is required in order to drive the unfavourable equilibrium position of this reaction towards ATP (c.f. Section 1.7)

As Fig. 4.7 shows, if chloroplasts are isolated from a leaf and illuminated in the presence of NADP, electron transport occurs from H_2O (which becomes oxidised) to NADP (which becomes reduced). At the same time oxygen is released and a proton gradient or ΔpH develops across the thylakoid membrane of the chloroplast. This is largely a consequence of the fact that, when a molecule of plastoquinone is offered an electron (e^-) by excited chlorophyll it is simultaneously obliged to take a convenient hydrogen ion (H^+) to keep this electron company. Moreover, it "picks up" (Section 1.2) this proton (H^+) from the outside of the thylakoid membrane in which it (plastoquinone) is located and then discharges a proton to the *inside* of the thylakoid space as it passes the electron on to the next carrier in the sequence. Accordingly, the space inside the membrane (the thylakoid space) becomes more acid (because it contains more protons or hydrogen ions) and the outside space (the stroma in Fig. 3.8b) becomes more alkaline (because it contains fewer protons and, consequently, more hydroxyl ions - see Eqn. 4.8). As this proton gradient builds up, it discharges through an ATPsynthase, an enzyme which catalyses the esterification of ADP (Eqn. 4.13). No one quite knows how this works but it is a bit like a head of water, built up behind a dam, discharging through a turbine which generates electricity. The ΔpH is created by light-driven electron transport and part of its energy is conserved as chemical energy, in the form of ATP,

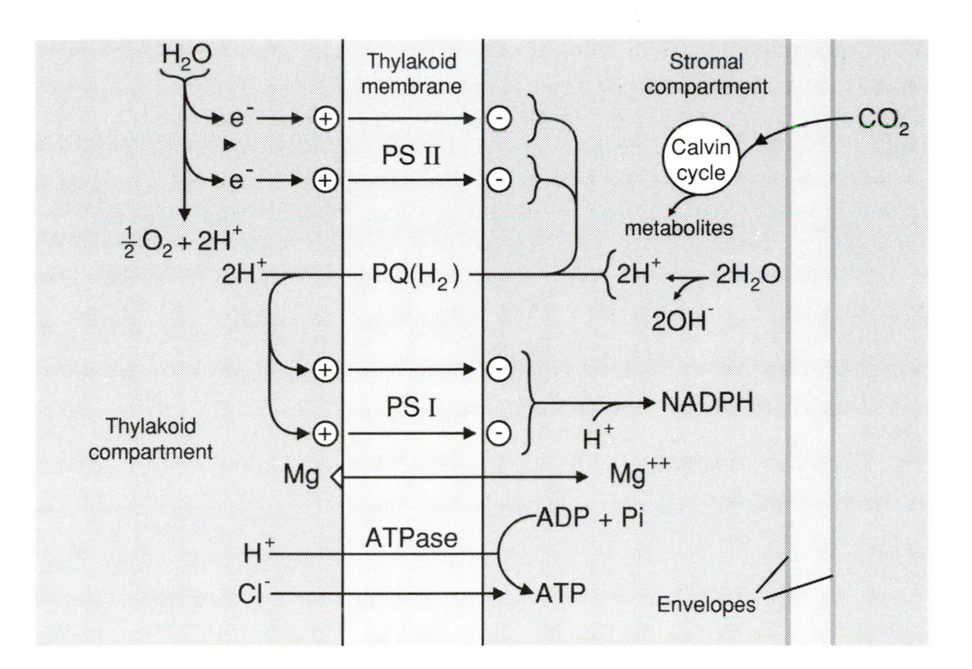

Fig. 4.7 Generation of ATP. The pigments and electron transport chain are housed in the thylakoid membrane which itself (see Chapter 3) lies within a protein gel (the stroma) enclosed within the envelopes (the external boundary of the chloroplast). On illumination, electrons from water are donated (top left) to chlorophyll in PSII excited by the absorption of photons. The transfer of electrons to PSI via plastoquinone (PQ) requires protons. These are taken up from the thylakoid compartment and released into the stroma. The stromal pH rises and Mg^{2+} moves as a counter-ion. ATP is formed as protons are discharged through the ATPsynthase. Electrons which have run "down" the thermochemical gradient which bridges the two photosystems are finally boosted "upwards" to NADP by light excitation of PSI. $NADPH_2$ and ATP are utilised in CO_2 fixation by the Calvin cycle in the stroma.

as the ΔpH discharges through the ATPsynthase. This concept is in accord with the "chemiosmotic hypothesis" for which Peter Mitchell was awarded the Nobel prize. Linear electron transport of 2 electrons from water to NADP is believed to generate 1.3 ATPs. This is consistent with an $H^+/2e$ ratio of 4 and an H^+/ATP ratio of 3. "Photophosphorylation" as it is called (i.e. light-driven ATP formation by chloroplasts) was first demonstrated by Arnon and Whatley. They proposed several varieties (see Sections 4.4 and 4.5). That which caused the greatest surprise to the scientific community at the time was "non-cyclic photophosphorylation" i.e. ATP formation associated with the linear transport (Section 4.4) of electrons from water to NADP. At the time this seemed as unlikely then as an English tennis player winning at Wimbledon or Margaret Thatcher being awarded an honorary degree by the University of Oxford. A generation of biochemists had become accustomed to the concept that *oxidation* (by molecular oxygen) of *$NADH_2$* generated in respiration, led to the formation of ATP from ADP and Pi. Now they were suddenly and unexpectedly asked to accept the fact that water could seemingly run up-hill as ATP was generated during the *reduction* of NADP and the *evolution* (rather than the uptake) of oxygen. Genius, it is sometimes said, is the recognition of the obvious before it becomes obvious to everyone else. It was obvious to Robert Hill, but not to a host of lesser mortals, that water could not run up-hill of its own accord but that it could be lifted up-hill twice in succession and allowed to run "down-hill", in between, doing work as it did so. Hence the Z-scheme. Of course Hill did not get there unaided and would no doubt have been the first to admit his debt to many other contributions (of which that of Robert Emerson's "enhancement" was perhaps the most important and most relevant). Even so the concept of the Z-scheme (which remarkably failed to impress the Nobel Prize Committee) pulled together all manner of things which have, at least in this writer's eyes, done more to advance our understanding of the photochemical events of photosynthesis than anything else this century. Amongst these is the simple notion that the reoxidation of Q_A by PSI is equivalent to the "down-hill" passage of electrons which occurs when $NADH_2$ is oxidised in respiration. Arnon and Whatley's non-cyclic photophosphorylation did not, after all, fly in the face of thermodynamics. Once electrons had run "down-hill" from the top of PSII to the bottom of PSI, generating a proton gradient *en route*, they were lifted to the top of PSI, and hence to NADP, by light excitation of chlorophyll essentially similar to that which had lifted them to the top of PSII in the first place. Light excitation of chlorophyll is, in the end of course, responsible for all biological "lifting". How do you suppose $NADH_2$ became reduced in the first place? Quite right, by accepting electrons from the break-down products of sugars, or whatever, first formed in photosynthesis. This is why "excited state one", the red excited state of chlorophyll is where it all starts. This winds the biological mainspring. This is the point of transduction of light energy into electrical energy. The subsequent conversion of electrical energy into chemical energy is all "down-hill" whether it occurs in the plant or in the tissues of animals which have eaten plants.

Electron transport through the Z-scheme of the sort that we have so far discussed is always associated with the generation of a proton gradient and this is "coupled" to electron transport. The coupling may be *"flexible"* in the sense that although a given proton gradient will only be able to drive a certain amount of ATP formation it may give rise to less than the maximum. Experimentally, electron transport may be uncoupled from ATP formation in various ways and electron transport then runs faster, just as a locomotive uncoupled from the wagons of a train could run faster. On the other hand, electron transport by isolated thylakoids also runs faster when it is coupled to ATP formation (i.e. in the presence of ADP + Pi). It is less easy to see, at first glance, why both coupling and uncoupling can accelerate electron transport. The reason is, that as a proton gradient is established, it exerts a "back-pressure" on electron transport. It becomes more difficult to release a proton into a space already containing a large number of protons. Both ATP formation and uncouplers discharge the proton gradient.

4.8 Assimilatory Power

Together, ATP and the reduced form of NADP, constitute *"assimilatory power"*. As Arnon, Allen and Whatley showed in a classic experiment, assimilatory power, generated in the light, can reduce carbon dioxide to CH_2O in the dark. In other words photosynthesis can be divided experimentally into two parts. The "light" part (the photochemistry), in which the oxygen is evolved, is like the Hill reaction (Eqn. 4.3) except that it also includes ATP formation (Eqn. 4.9). The "dark" part (the so-called *"dark biochemistry")* in which ATP and $NADPH_2$ are consumed (c.f. Section 3.10), is the conversion of carbon dioxide to carbohydrate. Similarly, the photochemical events, which lead to the generation of assimilatory power, are primarily associated with the thylakoid membranes (within the chloroplast) which house the photosynthetic pigments. The "dark" events (so called because they can be "driven" in the dark by assimilatory power even though they normally occur only in the light) are located in the stroma (Fig. 3.8b). The stroma is a gel of soluble protein in which the thylakoids are embedded and contains all of the enzymes (biological catalysts) needed to convert carbon dioxide to CH_2O. At the risk of

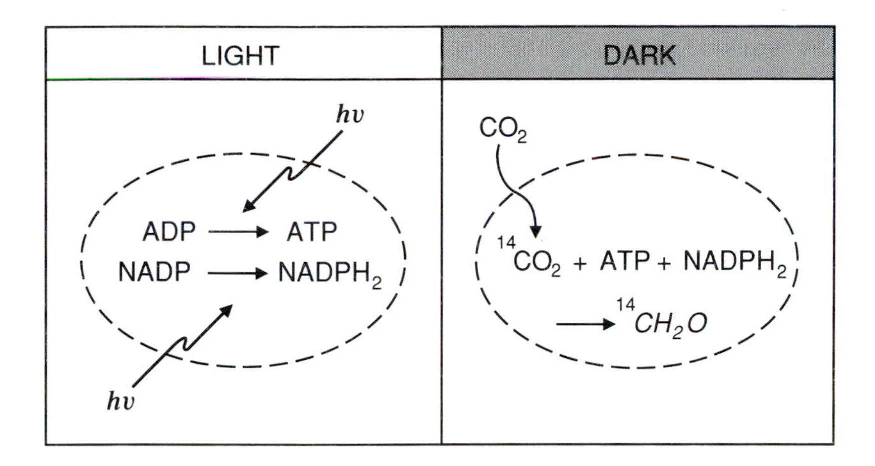

Fig. 4.8 Arnon, Allen and Whatley's Experiment. Experimental separation of photosynthesis into photochemistry (left) and "dark biochemistry" (right). At left, illuminated thylakoids give rise to ATP and $NADPH_2$. At right, enzymes of the chloroplast use ATP and $NADPH_2$ to reduce (radioactive) CO_2 to CH_2O in the dark.

tedious repetition, it is probably worth noting once again that although the "dark biochemistry" did not require light in this experiment (with isolated plastids in a test tube) the corresponding process in leaves occurs in the light and indeed some of the enzymes concerned are only fully operational when illuminated. In short, the "dark biochemistry" (Chapter 5) actually occurs in the light and, in that respect, is just as much a part of photosynthesis as the photochemical events which lead to the generation of assimilatory power.

4.9 Utilisation of Assimilatory Power

All plants consume ATP and $NADPH_2$ in a process usually referred to as the Calvin cycle. This will be described in more detail in Chapter 5. In outline, the cycle can be divided into three parts; carboxylation, reduction and regeneration. A "carboxylation" is the addition of carbon dioxide to a pre-existing acceptor in such a way that a "carboxylic" acid is formed. This is then reduced (Section 1.2), to the level of carbohydrate (CH_2O) by hydrogens from $NADPH_2$. Some molecules of this newly formed product are used immediately to regenerate the carbon dioxide-acceptor.

In all, there are 13 reactions in the Calvin cycle. Most occur three times in order to produce one molecule of product (a triose, or 3-carbon, sugar) from three molecules of carbon dioxide. This inevitably demands a somewhat complicated reaction sequence but, at least in outline, its principal features are readily understood (Fig. 4.9). In order to produce one molecule of triose phosphate from three molecules of CO_2 and one molecule of inorganic phosphate (Pi) the cycle must turn three times and is therefore best written in triplicate (Fig. 5.2). To do this it requires nine molecules of ATP and six molecules of $NADPH_2$. All of the $NADPH_2$ and six out of the nine ATPs are consumed in the only reduction (Section 1.2) sequence in the cycle - the conversion (Section 5.4) of PGA to triose phosphate. The remaining three ATPs drive the last stage (Section 5.14, equation 5.12) in the regeneration of the carbon dioxide-acceptor.

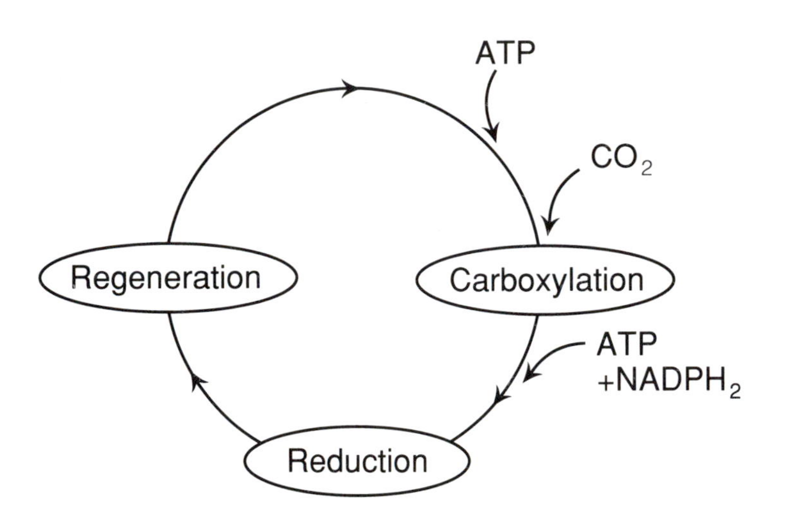

Fig. 4.9 The Calvin cycle in outline. Most of the assimilatory power (ATP + $NADPH_2$) is used in the reduction of the first formed product (PGA) of CO_2 fixation (Chapter 5), the remainder in the regeneration of the CO_2-acceptor. Product not used in regeneration is used in other metabolic processes such as starch and sucrose synthesis (Section 4.10).

4.10 Where It All Ends.

I know that I shall try the patience of some of my readers to the point of exasperation unless I am able to offer them an occasional glimpse of the common thread which strings this strange hotch-potch of subjects together. I hope, of course, that each chapter will stand on its own feet just as I could reasonably claim that John Steinbeck's "Pastures of Heaven" makes good reading even if it is regarded as a collection of short stories without a common factor. So I would be content with that at least, but let me say that what I am attempting to do is to trace the conversion of solar energy into chemical energy and that I hope finally to progress into ways that Man uses, or misuses, that energy. So far, we have seen how light drives electron transport and how electron transport produces the assimilatory power which, in turn, drives carbon assimilation. (More properly ATP and $NADPH_2$ are *consumed* by carbon assimilation but that, you might reasonably complain, is a bit of academic hair-splitting). In Chapter 5 we shall look into all of this in more detail but we might begin to wonder where it all ends and certainly *this* part of the tale ends with triose phosphate, sucrose and starch. These are the "end-products" of photosynthesis. More about sucrose and starch later. For the moment let us just note that they are both synthesised from triose phosphate (a 3-carbon sugar phosphate) and, since sucrose contains two hexose (6 carbon) units, and since starch contains many hexose units, they must be formed by a process of condensation which involves, at some stage, the release of Pi. Starch, when it is made in a leaf, is a short-term (mostly overnight) storage product and it is made in the chloroplast (hence starch pictures - Section 3.12). Sucrose, on the other hand, is made from triose phosphate in the cytosol (Fig. 4.10) and that brings us to a philosophical consideration of where photosynthesis ends. We have concluded that the "dark biochemistry" starts at the "top" of PSI (Section 3.10). We can similarly conclude that it finishes with sucrose, even though the reactions which make sucrose from triose phosphate occur in the cytosol rather than in the chloroplast (Section 4.10). Certainly, if we do things to a leaf which affect this part of the "dark biochemistry" we can see an impact on early photochemical events, like fluorescence, within seconds. That brings us to the question of how the triose phosphate finds its way into the cytosol from the stroma.

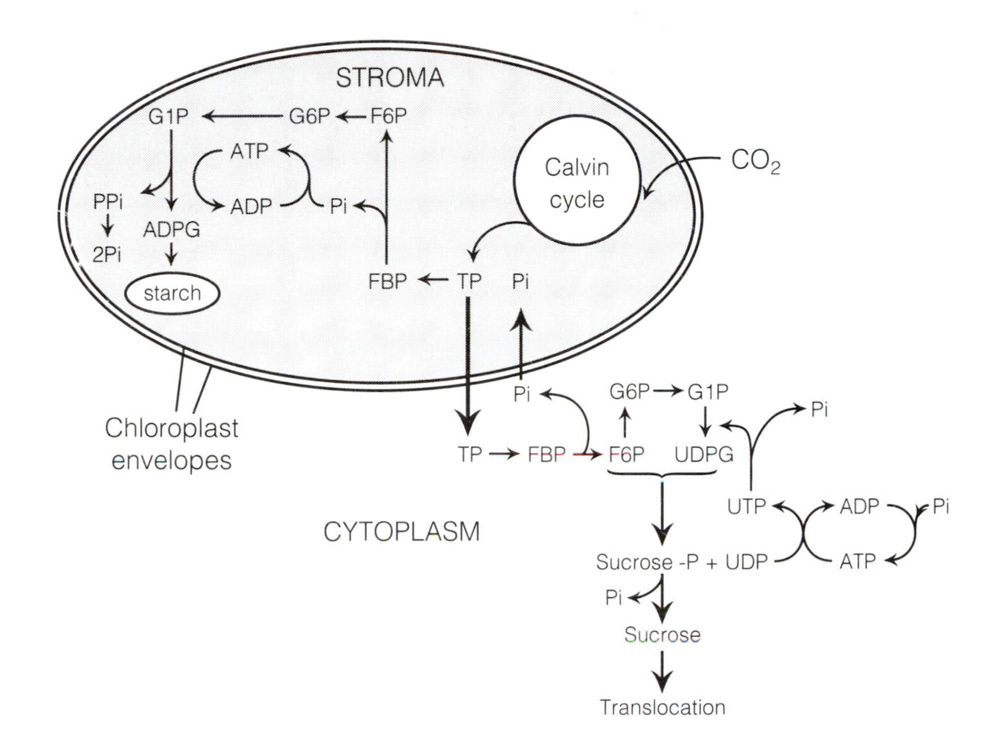

Fig. 4.10 Starch and sucrose synthesis. Both starch and sucrose are made of 6-carbon units. In turn these are synthesised from triose phosphate (TP). Starch is made in the stroma of the chloroplast, sucrose from TP exported through the phosphate translocator in exchange for Pi. Release and recycling of inorganic phosphate (Pi) during sucrose synthesis allow this process to continue.

4.11 The Phosphate Translocator

Isolated chloroplasts have figured large in my own contributions to photosynthesis. As you already know, (Section 4.1) Robert Hill's famous chloroplasts would evolve O_2 when supplied with an artificial electron acceptor but would not fix CO_2. Arnon and Whatley's chloroplasts were good at making "Assimilatory Power". They also fixed $^{14}CO_2$. (Fig. 4.3) but in such small amounts that it was stretching credibility to assume that they were fully functional. My own involvement came from my association with Robert Hill (my Ph.D. examiner) and my boss in my first job, Charles Whittingham. He suggested that improved performance by chloroplasts would be a useful enterprise for a young man to embark upon. Robin Hill (as ever) pointed me in the right direction and, to someone who had learned to prepare record-breaking mitochondria in Harry Beevers' laboratory at Old Purdue, it seemed to be no great problem. Of course it was more difficult than it seemed. There were so many variables that a logical approach would have taken too long and changing several factors at a time, although desperately unscientific, was the only way forward. In the end it worked and although, to some extent, my

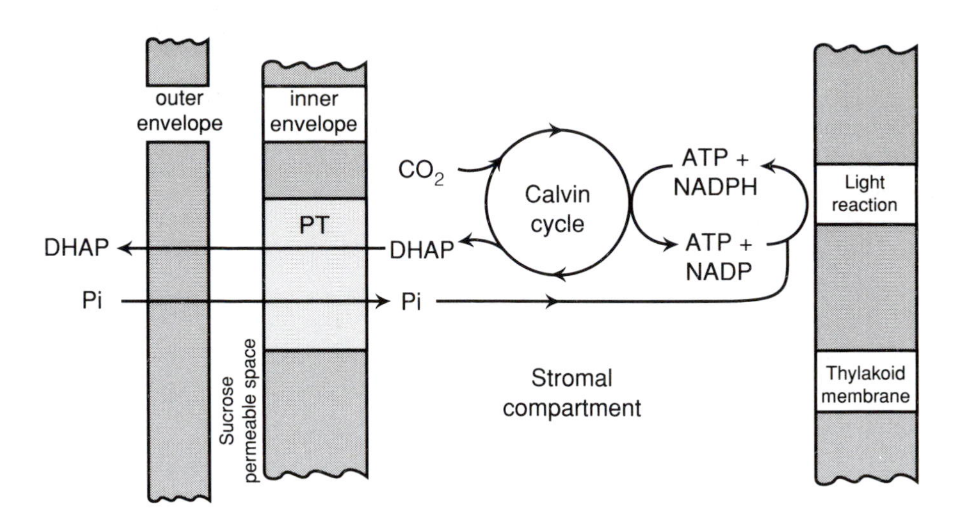

Fig. 4.11 The Pi Translocator. This is located in the inner envelope and permits a strict stoichiometric exchange of external Pi with internal triosephosphate (dihydroxyacetone phosphate or DHAP). Movements may occur in either direction according to circumstance (after Heldt).

colleagues and I had some of our thunder stolen at the eleventh hour by Dick Jensen in the United States (who used our techniques plus a magic component called inorganic pyrophosphate) the first achievement of rates by isolated chloroplasts similar to those attained by intact leaves was a matter for some rejoicing. Of course when you have succeeded in isolating fully functional chloroplasts for the first time there is then the problem of what to do with them. Happily the chloroplasts

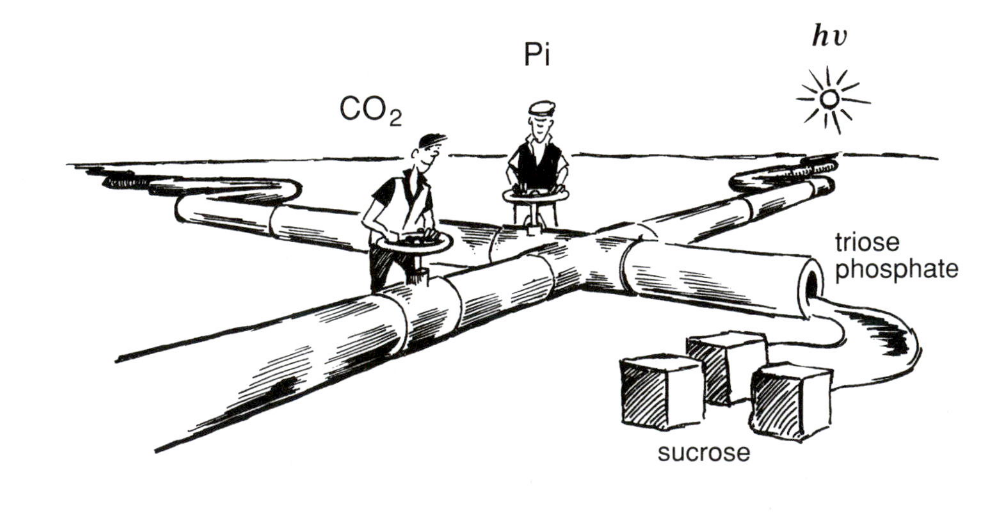

themselves nudged us in the right direction. Firstly, they would not make sucrose, which everyone at that time regarded as the end-product of photosynthesis. Secondly, they responded very readily to PGA and triose phosphates (which were not supposed to cross membranes) and finally they required Pi to function but were inhibited by the same compound in fractionally larger amounts - an inhibition that was reversed by triose phosphate and PGA. The reason for the failure to make sucrose was evident.

THE PHOSPHATE TRANSLOCATOR

It had to be made somewhere else. The trouble was that the enzymes concerned in sucrose synthesis were found in chloroplasts prepared by non-aqueous techniques and it was a long time before these enzymes could be accepted as contamination from the cytosol. The Pi requirement was easier to understand because there was a clear requirement for one molecule of Pi and three molecules of CO_2 in the synthesis of one triose phosphate. The inhibition by larger concentrations of Pi and its reversal by triosephosphate or PGA was perplexing but finally, in 1970, my colleagues Carl Baldry, Chris Bucke, Bill Cockburn and myself were bold enough to come up with the following

"if sugar phosphates are exported from the chloroplast there must be a corresponding import of phosphate (in some form) if steady-state photosynthesis is to be maintained. .... A direct obligatory exchange between orthophosphate (outside) and sugar phosphate (inside) could account for the inhibition of photosynthesis by orthophosphate and its reversal by sugar phosphates."

Meanwhile, in the spring of that year (1970), a German colleague, Hans Heldt, visited Imperial College (where this work had been done) and discussed with me his use of "centrifugal filtration" in examining the movement of ATP and ADP across the chloroplasts envelope. I suggested, on the strength of our results, that he looked, in the same way, at the transport of PGA and Pi. He did and the phosphate translocator grew from hypothesis into physical reality. So what does the phosphate translocator do? It facilitates an exchange across the chloroplast envelope, on a one to one basis, of internal triosephosphate with external Pi. This is its principal function. The chloroplast is not an independent entity living in a protected environment of its own within the leaf. It has to maintain itself, of course, but it is there to supply the plant with the wherewithal for the plant's existence. If photosynthesis is to continue unchecked and, if internal cycling of Pi is insufficient to maintain rates of synthesis, then 1 molecule of Pi *must* be imported for every triosephosphate exported. In other words, photosynthesis is a cellular affair. It is not

Photosynthesis and phosphate ***"a cellular affair"***

a process which chloroplasts can achieve unaided for very long. This is why photosynthesis will virtually cease in isolated chloroplasts as external Pi is consumed and diminishes and why three molecules of O_2 are then evolved for every Pi added. This is why Pi, in relative excess, also inhibits. Clearly if too much triosephosphate is pulled out of the chloroplasts because there is too much Pi competing for entry, depletion of Calvin cycle intermediates will diminish replenishment of the CO_2 acceptor. Competition is central to this last statement because if we regard the translocator as a door controlled by a demon (Section 3.13), intent on regulating one-to-one exchange, regulation will only be maintained if competitive pressures are maintained. Thus Pi competes with C3 intermediates and *vice versa*. When internal triose phosphate is in relative excess, and Pi is freely available externally, all is well but this implies a fine balance between too little and too much cytosolic Pi.

All of this has implications for a complex system of regulation which is still not fully understood. Relatively speaking, there is an immense pool of Pi within the reservoir of the central vacuole but this is not free to enter the cytosol, otherwise inhibition by excess Pi would rapidly follow. This leaves sucrose synthesis (from triose phosphate in the cytosol) as the main mechanism for recycling Pi. Fine regulation of enzymes concerned in sucrose synthesis, including the involvement of an "effector" called fructose 2,6-bisphosphate, serves to allow sucrose synthesis to proceed at a rate which matches the availability of triose phosphate for export.

[According to present understanding, fructose 2,6-bisphosphate ($Fru2,6P_2$) works by 'sensing' the availability of triose phosphate. Triose phosphate increases the amount of substrate available to the FBPase (the enzyme in Fig. 4.10 which splits FBP into F6P and Pi), but decreases the concentration of $Fru2,6P_2$ (because it inhibits the activity of F6P,2-kinase, the enzyme which synthesises $Fru2,6P_2$). This decrease, in turn, releases an inhibition of FBPase. This has a critical effect on sucrose synthesis, because FBPase catalyses the first irreversible reaction in the cytosol.]

4.12 The Red Drop

This section might seem to those of my readers already knowledgeable about photosynthesis, to have been added as an afterthought and out of context. In a way they would be right but it is difficult, as we have already seen at the beginning of this chapter, to avoid problems of this nature while striving for simplicity. So here I have erred by neglecting to afford a proper recognition to a key observation which influenced the development of the whole concept of the Z-scheme and its formulation by Robert Hill and Fay Bendall. Now is my opportunity to put this right but, in order to do that, we must forsake carbon assimilation momentarily (before returning to it in more detail in Chapter 5) and return to the photochemistry.

Robert Emerson made a number of important contributions to photosynthesis before his life was tragically ended when an aircraft in which he was travelling crashed into the Hudson River. One of these was his recognition of the *"red drop"* and the associated phenomenon of *"enhancement"*. Fig. 4.13 shows the absorption spectrum of chlorophyll. It is also possible to derive an "*action* spectrum" for photosynthesis to see how effectively different wavelengths of light drive photosynthesis. Fig. 4.12 shows such a comparison based upon equal numbers of incident photons at each wavelength. On *this* basis, the action spectrum closely follows the absorption spectrum. This is because red and blue photons are more readily absorbed by chlorophyll and if a leaf is offered equal numbers of blue, green

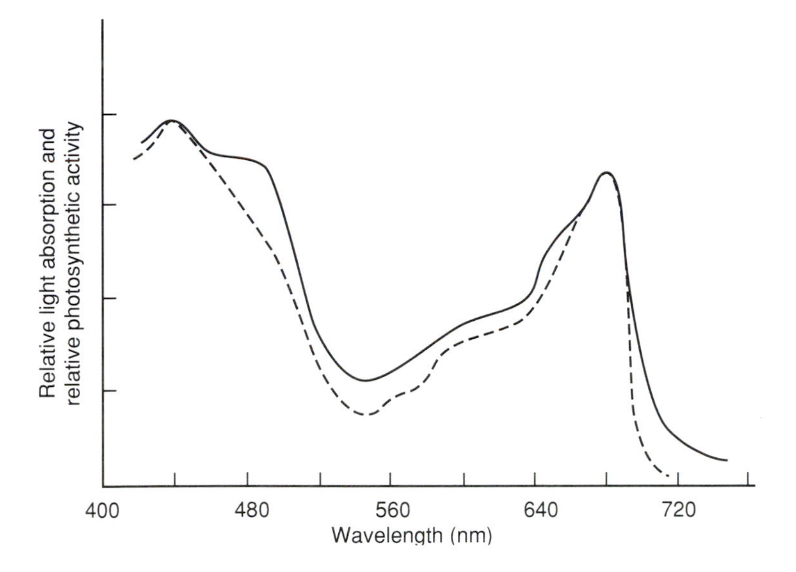

Fig. 4.12 Absorption and action spectra compared. Solid line, absorption spectra for *Ulva*. Broken line, relative photosynthetic rates for equal numbers of incident quanta at each wavelength (after Haxo and Blinks).

and red photons, fewer green will be absorbed. Thus, photosynthesis in weak green light will be slower than in red or blue light at equal photon flux densities. Fig. 4.13, on the other hand, shows an action spectrum *based on quantum yield* (Section 6.6) at different wavelengths. In other words, it compares effectiveness of red, blue and green photons once absorbed. We might expect such an action spectrum to be a horizontal line if photons were equally effective over the whole of the range of

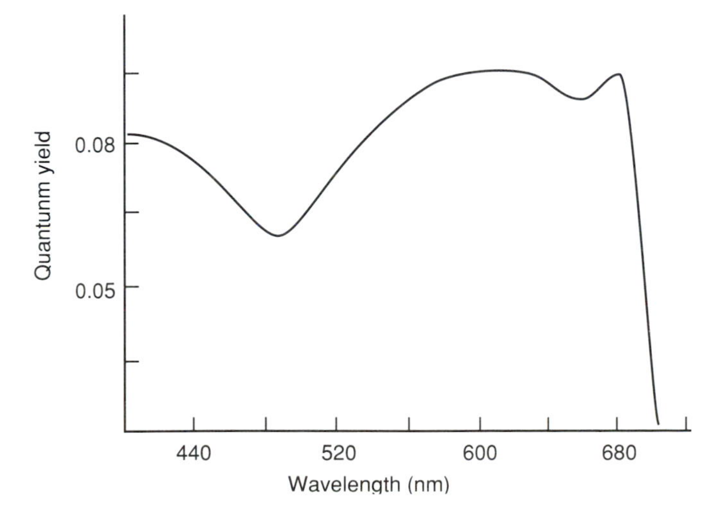

Fig. 4.13 Action spectrum of O_2 evolution by *Chlorella*. Maximum quantum yield as a function of wavelength (after Emerson and Lewis). Note the dramatic fall in yield (the "red-drop") at wavelengths above 680 nm.

wavelengths absorbed by chlorophyll since blue photons as well as red give rise to "excited state I" (Section 2.8) and even those few green photons which are absorbed will do the same. The fact that there is a big dip (Fig. 4.13) at about 450 nm may be due to absorption of photons by carotenoids and inefficient transfer of excitation energy to chlorophylls. Emerson was much more intrigued by the abrupt drop in efficiency in red light at wavelengths above 700 nm. The absorption spectrum indicates that there is still appreciable absorption of light at these wavelengths. The red drop suggests that these long wave photons are not used to best effect. This led Emerson to mix red light with weak blue light. Such mixtures supported more photosynthesis than the sum of both applied singly and the "extra" photosynthesis resulting from adding red light between (700 and 730 nm) to blue light is now known as "enhancement". We can now explain this in terms of the Z-scheme - a concept which this observation did much to create. Excitation energy from short wave (higher energy) red photons or blue photons can be readily transferred "down-hill" to chlorophyll molecules in both photosystems but "up-hill" transfer of excitation energy from red photons at 700 nm above is difficult. Accordingly, light at, say, 680 nm is more effective in operating both photosystems than light at 700 nm which favours PSI. Similarly a mixture of both short and long wave photons is better than either above. Hence "enhancement" and the inference, that came to be drawn, that two photosystems must be involved. All of this, of course, bears on Emerson's long-standing controversy with Warburg (Section 6.7) because a quantum efficiency of 8 is implicit in the Z-scheme which requires input of 4 photons in each of its two photosystems in order to drive linear transport of electrons from water to NADP.

4.13 Summary

In outline, and in principle, photosynthesis can still be described in terms of a classic experiment (Fig. 4.8). Light energy ($h\upsilon$), absorbed by chlorophyll initiates electron transport (i.e. light energy is converted into electrical energy at this stage). Electron transport, in turn, generates chemical energy as assimilatory power (ATP and $NADPH_2$). In linear (non-cyclic) electron transport, electrons from water are lifted uphill, to NADP, by the combined effort of photons absorbed by each of the two photosystems (PSII and PSI) which constitute a major part of the Z-scheme. In cyclic electron transport there is no consumption of water or generation of $NADPH_2$. On the contrary, electrons simply cycle round PSI. However, both cyclic and linear electron transport generate a proton gradient which discharges through an ATPsynthase driving the formation of ATP from ADP and inorganic phosphate. In effect, ATP and $NADPH_2$ are the end products of the photochemistry. They are used by the *"dark biochemistry"* to join carbon dioxide to an acceptor and reduce it to CH_2O. Triose phosphate produced in carbon assimilation (which is considered more fully in Chapter 5) may be used to form starch in the stroma of the chloroplast or sucrose in the cytosol of the cell. Export of triose phosphate from the chloroplast occurs via the phosphate translocator. Inorganic phosphate released by sucrose synthesis in the cytosol is recycled, entering the stroma in exchange for exported triose phosphate. Once again this inorganic phosphate is combined with ADP to give ATP in the reactions driven by electron transport. Once again ATP is consumed in carbon assimilation so that the whole process of energy transduction (light energy $\rightarrow$ electrical energy $\rightarrow$ chemical energy) can continue indefinitely.

Chapter 5

The Dark Biochemistry

Fixing carbon

Doubling up with bootstrings

Necessary evils?

Virtue or necessity?

Greenhouses within plants

C3, C4 and CAM

5.1 Photosynthetic Carbon Assimilation

Photosynthesis, as we have seen, is a process of energy transduction in which light-energy is first of all converted into electrical energy and then into chemical-energy. In the first stage, light energisation of chlorophyll initiates some cyclic electron transport but principally linear electron transport from water to NADP which, accordingly, becomes reduced to $NADPH_2$ (see Chap 4). ATP is also formed during electron transport by addition of Pi to ADP, thereby conserving another fraction of this electrical energy in a chemical form. Together ATP and $NADPH_2$ constitute "assimilatory power" (Section 4.8). This assimilatory power is then used in photosynthetic carbon assimilation (Fig. 5.1). There are three main forms of carbon assimilation which are referred to, for reasons which will become clear, as C3, C4 and CAM. In a sense they are not photosynthetic processes at all, because all of the individual reactions can be *made* to work in the dark and indeed they are sometimes referred to, often rather dismissively, by photochemists and biophysicists, as *"the dark biochemistry"*.

PHOTOSYNTHETIC CARBON ASSIMILATION

[There is a sort of hierarchy amongst those who work in photosynthesis, which is such a vast subject that it embraces all manner of scientific disciplines and occupations. At one end there are the physicists, the biophysicists and the photochemists. At the other end there are the farmers, the foresters and the horticulturists. By the very nature of things, concepts in physics tend to be more difficult to grasp than concepts in farming. Relativity, or quantum mechanics, is less easily grasped than the need to plough, fertilise or spray. In general we all tend to respect skills which we perceive as useful but do not understand. Sadly, those to whom we afford our respect do not always respond with due humility. All of us must surely have come across the medical practitioner who addresses his patients like a school teacher explaining things in a kindergarten. In this hierarchy, therefore, the biophysicist is undoubtedly perceived as rather superior to the agronomist even though agronomy (like ecology) is an immensely complicated subject. In this context there was also the classically slanderous remark, attributed to C. P. Darlington (a former Professor of Botany at Oxford) who, paraphrasing Oscar Wilde, described ecologists as *"the incompetent in pursuit of the incomprehensible"*.]

So the "dark biochemistry" may not be really ***photo***synthesis (although it must always be remembered that, even though it can be made to go in the dark, it normally occurs in the light and involves many enzymes which are light-activated). That is not to say that it is in any way less important than the photochemistry or that its elucidation has been anything other than a triumph for plant biochemists. Central to all forms of carbon assimilation is the Calvin cycle - named after Melvin Calvin who was honoured by the Nobel Prize for its elucidation. It is also sometimes referred to as the Benson-Calvin cycle by those who feel that the Nobel committee might have been more generous to Andy Benson, the second most senior investigator in the team that Calvin then led. There are also more neutral names like *"the reductive pentose phosphate pathway"* which reminds us that this is a reductive process (Section 1.2) in which carbon dioxide becomes CH_2O, and that 5-carbon sugar phosphates (pentose phosphates) are key intermediates.

C3 PHOTOSYNTHESIS

5.2 The Calvin Cycle

In concept, carbon assimilation, via the Calvin cycle, is simple. It involves the addition of a molecule of carbon dioxide to an acceptor molecule. The newly formed addition compound then undergoes a series of changes (Figs. 5.2 and 5.3) including reduction, which are designed to form (and set aside) a product. At the same time, the acceptor is regenerated so that the process might continue.

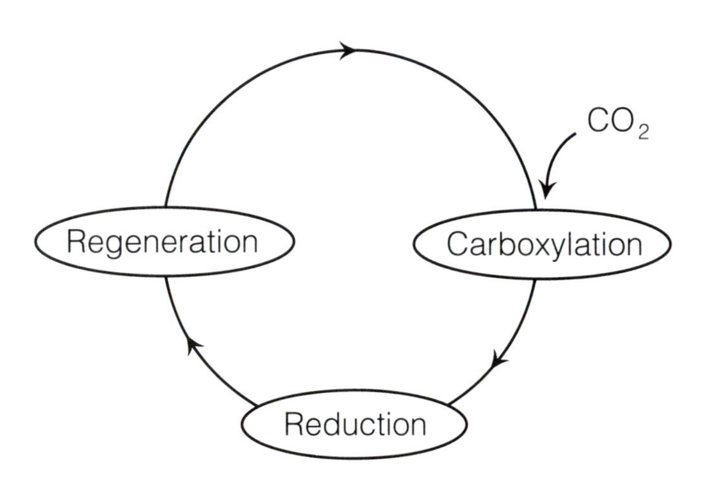

In actuality, the Calvin cycle, is rather more complicated than this but not as complex as it is sometimes portrayed. The CO_2 acceptor is a sugar phosphate, containing five atoms of carbon (i.e. it is a pentose) In addition it contains two atoms of phosphorous. It is called rib***u***lose 1,5-bisphosphate.

[To a chemist this name indicates that the sugar is a ketose, like fructose (containing a C=O group), rather than an aldose, like glucose (containing a CHO group) - hence *ribulose* rather than *ribose*. The term 1,5-***bisphosphate*** means that there are two phosphate groups, one at either end of the molecule, attached to carbons 1 and 5 respectively.) Reflection would also tell the chemist that this is rather an unhappy structure, that it has been wound up by its very nature into a rather unstable configuration and that (given an opportunity, a catalyst, and a new partner) it would happily change into something else.]

This then is the CO_2 acceptor and, when offered carbon dioxide in the presence of the enzyme RUBISCO, it happily embraces the carbon dioxide and immediately splits into two new 3-carbon compounds. Thus, in arithmetical terms we have:-

$$\underset{\text{(the acceptor)}}{C5} + \underset{\text{(the } CO_2\text{)}}{C1} \rightarrow \underset{\text{(the product)}}{2C3} \qquad \text{Eqn. 5.1}$$

THE CALVIN CYCLE

This is a process of "carboxylation" because the carbon dioxide joins to the acceptor (ribulose 1,5-bisphosphate or RuBP for short) in such a way that a new carboxyl group is formed. Indeed, as a result of the addition of the carbon dioxide and the splitting of the 6-carbon intermediate so formed into two C3 compounds, *two* new carboxyl groups are produced, in the shape of two molecules of 3-phosphoglycerate.

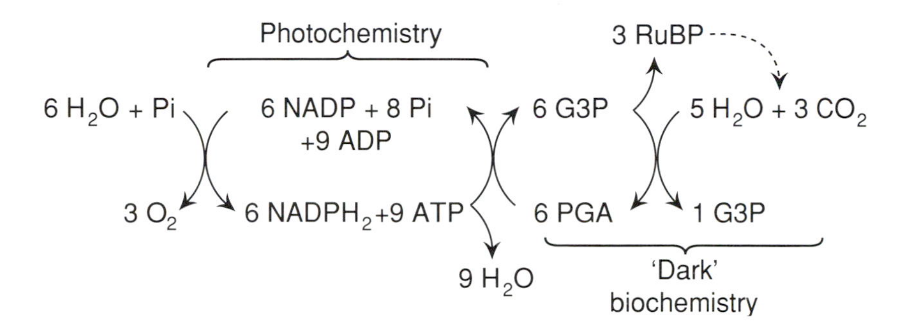

Fig. 5.1 Light and Dark Events in Photosynthesis The first product of photosynthetic carbon assimilation is 3-phosphoglycerate (PGA) a three-carbon acid. This is converted to a corresponding three-carbon sugar phosphate called glyceraldehyde 3-phosphate (G3P) in a reductive process which consumes two thirds of the ATP and all of the $NADPH_2$ generated by the photochemistry. The rest of the ATP is utilised in the regeneration of the CO_2 acceptor, ribulose 1,5-bisphosphate (RuBP) in a process in which five 3-C compounds (triose phosphates) are rearranged to give three 5-C carbon compounds (see text). The "assimilatory power" (ATP + $NADPH_2$) generated by photochemical events in the light is consumed in the "dark biochemistry" - so called because none of the biochemical reactions involved have an absolute requirement for light even though, in a photosynthetic context, they normally occur only in the light. Triose phosphates not used in regenerating the CO_2-acceptor undergo subsequent transformations to starch, sucrose etc. (Section 5.3).

These newly formed molecules of 3-phosphoglycerate (3PGA, or PGA, for short) are now reduced and it is at this stage that all of the $NADPH_2$ and most of the ATP which is consumed, by the Calvin cycle, is used. Once the PGA has been reduced (to triose phosphate - i.e. to a sugar phosphate containing 3 carbons) the next trick is to generate a 5-carbon acceptor molecule from 3-carbon precursor. Arithmetic demands that, if you wish to make 5-carbon compounds from 3-carbon compounds without loss, you must convert five threes into three fives

$$5\ C3 \rightarrow 3\ C5 \qquad \text{Eqn. 5.2}$$

and this is precisely what is achieved in the regenerative part of the cycle. In essence, therefore, we have carboxylation, reduction and regeneration. The latter may seem a little involved at first because it involves two types of reaction. The first joins (or condenses) two C3s to give one C6. This is a so-called aldol, condensation i.e. **ald**ehyde + alcoh***ol***. It is catalysed by an enzyme which is therefore called **ald*ol*** ase.

$$C3 + C3 \rightarrow C6 \qquad \text{Eqn. 5.3}$$

The second type of reaction transfers a C2 fragment from the top of the C6 to a C3, thereby forming a C5 and a C4

$$C6 + C3 \rightarrow C4 + C5 \qquad \text{Eqn. 5.4}$$

and another aldol condensation, this time involving a 4-carbon sugar phosphate

$$C3 + C4 \rightarrow C7 \qquad \text{Eqn. 5.5}$$

This is followed, in turn by a second reaction catalysed by transketolase in which a C2 unit is transferred (from the top of the newly formed C7 sugar) to a C3, to give two C5s

$$C3 + C7 \rightarrow C5 + C5 \qquad \text{Eqn. 5.6}$$

As stated above, the principal function of this repeated sequence (of condensation and transfer) is to rearrange five C3s to form three C5s.

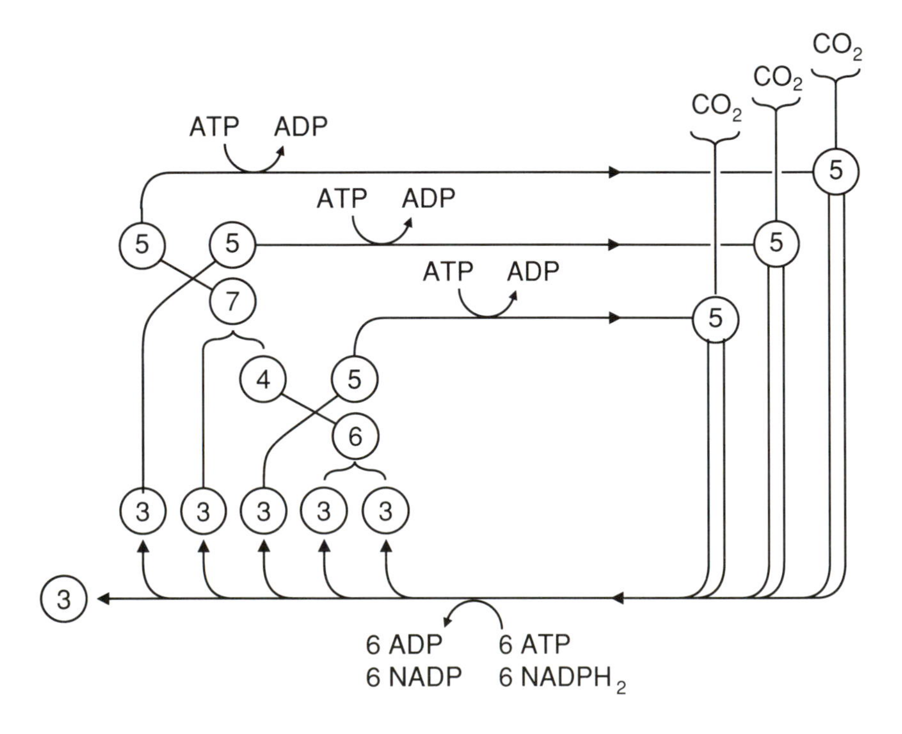

Fig. 5.2 The Calvin Cycle in outline showing how three molecules of CO_2 give rise to one molecule of 3-carbon product and how five C3 molecules are rearranged to give three C5 acceptors. The piece transferred is a 2-carbon **ketose** unit and the transfer is catalysed by an enzyme therefore called trans**ketolase.**

In striving for simplicity, emphasis has been put on three prime features of the Calvin cycle, i.e. carboxylation (of the acceptor), reduction (of the product) and regeneration (of the acceptor). The complete sequence is, however, just a shade more complex (Fig. 5.3) and includes aspects of equal importance.

As noted above, the second sequence of condensation and transfer (Fig. 5.3) (again catalysed by aldolase and transketolase respectively) involves the addition of a C3 or triosephosphate to the C4 molecule (erythrose 4-phosphate) created in the first sequence. The alert reader will already have noted that whether we condense two molecules of triosephosphate (two C3s), or one C3 and one C4, the end-product

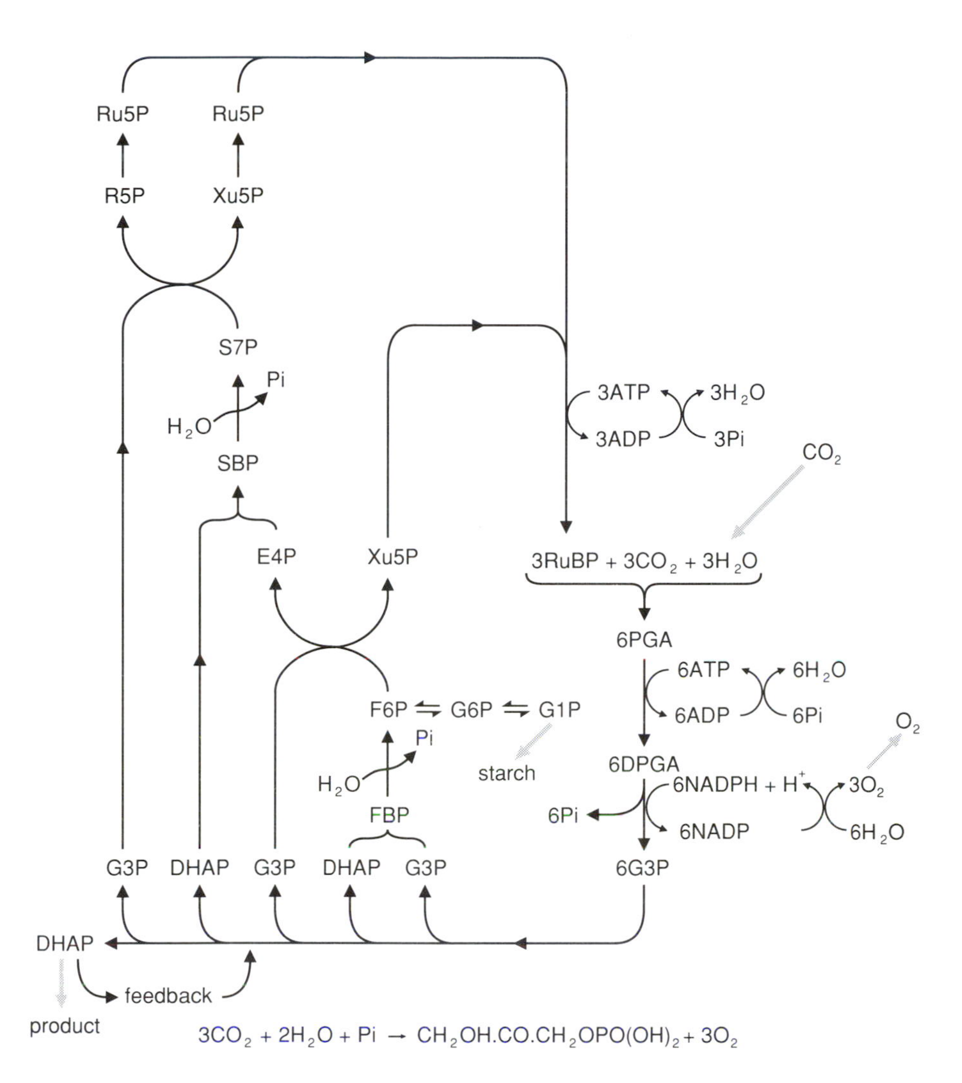

Fig. 5.3 Calvin Cycle in more detail. (c.f. simplified version in Fig. 5.2) In this version, the intermediates are named. Thus the triose phosphates are dihydroxyacetone phosphate (DHAP) and glyceraldehyde 3-phosphate (G3P). The tetrose phosphate is erythrose 4-phosphate (E4P), the pentose monophosphates are xylulose 5-phosphate (Xu5P), ribose 5-phoshate (R5P) and ribulose 5-phosphate (Ru5P); the hexose monophosphates are fructose 6-phosphate (F6P) and glucose 6-phosphate (G6P) and the heptose monophosphate is sedoheptulose 7-phosphate. Finally, the bisphosphates are 1,3 diphosphoglycerate (DPGA) which is also more properly called glycerate 1,3-bisphosphate (GBP), ribulose 1,5-bisphosphate (RuBP), fructose 1,6-bisphosphate (FBP) and sedoheptulose 1,7-bisphosphate (SBP) respectively. All of the left-hand side of the sequence is concerned with the regeneration of the CO_2 acceptor as 5C3s are converted to 3C5s. The reduction of the CO_2 fixation product (PGA) occurs at bottom right and is shown in still more detail in Fig. 5.4.

will, in each case, be a ***bis***phosphate (in these particular circumstances, a C6 and a C7 bisphosphate - i.e. C6 and C7 sugars, respectively, with a phosphate group at each end). Before the C2 transfer can follow, a phosphate group has to be removed from the end of the molecule which is to receive the C2 unit. This is accomplished, in each case, by a hydrolysis catalysed by the appropriate bisphosphat***ase*** e.g:-

$$\text{fructose 1,6-bisphosphate} + H_2O \rightarrow \text{fructose 6-phosphate} + \text{inorganic phosphate} \quad \text{Eqn. 5.7}$$

and

$$\text{sedoheptulose 1,7-bisphosphate} + H_2O \rightarrow \text{sedoheptulose 7-phosphate} + \text{inorganic phosphate} \qquad \text{Eqn. 5.8}$$

At first glance, these are relatively trivial reactions but they have an importance in regulation and they also constitute two points in the cycle at which inorganic phosphate (Pi) is released. A further six molecules of Pi are released in the reduction of PGA to triose phosphate so that, in the Calvin cycle as formulated in Fig. 5.3, eight molecules of Pi are released in all. Moreover, *and this is of crucial importance,* only five of the six molecules of PGA created by the initial carboxylation of three molecules of ribulose 1,5-bisphosphate

$$3C5 + 3 \times CO_2 \rightarrow 6C3 \qquad \text{Eqn. 5.9}$$

are consumed in the regeneration of the acceptor (Eqn. 5.2). The remaining C3 constitutes product. (Figs. 5.1 & 5.2)

As already noted, the "dark biochemistry" of CO_2 fixation owes its name to a classic demonstration by Arnon, Whatley and others that isolated chloroplasts can be made to generate ATP and $NADPH_2$ in the light and that these two compounds can then be consumed, in darkness, in biochemical reactions in which CO_2 is incorporated or "fixed". This is not to say that the reaction of carbon fixation normally occur in the dark. On the contrary CO_2 fixation by the Calvin cycle is a process which normally occurs in the light and many of the enzymes involved are "light-activated" i.e. they work better in the light than in the dark because of regulatory processes. Many of these are themselves reductive in nature and involve ferredoxin, a compound reduced by photosystem I in the light (Chap 4).

5.3 End Product and Autocatalysis

What happens to the C3 product of the Calvin Cycle ? As we have seen, two C3s can be condensed to give a C6 (a hexose). Similarly, after some alteration, two C6s (essentially fructose and glucose) can be condensed to give the disaccharide sucrose. This C12 compound, which we all know simply as "sugar", is one of the most important compounds in plant metabolism because it is a transport metabolite, i.e. it is in this form that most of the carbon (in most plants) is transported from the leaves to the stems, roots, flowers, fruits, storage organs and so on. At these sites it is this compound which is then used to provide the organic carbon component of all other plant products - amino acids, proteins, lipids (fats), structural materials such as cellulose and lignin (wood) etc., etc.

Some C6 units also undergo condensation reactions of various types to form straight-chain or branched-chain polysaccharides - such as starch which is composed of many C6 units. Starch (see starch pictures, Section 3.12) is an important, temporary (overnight), storage product in chloroplasts and a more permanent storage product in stems, roots, tubers, etc. In chloroplasts it is formed directly from C3s (via C6s). In other places (roots, tubers, fruits, seeds, etc.) it is formed from C6s released (by hydrolysis) from sucrose translocated to those sites from the leaves where it is formed.

END PRODUCT AND AUTOCATALYSIS

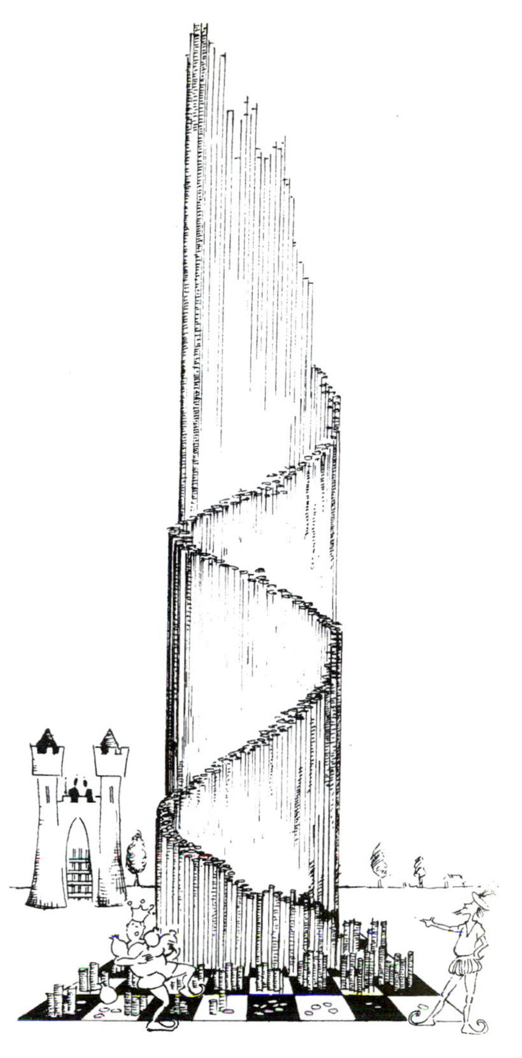

This then, is what carbon dioxide fixation or carboxylation is all about. The function of the Calvin cycle is to combine carbon dioxide with an acceptor molecule to form a product and, at the same time, to regenerate the CO_2-acceptor so that the cycle might continue to turn. It will be evident, either immediately or on reflection, that this is not enough. If the cycle only regenerated as much acceptor as it used there would be no capacity for growth. The Calvin cycle is, and must be, *"a breeder reaction"* with the capability of producing more CO_2-acceptor than is present in the first instance. This is called *"autocatalysis"* - a way of pulling yourself up by your own bootstrings. It works very simply because the C3 compounds which go on to form hexoses, sucrose or starch are precisely the same as those used to regenerate the C5 CO_2-acceptor (ribulose 1,5-bisphosphate). Of course the leaf can not get something for nothing. If it deflects potential C3 product into C5 acceptor it will be left with less to export from the chloroplast. Fortunately, *"doubling up"* is arithmetically very sound (or equally disastrous depending on circumstance).

[There is the familiar fable of the king who sought to reward one of his subjects by offering him half his kingdom. The man, no doubt a greedy mathematician, asked instead that the king should give him a chess board with one penny on the first square, two on the second and so on, thereby emptying the Treasury. There is an equally "infallible" system" for betting on horses - if you lose the first time simply put twice as much stake on a horse in the next race and so on. Perhaps you will not tire of the arithmetic before I did but even doubling up on the first 32 squares of the chessboard would increase the original one unit stake by six orders of magnitude.]

The Calvin cycle can, if it deflects all of its products into new substrate, double its RuBP for every 5 turns of the cycle. This is the basis of plant growth, and, indirectly, the basis of all animal life. *"All flesh is grass"* and all plants depend upon the autocatalytic aspect of the Calvin Cycle in order to grow.

5.4 The Reduction of PGA

Having looked at the cycle as a whole, we can now take a second look, in more detail, at some of its salient features. Much attention was paid in Chapter 1 to the nature of oxidation and reduction as the source of biological energy. The reduction of PGA, the C3 compound formed from RuBP by carboxylation, is the only reductive step in photosynthetic carbon assimilation. The hydrogen donor is the $NADPH_2$ formed as a result of linear transport of electrons from water (Chap 4) so that, in effect, the hydrogen from water is passed, via NADP, to PGA in order to produce a chemical compound which is basically a form of CH_2O. The reduction of PGA (Fig. 5.4) occurs in two stages. In the first stage PGA is phosphorylated at the expense of ATP to give diphosphoglycerate or, more properly, glycerate 1,3-bisphosphate (GBP)

$$PGA + ATP \rightarrow GBP + ADP \qquad \text{Eqn. 5.10}$$

This reaction is energetically unfavourable and is pushed, as though uphill, by the free energy of hydrolysis of the ATP molecule, by the 7 Kcal or so of energy *which would have been released had the ATP been hydrolysed to ADP and Pi* (Chap 4). Once the worst is over, once GBP has been formed, the second stage of the reaction (the actual reduction) is energetically downhill as the newly formed GBP is reduced to glyceraldehyde 3-phosphate (G3P)

$$GBP + NADPH_2 \rightarrow G3P + NADP + Pi \qquad \text{Eqn. 5.11}$$

It should be noted, in passing, that this reaction, like those catalysed by the bisphosphatases, is one which releases Pi. In fact, since all six molecules of PGA (produced by the carboxylation of three molecules of RuBP in Fig. 5.1) are reduced to G3P, six molecules of Pi are released at this stage in the cycle. Together with one Pi from each of the two bisphosphatase reactions, this makes a grand total of 8

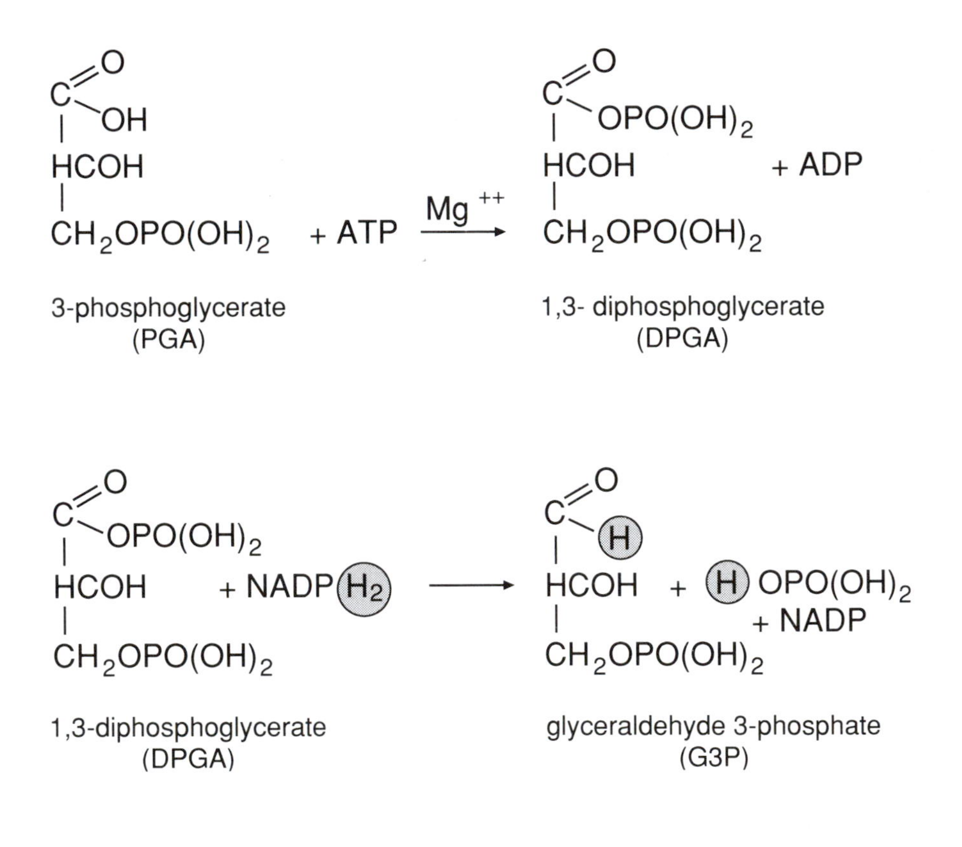

Fig. 5.4 Reduction of PGA to triose phosphate (Equations 5.10 and 5.11 in more detail). This reduction is also shown at the bottom right of Fig. 5.2.

molecules of Pi. On this basis we can start to see what goes in and out of the cycle, in terms of energy and molecules and what additional implications should be considered.

If we represent the Calvin cycle as we have in Figs. 5.2 and 5.3 we see that 3 molecules of RuBP are joined to 3 molecules of carbon dioxide to form 6 molecules of PGA, all of which are reduced to G3P. Of these, 5 are used to regenerate the CO_2-acceptor, RuBP (Eqn. 5.2) and, in this process, 3 molecules of ATP are consumed in the final stages in which 3 molecules of ribulose 5-phosphate (Ru5P) are converted to RuBP

$$3Ru5P + 3ATP \rightarrow 3RuBP + 3ADP \qquad \text{Eqn. 5.12}$$

This again is an example of how energy is utilised or transferred. The conversion of Ru5P to RuBP is a "winding-up" process. The free energy of hydrolysis of ATP (Section 4.6) is, in a sense, consumed in a process in which the relatively stable Ru5P molecule is twisted into the relatively unstable RuBP configuration. This makes RuBP ripe for carboxylation and a return to stability (by way of stable end-products). Of the nine ATP molecules used in the Calvin cycle (as represented in Figs. 5.2 and 5.3), three are used for this purpose and six in the phosphorylation of PGA to GBP, prior to its reduction to G3P; a reaction (Eqn. 5.11) in which all six molecules of $NADPH_2$ are consumed.

5.5 Energy Inputs

The ways in which ATP and $NADPH_2$ (Arnon's "assimilatory power. - Section 4.8) are consumed in the Calvin cycle are outlined above. We can also take the opportunity to do a little arithmetic along these lines. Fig. 5.3 portrays the Calvin cycle in a way which makes it easier to follow the "sugar phosphate shuffle" in which five molecules of C3 product are converted into three molecules of the C5 sugar phosphate which is the CO_2-acceptor. Represented in this way, Fig. 5.3 calls for the consumption of three molecules of carbon dioxide (fixed and converted into CH_2O). We can put values on the energy inputs involved but it should be noted that the values used here are approximate rather than precise because we are seeking to arrive at "ball-park" figures, and that precision, in these regards, depends on detail which is not readily available to us.

[We can burn sucrose in a calorimeter and arrive at a precise value of the heat released but fuel consumption by a vehicle is vastly influenced by the way it is driven and by external conditions such as head winds. Similarly free energy changes in metabolism depend on things like the prevailing Mg^{++} concentration etc.]

With these provisos in mind, we can put a figure of 7 Kcal or 29 Kjoules on ATP and 52 Kcal (or 218 Kjoules) on $NADPH_2$ (Fig. 4.2). This gives us a total of 125 ($3 \times 7 + 2 \times 52$) Kcal or 523 Kjoules for 3ATP + $2NADPH_2$. In terms of our energy approximation, this offers reasonable agreement in the sense that we can ascribe a value of 672/6 = 112 Kcal or 470 Kjoules to CH_2O on the basis of calorific burning of glucose, $C_6H_{12}O_6$ (which gives us 672 Kcal or 2822 Kjoules). In other words, the Calvin cycle is consuming about 125 Kcal or 525 Kjoules in order to produce one molecule of product (G3P) with a structure not too different from CH_2O. Any attempt to look into this more closely would take us down the path illuminated by C.P. Snow (Section 1.8) into the esoteric world of entropy which is largely beyond the scope of this book. We can, however, find comfort in the fact that the Calvin cycle is not only energetically feasible but also extremely efficient in terms of energy locked into product per unit of energy used in synthesis.

5.6 Stoichiometries

This is a difficult word to spell and indeed, one which is impossible to spell to the universal satisfaction of scientific journals in both the U.K. and the United States. Its meaning is less difficult. The stoichiometry of a reaction is simply the numerical relationship between the number of molecules involved. For example, in the reaction:-

$$A + B \Leftrightarrow C + D \quad \text{Eqn. 5.13}$$

there is a one-to-one stoichiometry, i.e. one molecule of A reacts with one molecule of B to give one of C and one of D. Conversely, two molecules of NADP are

$$2NADP + 2H_2O \rightarrow 2NADPH_2 + O_2 \quad \text{Eqn. 5.14}$$

involved when NADP serves as an oxidant in the Hill reaction in order to bring about the release of one molecule of O_2 from two molecules of water.

In Figs. 5.2 and 5.3 the overall stoichiometry is as follows. First of all, in order to generate the necessary assimilatory power (6 molecules of $NADPH_2$ + 9 molecules of ATP) we have

$$\begin{aligned} 6NADP + 6H_2O + 9ADP + 9Pi \rightarrow \\ 6NADPH_2 + 9ATP + 9H_2O + 3O_2 \end{aligned} \qquad \text{Eqn. 5.15}$$

This equation immediately poses problems which require explanation. It follows from Eqn. 5.15 that six molecules of water need to undergo photo*lysis* in order to provide the hydrogens for the reduction of the NADP. This will release three molecules of oxygen. The source of the nine molecules of water on the right of the equation is less obvious until it is remembered that the condensation of ADP and Pi releases one molecule of water per ADP esterified (Eqn. 5.16). The assumption, implicit in Eqn. 5.15 that linear electron transport (of $6H_2$) from water to NADP can produce nine molecules of ATP *is* only an assumption. Measurements put this particular stoichiometry at between one and two ATP per $2e^-$ (or, in present terms, 2H) transported but, in fact, the stoichiometry or the coupling might be "flexible" within these limits. Stoichiometry in a strictly chemical reaction is invariable. A molecule of water is comprised of 2 hydrogen atoms and one oxygen atom (i.e. 1/2 of an oxygen molecule, O_2). There is no way that more or less could be accommodated. Conversely, if we can picture ATP formation from ADP and Pi as driven by a proton gradient, discharging through an ATPsynthase, the situation is somewhat different. The stoichiometry of the reaction:

$$ADP + Pi \rightarrow ATP + H_2O \qquad \text{Eqn. 5.16}$$

remains invariable but how much ATP is formed by the discharge of a given number of protons through the ATPsynthase might conceivably be as flexible, within limits, as the amount of electricity generated by the discharge of water through a turbine. We are no longer talking solely in terms of chemical stoichiometry but more in terms of mechanical efficiency. Burning a litre of a given sample of petrol (gasoline) in a calorimeter will produce a precise number of kilocalories. Burning the same fuel in an internal combustion engine will produce a smaller amount of mechanical work depending upon the efficiency of the engine and the nature of the coupling (the transmission) between the engine and what is being driven by the engine. If we think in terms of motor vehicles, the number of miles travelled per unit of fuel consumed will also depend to some extent upon the driver, i.e. on the way in which the operation of the machine is regulated. In biological situations, the efficiency of energy transduction, will also vary, within thermodynamic limits, according to regulatory mechanisms within the cell.

If we can accept that Eqn. 5.15 is both accurate in regard to its chemical stoichiometry and reasonable in regard to how much ATP can be generated by the linear transport of two electrons from water to NADP, we can proceed. As we see in Eqn. 5.17,

$$\begin{aligned} 3CO_2 + 6NADPH_2 + 9ATP \rightarrow \\ 1G3P + 6NADP + 9ADP + 8Pi \end{aligned} \qquad \text{Eqn. 5.17}$$

the conversion of three molecules of carbon dioxide into one molecule of G3P requires the utilisation of nine molecules of ATP and six of $NADPH_2$. If we now add Eqn. 5.15 and Eqn. 5.17 together, and cancel terms which appear on either side of the new combined equation, we ought to arrive at the overall equation for photosynthesis (Eqn. 5.22). In that equation, however, we have used CH_2O to described the product and the *real* product of the Calvin cycle is a triose **phosphate** rather than a triose ($C_3H_6O_3$). To make things balance, for the purpose of this exercise, we therefore need to introduce a hypothetical hydrolysis of our triose phosphate to yield free triose:-

$$G3P + H_2O \rightarrow C_3H_6O_3 + Pi \qquad \text{Eqn. 5.18}$$

Adding all three equations together (and cancelling) we would now arrive at something quite unsatisfactory, i.e.

$$3CO_2 \rightarrow C_3H_6O_3 + 3O_2 + 2H_2O \qquad \text{Eqn. 5.19}$$

In this nonsense equation, we have three molecules of water too few on the left and two too many on the right. The two on the right are easily disposed of because we have neglected to add the two molecules of water to the left-hand side which were needed to hydrolyse the two bisphosphates (Eqns. 5.7 & 5.8) during the regeneration of the CO_2-acceptor. What is worse, we have also forgotten that the actual carboxylation also consumes water and should be written

$$3RuBP + 3CO_2 + 3H_2O \rightarrow 6PGA \qquad \text{Eqn. 5.20}$$

$CH_2OPO(OH)_2$ | $C{=}O$ (CO_2 →) | $HCOH$ | $CH_2OPO(OH)_2$ $+ H_2O \longrightarrow$

RuBP

$CH_2OPO(OH)_2$ | $HCOH$ | ⓒOOH + COOH | $HCOH$ | $CH_2OPO(OH)_2$

2 x PGA

With a sigh of relief we can therefore add five molecules of water to the left hand side of the combined equations so that, on cancelling, we now arrive at the more or less respectable

$$3CO_2 + 3H_2O \rightarrow C_3H_6O_3 + 3O_2 \qquad \text{Eqn. 5.21}$$

which, on dividing by three gives

$$CO_2 + H_2O \rightarrow CH_2O + O_2 \qquad \text{Eqn. 5.22}$$

although we should bear in mind that, while this is stoichiometrically correct, it carries unfortunate implications (see 4.1) which can be avoided by adding a molecule of water to both sides to give

$$CO_2 + 2H_2O \rightarrow CH_2O + O_2 + H_2O \qquad \text{Eqn. 5.23}$$

No doubt there will be those amongst my readers who find this playing about with equations somewhat trying and perhaps out of place in a simple description of carbon assimilation. If, however, we are to accept that photosynthetic carbon assimilation is indeed brought about by the thirteen reactions of the Calvin cycle (not all of which have been mentioned here), it is important that these partial reactions actually do add up to the overall equation for photosynthesis. They do, i.e., if we were to repeat this exercise in full detail (by adding up all the equations in Fig. 5.2) we would arrive at precisely the same end point. We can, therefore, conclude that the Calvin cycle is an adequate representation of *this aspect* of carbon assimilation and that the arithmetic concerning its energy requirements (2 $NADPH_2$ + 3 ATP per CO_2 fixed, as in Eqn. 5.17) is soundly based.

PHOTORESPIRATION

5.7 Oxygenation

The reason for the italics in the last sentence, on page 97, is that this is not, by any means, the whole story. If, instead of having the situation in which

$$C5 + CO_2 \rightarrow 2 \times C3 \qquad \text{Eqn. 5.24}$$

we have

$$C5 + O_2 \rightarrow C3 + C2 \qquad \text{Eqn. 5.25}$$

both the overall stoichiometry and the energy requirements become quite different. When this happens a molecule of RuBP is consumed (by oxygenation) thereby denying the possibility of carboxylation of this particular molecule and giving rise to a C2 compound (phosphoglycolate) which is obviously not part of the Calvin cycle as we have represented it in Figs. 5.2 and 5.3. For this reason we sometimes prefer more complex figures (Fig. 5.5) which seek to illustrate both processes occurring side by side.

5.8 Virtue or Necessity?

Oxygenation is the first step in a very complex series of reactions which are collectively referred to as "photorespiration" but, before we plunge headlong into this morass, into this slough of despond, perhaps we can pause and ask what, in broad terms, is happening. What now seems to be the most credible story is this. Before the Proterozoic Era, the global atmosphere contained ample carbon dioxide but little

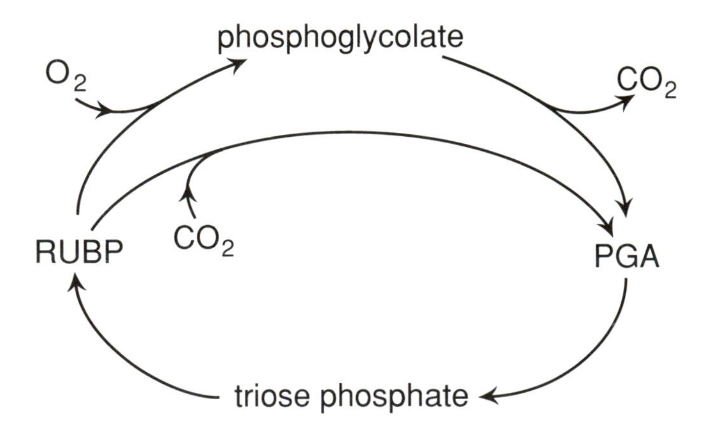

Fig. 5.5 Photorespiration. In C3 species in normal atmosphere the Calvin Cycle is depleted when oxygenation of RuBP gives rise to phosphoglycolate. Some of the carbon which leaves the Calvin Cycle as phosphoglycolate is "scavenged" and returned as PGA but this involves loss of carbon dioxide and of ammonia.

or no oxygen (Fig. 9.14). The evolution of green cyanobacteria and other plants containing the Calvin Cycle, with RUBISCO at its heart, changed all this.

VIRTUE OR NECESSITY?

Carboxylation proceeded apace, about one fifth of atmospheric nitrogen was gradually displaced by oxygen and vast quantities of CH_2O were locked away as coal and oil. Oxygenation of RuBP (Eqn. 5.25) during periods of high atmospheric CO_2 would have been relatively trivial because oxygen competes only poorly with carbon dioxide for the attention of RUBISCO and indeed is largely suppressed in high concentrations of carbon dioxide or if the oxygen concentration around a leaf is (experimentally) decreased to about 2%. That oxygenation occurs at all, seems to be an inevitable consequence of the nature of the enzyme and the reaction. Whether or not photorespiration serves an essential or a useful purpose is still a matter for argument and conjecture. There is little doubt that it is not entirely deleterious and a number of researchers have been quick to draw attention to possible or probable advantages but, at the same time, it unquestionably diminishes the rate and extent of carbon dioxide fixation on which the plant depends. Although the plant may have (in an evolutionary manner) contrived to make a virtue out of a necessity, many plants grow faster if photorespiration is suppressed by high carbon dioxide or low oxygen. In other words, even if photorespiration serves some useful purpose in normal circumstances it is clearly not *essential* in more favourable conditions.

5.9 A Necessary Evil

Of course we do not live in a perfect world. In low oxygen or high carbon dioxide, photorespiration is diminished and growth proceeds a pace. There is no doubt that, in the absence of other constraints, plants grow better if they are not allowed to photorespire. Seen from this stand-point oxygenation of ribulose

"Plants cannot move into the shade...."

bisphosphate can be regarded as an inevitable consequence of the characteristics of RUBISCO. But there may be more. In their natural environment plants are more often under stress than not and one sort of stress can lead to another. Drought is commonplace. Stomata close in response to water deficit. Closed stomata limit water loss by transpiration. At the same time closed stomata limit free diffusion of carbon dioxide. High light, combined with low carbon dioxide, is one of a number of conditions which can lead to photoinhibition and there is much contemporary preoccupation with mechanisms which facilitate safe dissipation of excitation energy. Plants can not move into the shade if they are uncomfortable although some do their best by adjusting the inclination of their leaves in order to intercept less light. Failing this, and failing safe dissipation of excitation energy, the initiation of electron transport which is unable to bring about "useful" work (carbon assimilation) can result in destructive photoinactivation of the photochemical apparatus. Heber argues that photorespiration serves a useful function in these circumstances by allowing a degree of linear electron transport (Section 4.4) to NADP. This in turn generates a useful proton gradient and poises cyclic electron transport. All of these combine, in ways not properly understood, to contain photoinhibitory damage.

"As early as 1972, Osmond and Bjorkman have suggested that photorespiration is involved in the protection of the photosynthetic apparatus against photoinactivation. Light absorbed by the pigment system of chloroplasts can give rise to the formation of highly reactive chemical species, if its energy is not used to reduce NADP or to form a transthylakoid proton gradient. Particularly under high intensity illumination, production of oxygen radicals and similarly dangerous species needs to be checked in order to enable available scavenging and detoxification mechanisms to protect the photosynthetic apparatus. For this reason, NADP needs to be available as a highly competitive electron acceptor under all illumination conditions to prevent or reduce the formation of oxygen radicals and singlet oxygen. In proper balance with NADPH, it poises the electron transport chain so as to facilitate cyclic electron flow which increases the transthylakoid proton gradient under high intensity illumination. The proton gradient has different functions. It permits synthesis of ATP controls photosystem II activity and initiates reactions

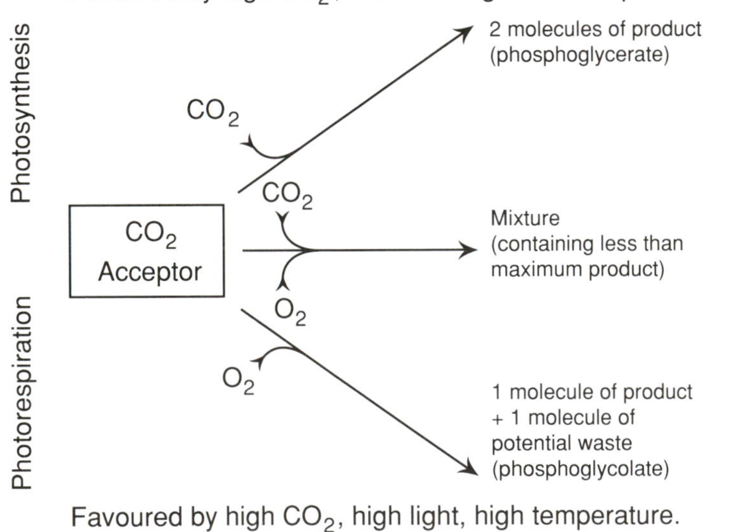

Fig. 5.6 Distribution of newly fixed carbon between photosynthesis and photorespiration according to conditions.

capable of rapidly degrading excess light energy into heat thereby contributing to protection of the photosynthetic apparatus. Thus photorespiration is protective by permitting the formation of a large proton gradient and by preventing the accumulation of NADPH particularly when stomatal closure limits the diffusion of atmospheric carbon dioxide into leaves, as it regularly does under water stress. There is the question how the increased proton gradient made possible by photorespiratory carbon turnover controls photosystem II activity and facilitates non-destructive dissipation of excess excitation energy. In this area, we know little about molecular mechanisms. Future research will have to concentrate on this aspect of photoprotection. So far, however, it is clear that photorespiration must be considered, from a human point of view, as a necessary evil. The emphasis is on 'necessary'. In C3 plants, photorespiration decreases productivity. This, however, appears to be a small price if weighed against the protection it provides when plants are exposed to excess irradiation particularly during periods of water stress" -Heber

5.10 All is not Lost

Carnegie, Gulbenkian, Rockefeller and Nuffield are renowned for their philanthropy but there are rich people who are better known for their meanness and it can be argued that only the mean can become wealthy in the first place. Plants have little choice. In a highly competitive world they can not afford to squander hard-won riches. Faced with a loss of newly fixed carbon by oxygenation of ribulose bisphosphate it seems that evolution favoured the development of a scavenging system which claws back carbon which would otherwise be lost as carbon dioxide. Indeed understanding of this process came, in part, from experiments with inhibitors which blocked steps in the glycolate pathway and from mutants which lacked key steps. These it might be supposed would have been beneficial but, instead, they were deleterious. The reason is not hard to see (Figs. 5.7, 5.8). What causes the damage is the oxygenation of RuBP. Thereafter a scavenging process comes into operation which converts a significant fraction of carbon (which would otherwise be lost) into a form in which it can re-enter the Calvin Cycle. Phosphoglycolate which is formed

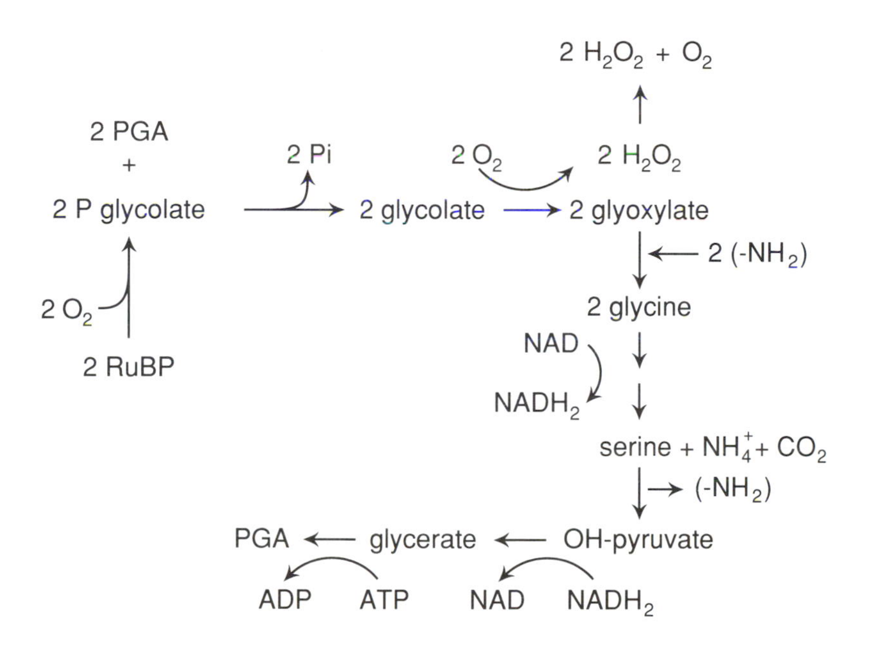

Fig. 5.7 Production and Utilisation of Glycolate. Oxygenation of RuBP produces one molecule of PGA and one molecule of phosphoglycolate (whereas carboxylation of RuBP gives rise to two molecules of PGA). A complex scavenging mechanism (see also Fig. 5.7) can convert half of the phosphoglycolate back to PGA but this conversion involves loss of carbon dioxide and ammonia

by oxygenation of RuBP contains only two carbons unlike the phosphoglycerate formed by carboxylation which contains three. Reactions which occur outside of the chloroplast, in mitochondria and peroxisomes, contrive to avoid total degradation of these 2-carbon molecules, ensuring that two molecules of phosphoglycolate (C2) eventually give rise to one molecule of 3-phosphoglycerate (C3) which can once again participate in the carbon flux of the Calvin Cycle i.e.

$$2 \times C2 \rightarrow 1 \times C3 + CO_2$$

Oddly enough, the conversion of two C2 compounds to one C3 and one CO_2 is even more complex than the conversion of five C3s to three C5s in the Calvin Cycle. However, while the complete photorespiratory cycle starts and finishes in the chloroplast, some of its reactions occur in other organelles and some of the intermediates may be diverted into other metabolic pathways. One very important aspect which should not be overlooked is that amino groups are first of all introduced (in the conversion of serine to hydroxypyruvate). These transaminations (catalysed by transaminases or aminotransferases) simply involve transfer of amino groups in and out of the pathway, but if you look carefully at Fig. 5.5 you will see that two amino groups come in (as two glyoxylates becomes two glycines) but only one goes out (as one serine becomes one hydroxypyruvate).

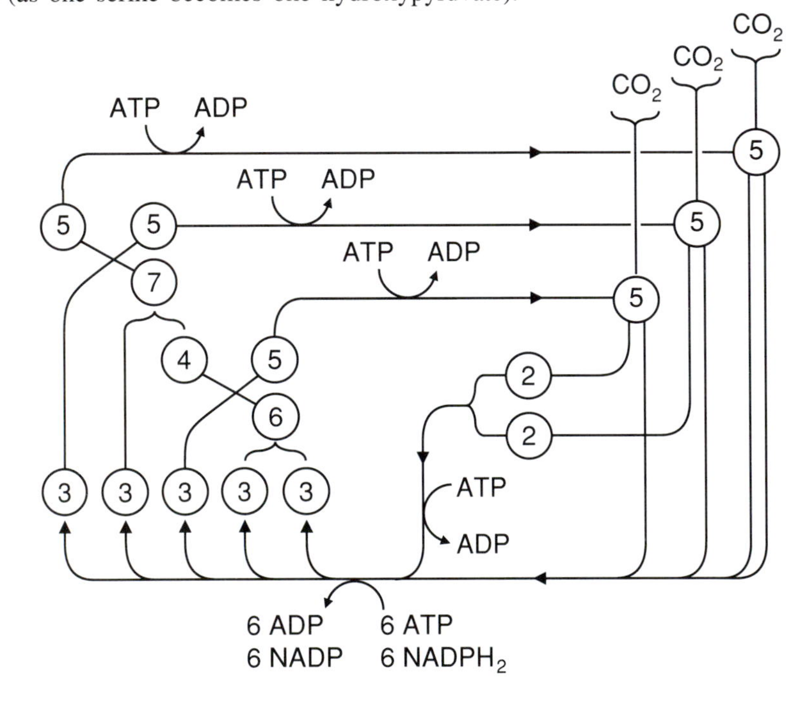

Fig. 5.8 Maximal oxygenation of RuBP in the Calvin Cycle (c.f. Fig. 5.2 which shows no oxygenation). The above situation occurs only at the carbon dioxide compensation point (e.g if a leaf has been placed in a closed chamber and allowed to photosynthesise until it can diminish the surrounding carbon dioxide no further). C4 species can take the carbon dioxide concentration down to near zero but, because of photorespiration, C3 species can only diminish the carbon dioxide to about 50 ppm. At this point there is no net product, two out of every three molecules of RuBP are oxygenated (because the carbon dioxide concentration is so low that it cannot offer more competition) and release of newly fixed carbon dioxide from glycine (Fig. 5.6) maintains the carbon dioxide concentration at this low steady-state level.

This leaves one amino group unaccounted for and, in fact, *it is released as free ammonia* as two glycines become one serine. Accordingly, one molecule of ammonia is released for every molecule of "re-formed" PGA which re-enters the Calvin Cycle. The potential loss of ammonia from the leaf is an aspect of photorespiration which is of considerable significance.

C4 PHOTOSYNTHESIS

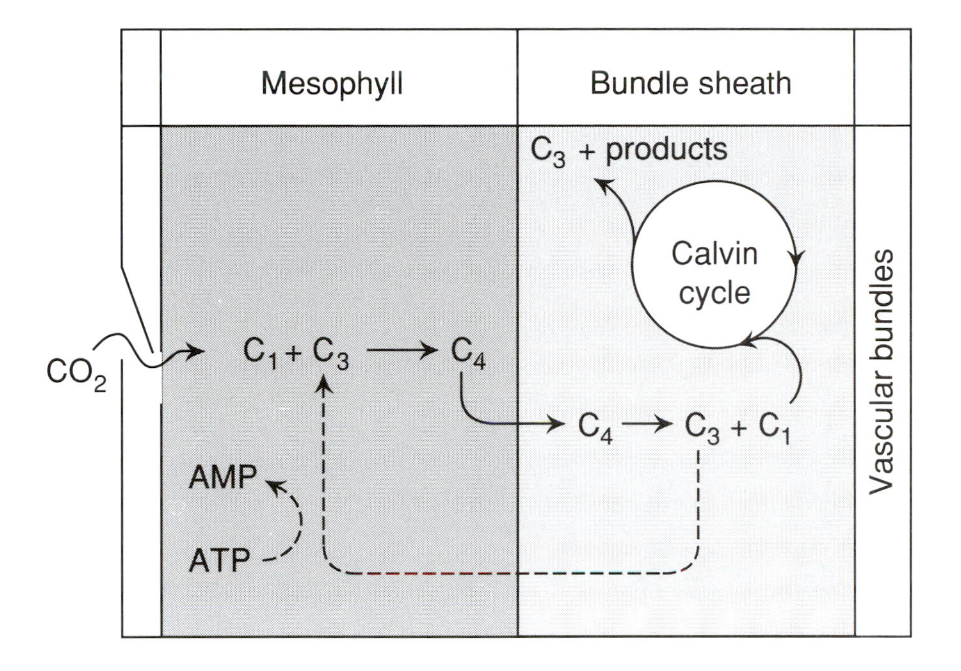

Fig. 5.9 C4 Photosynthesis This stark figure is an attempt to encapsulate the essentials of C4 photosynthesis. This is a process of carbon assimilation which embraces the Calvin Cycle but also involves additional reaction and a requirement for a specialised physical structure (Kranz anatomy). The three boxes represent this structure in which, in actuality, the mesophyll and bundle-sheath compartments are arranged in more or less concentric circles around the central vascular tissue or "bundle" (c.f. Fig. 5.10). Carbon dioxide enters the mesophyll via stomata and, in the cytosol of these cells, is fixed by PEP carboxylase, creating a four-carbon acid called oxaloacetic acid (OAA) from CO_2 and the three-carbon phospoenolpyruvate (PEP). Following reduction to malate, or transamination to aspartate (neither shown in this outline diagram), the newly formed C4 acid is physically transported into the inner bundle-sheath compartment where it is decarboxylated, releasing CO_2 and another C3 acid. Depending on the species, this C3 acid is either PEP or pyruvic acid. In either case these are transported back to the mesophyll and, if pyruvic acid is involved, this is converted back to PEP in a curious reaction, catalysed by pyruvic phosphate dikinase, consuming ATP and releasing AMP and inorganic pyrophosphate (PPi). Putting aside these details, the entire process constitutes a CO_2 pump in which this complex sequence of carboxylation, decarboxylation and transport creates a living CO_2 enriched greenhouse. Within this environment, the Calvin Cycle works to advantage.

5.11 C4 Plants

C3 plants (such as wheat, potatoes and most other temperate crop species and members of the native temperate flora) fix carbon dioxide in the light almost exclusively by the unaided mechanism of the Calvin Cycle (Fig. 5.2, 5.3). They are called "C3 plants" because the product first formed, by the carboxylation of ribulose 1,5-bisphosphate (RuBP), is 3-phosphoglycerate (PGA), a molecule containing *3* atoms of carbon. There is, however, a second major category of higher plants (C4) in which acids, containing *4* atoms of carbon are the initial products of carbon dioxide fixation (Fig. 5.9). It should not be supposed that these plants (such as sugar cane, maize and many sub-tropical grasses) somehow contrive to manage without the Calvin cycle. On the contrary, they are also *entirely dependent* on it but they have an *extra* piece of metabolic and structural machinery which makes them more effective in certain circumstances.

Just as the names of Benson and Calvin will always be associated with the carboxylation of RuBP and the reaction which lead to the regeneration of RuBP, so are the names Hatch, Slack and Kortschack linked to C4 photosynthesis. The first observations were those of Kortschack, in Hawaii, although it should be noted that, at about the same time, there was independent work (by Karpilov and his colleagues) in the Soviet Union which was almost entirely overlooked in the West and which did not lead, as did Kortschack's, to final elucidation of the pathway involved. Kortschack worked for the Hawaiian Sugar Planters Association and, in the 1950's, attempted to repeat, with sugar cane, the same sort of experiments that Calvin and his colleagues had carried out with *Chlorella*. To be precise, he exposed leaves of sugar cane to radioactive CO_2 ($^{14}CO_2$) for short periods of time in the light and then, after extraction, ran two-dimensional paper chromatographs of the products of $^{14}CO_2$ fixation. To his surprise, although 3-phosphoglycerate (PGA) appeared in a relatively short time, the very first products were four-carbon acids (malate and aspartate). These he correctly assumed to be both derived from a common four-carbon precursor, oxaloacetate. News of this striking difference between the behaviour of sugar cane, *Chlorella*, and those higher plants which Benson, Calvin, Gaffron and others had worked on in the United States, soon travelled to the Colonial Sugar Refining Company in Queensland, Australia where Hal Hatch and an English colleague, Roger Slack, were persuaded, by Kortschack's results, to look more deeply into the underlying biochemistry. Over the next few years Hatch, Slack and their colleague Hilary Johnson made a number of very important contributions which led to the concept of "C4 photosynthesis". As in the work by Benson and Calvin on *Chlorella*, initial suggestions were made which subsequently proved to be untenable but, by 1970, when a seminal conference on this subject was held in Canberra, everything had more or less fallen into place and the main aspects of the C4 pathway had been defined. One important technique employed by Hatch and Slack was the "pulse-chase" experiment. This involved a brief exposure to $^{14}CO_2$ followed by a longer period of illumination in which the $^{14}CO_2$ was flushed from the reaction chamber and replaced by $^{12}CO_2$ When this was done, radioactivity (which initially appeared in malate and aspartate) soon left these compounds only to reappear in substances, such as sucrose, which are regarded as typical end-products of photosynthesis. Accordingly, it seems clear that the transient appearance of radioactivity in malate and aspartate must involve a process of carboxylation, decarboxylation and refixation. Hatch and Slack identified several enzymes of this novel C4 pathway and provided preliminary evidence for the distribution of these enzymes between different compartments in the leaf. They also showed that several other species, such as maize, behaved in a broadly similar way to sugar cane.

5.12 Greenhouses Within Plants

Before we go further into the subject of C4 photosynthesis, we might well ask what it is all about. This touches on many aspects of carbon dioxide metabolism which we have considered in other chapters and in other sections. There we have seen that primaeval atmospheres contained high concentrations of carbon dioxide and little oxygen. Photosynthesis, as it evolved, changed all of this. Carbon dioxide was taken from the atmosphere and replaced, on a molecule to molecule basis, by oxygen.

In consequence, contemporary global atmospheres contain about 21% oxygen and very small quantities of carbon dioxide (although carbon dioxide taken up from the atmosphere must have been constantly replenished by equilibration with oceanic reserves and by volcanic activity etc). Within recent history, carbon dioxide has been even lower in concentration than it is at present but although the concentration is continuing to rise remorselessly, as a result of Man's activities, it is still far too low (in the 1990's) to saturate photosynthesis. The relationship between carbon dioxide concentration and rate of photosynthesis is broadly similar to the relationship between light intensity and rate of photosynthesis (Chap 6). Although near saturation is reached at low concentrations of carbon dioxide (below 1000 parts per million),

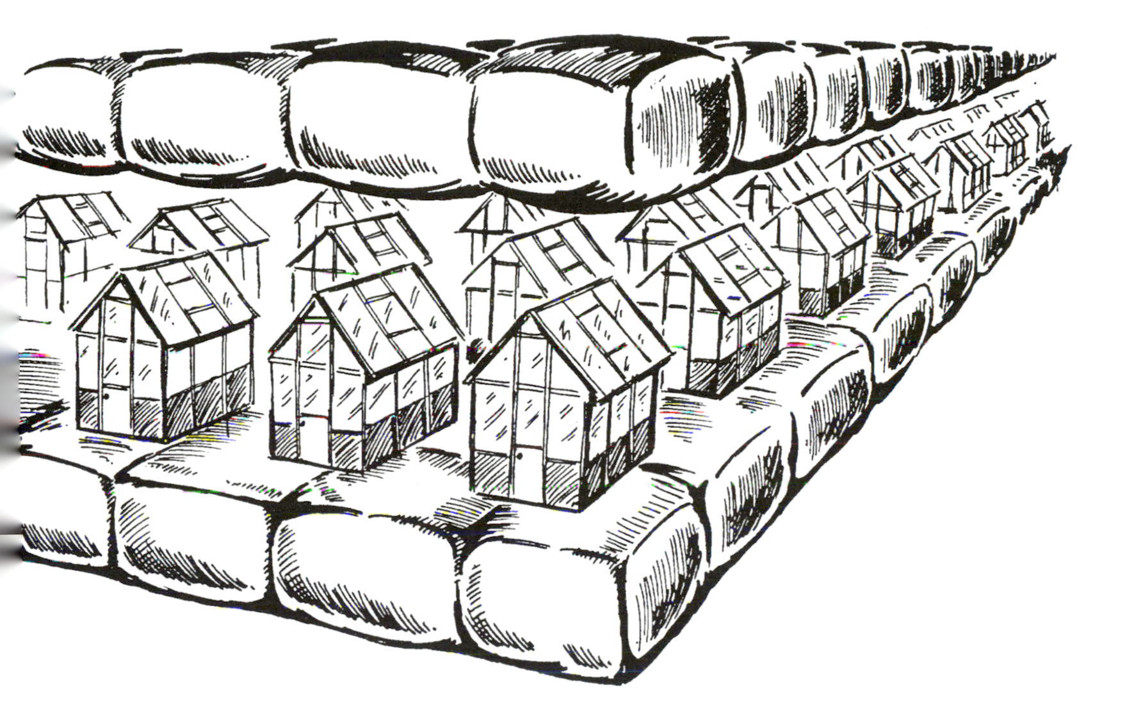

atmospheric carbon dioxide still lies well within the linear part of the relationship. In other words if, in an experiment, we double the concentration of carbon dioxide surrounding a plant we might expect something like a doubling in the rate of photosynthesis. Indeed, if we were to carry out such an experiment with a C3 plant this is what we would observe (Fig. 9.19). If, on the other hand, we used a C4 plant we would find very little response to *augmented* carbon dioxide (i.e. to increases in carbon dioxide above the ambient concentration of about 350 ppm - Fig. 9.19). For many years, tomato and cucumber growers in Western Europe have used carbon dioxide enrichment to increase the productivity of these glasshouse crops. C4 plants, it turns out, had the idea first in the sense that they have their own internal carbon dioxide-enriched greenhouses in which C3 photosynthesis can proceed at an enhanced rate. As we have already noted, C4 photosynthesis is not an alternative to C3 photosynthesis but rather an addition to it. What is added is, in effect, a carbon dioxide-enriching mechanism which captures carbon dioxide at low concentrations in an outer compartment (the mesophyll compartment) of C4 leaves and releases it, at a higher concentration, in an inner compartment (the bundle sheath compartment) which houses the Calvin Cycle. This, of course, smacks a little of alchemy and tampering with the laws of thermodynamics. In actuality, C4 plants do not get something for nothing, any more than the rest of us. Their carbon dioxide enriching mechanism requires an energy input and, for this reason, C4 plants are less efficient in terms of energy usage, than C3 species. Nevertheless, this additional expenditure of energy brings obvious advantages in its wake and others which will become more clear as we look into this process more deeply (Section 5.16). Even so, you may already have begun to wonder why, if one carbon dioxide-fixing mechanism proves to be inadequate, tacking on an additional carbon dioxide-fixing mechanism (as part of a carbon dioxide enriching system) can make the whole process more effective. The answer lies in the intrinsic characteristics of the enzymes concerned. Ribulose bisphosphate carboxylase (Rubisco), the carbon dioxide-fixing enzyme at the heart of the Calvin cycle, evolved during a period in geological time when oxygen was virtually absent. The nature of the enzyme and the reaction is such, however, that both carbon dioxide and oxygen will serve as substrates (Section 5.7). The two gases compete for an active site on the enzyme. If carbon dioxide wins, two molecules of PGA are formed and then consumed in the subsequent reactions of the Calvin cycle and in the eventual production of sucrose and starch. If oxygen wins, i.e. if oxygenation of RuBP occurs (Section 5.7), one molecule of PGA and one molecule of *phosphoglycolate* are formed. Such oxygenation constitutes the first step in photorespiration (5C). In C3 plants, in atmospheres containing normal concentrations of carbon dioxide and oxygen, a significantly large fraction of RuBP is consumed in oxygenation rather than in carboxylation. In the internal, carbon dioxide-enriched, greenhouse in C4 plants this first step in photorespiration is largely suppressed. To the extent that the reaction catalysed by Rubisco is susceptible to oxygenation it is an inefficient carboxylating mechanism. In contrast, the carboxylating enzyme which plays the essential role in the carbon dioxide enriching system in C4 plants (Eqn. 5.26 is not affected by oxygen and becomes saturated at very low levels of carbon dioxide. This enzyme is called phosphoenolpyruvate (PEP) carboxylase. It is a widely distributed enzyme which is present in C3 as well as C4 plants and one which is responsible both for dark fixation in Crassulacean Acid Metabolism (Section 5.12) and an important feature of "dark respiration". First described by Axelrod and Bandurski in the 1950's, it carboxylates one molecule of phosphoenolpyruvate to yield one molecule of oxaloacetate and one molecule of orthophosphate.

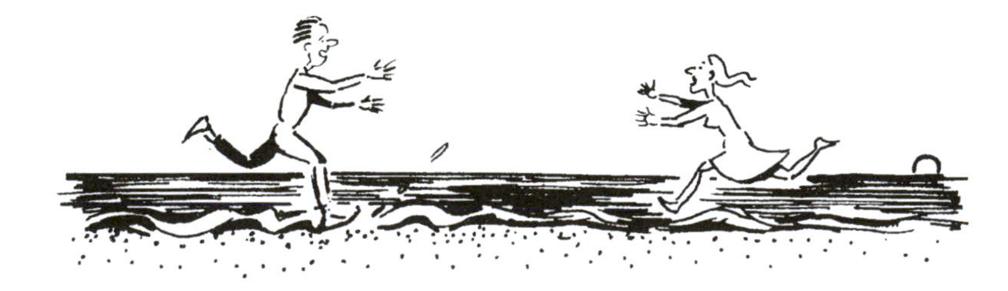

"extremely high affinity"

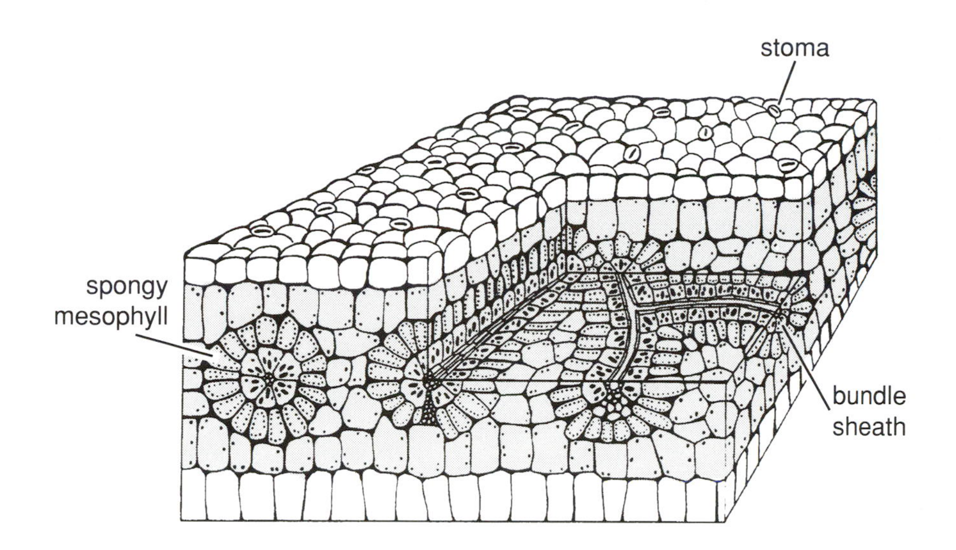

Fig. 5.10 Kranz Anatomy C4 photosynthesis is based on both biochemistry and structure. The preliminary carboxylation by PEP carboxylase occurs in the mesophyll cytosol. The Calvin cycle is restricted to the bundle-sheath chloroplasts. The term "Kranz" comes from a wreath, worn on the head, which the concentric layers of tissue in C4 leaves were thought to resemble. In this transverse section through a leaf the vascular tissue at the centre is surrounded by the bundle sheath which contains large chloroplasts. This, in turn, is embedded in a spongy mesophyll containing many small chloroplasts.

In C4 plants, this enzyme with its extremely high affinity for carbon dioxide,is situated in an outer (mesophyll) compartment of the leaf (Fig. 5.10). Now we can begin to see how the whole thing functions and how the whole system depends on enzymes with different characteristics housed in different compartments. PEP carboxylase captures carbon dioxide from the atmosphere in the outer, or mesophyll, compartment of the leaf. The immediate end product of this carboxylation (oxaloacetate) gives rise to malate and aspartate. Either, or both, of these acids are translocated into an inner compartment of the leaf where they are decarboxylated. The inner, or bundle-sheath compartment (the carbon dioxide enriched greenhouse), is not readily permeable to carbon dioxide. The internal carbon dioxide concentration is therefore increased, thus facilitating RuBP carboxylation and the reactions of the Calvin cycle (Figs. 5.2, 5.3).

CRASSULACEAN ACID METABOLISM

5.13 Came you not from Newcastle ?

It will have become readily obvious that this is a book which takes a great many liberties with style and form and opinion. It will, therefore, come as no surprise that it can also be blatantly self-indulgent so, if you wish to avoid an exercise in personal nostalgia (intertwined with a little of the history of Crassulacean Acid Metabolism), now is the time to skip quickly to the next section where you will find the substance of that aspect of photosynthesis. Otherwise, you will have to endure the fact that I am entirely incapable of writing about this particular subject without going back to my youth and, what is worse, sharing it with you.

CHAPTER 5 - THE DARK BIOCHEMISTRY

As one of a number of research students (at Newcastle in the 1950's) I did my Ph.D on the enzymic basis of Crassulacean Acid Metabolism without even knowing what CAM achieved. Such an awful lack of understanding is not entirely without precedent amongst research students but, in my own defence, I should hasten to add that none of us, at that time, knew what it was about. Even so, we had flocked to study it in some numbers, because CAM was the main area of interest of our Professor, a famous, and gently eccentric Welshman, called Meirion Thomas, who made such an impression on his students that very few chose to do anything other than CAM for their research. So I was not to turn to photosynthesis, or so I thought, until I first met my Ph.D. examiner in 1956. This turned out to be Robert Hill, of chloroplast fame, and it was he who was to steer me in that direction. But surely, you might ask, is not Crassulacean Acid Metabolism a form of photosynthesis ? And, of course, you would be right but we did not regard it as such at the time because, often quite literally, we were working in the dark. We were interested to know more about a process in which certain plants fixed carbon dioxide into organic acids in the dark. Many, but not all of these, were members of the family *Crassulaceae*. Photosynthesis was not our concern and indeed, precious little was known about it at that time. Calvin and his colleagues were scarcely embarked on their work, Arnon and Whatley's classic experiments (Section 4.8) were yet to be done. The Z-scheme (Section 4.3) and C4 photosynthesis (Section 5.10) were still undreamed of. It was Meirion Thomas (along with students, then unknown but soon to become renowned, like Harry Beevers, Stan Ranson, John Brown and Bill Bradbeer) who clearly established that the accumulation of malic acid, during *"dark acidification"* in CAM plants, involved carbon dioxide fixation. In other words, carbon dioxide in CAM plants was as much a substrate by night as it was in C3 and C4 species by day.

The discovery of dark acidification by Benjamin Heyne came long before. He tasted a CAM plant and found it more acid in the morning than in the evening. He wrote the following in 1813 but sadly failed to say why he had ever felt inclined to taste these unlikely leaves in the first place or to repeat this unpleasant experience later in the day.

A.B. Lambert, Esq., Vice President of the Linnean Society.

Dear Sir,

I had an opportunity some time ago of mentioning to you a remarkable, deoxidation of the leaves of a plant in day-light. As the circumstance is in itself curious, and throws great light on the opinion of those celebrated philosophers who have written on the subject, I will state it shortly in this letter, which if you please, you may in extract, or in any other way you think proper, lay before the Society.
The leaves of the Cotyledon calycina, *the plant called by Mr. Salisbury* Bryophyllum calycinum, *which on the whole have an herbaceous taste, are in the morning as acid as sorrel, if not more so. As the day advances, they lose their acidity, and are tasteless about noon; and become almost bitterish towards evening. This is the case in India, where this plant is pretty generally cultivated in our gardens and it remains to be seen if the same takes place in the hot-houses in England, where it has been lately introduced.*

I have seen this plant but once in this country, and that was at Mr. Loddiges', at Hackney, about twelve o'clock in the day-time, when I found it quite tasteless. The distance of that place from my habitation has hitherto prevented me from attending to it at an earlier hour in the morning, I have, however, but little doubt it will be found as acid as I have described it to be in India.

"Der Kaktusliebhaber"
- with acknowledgements to Spitzweg

CAME YOU NOT FROM NEWCASTLE?

I need scarcely observe, that the acidity which these leaves possess in the morning cannot be ascribed to anything else than to the oxygen which the plant has absorbed during the darkness of the night, or which has been transferred from other constituent principles of the plant during that period. I think it has been absorbed, as it is so loosely united at its base, that even the light of day has an immediate effect of disengaging it again.

Both Priestly and Ingenhousz have concluded, from numerous experiments, that all plants exhale vital air in the day-time, and fix air or carbonic acid gas during the night; but these conclusions have been called in question by some, from the various results of experiments since made on this subject. What I have now related is therefore not destitute of interest, as it seems incontrovertibly to establish the theory of these celebrated philosophers.

I was in the hopes of learning something new or pertinent on this interesting subject in Sprengel's work on the Structure and Nature of Plants; but, to my great disappointment, there is nothing to be found but what has been advanced by the two philosophers mentioned, and by Saussure and Sennebier in later times. Sprengel expatiates much on the exhalation and absorption of carbonic gas, and only once mentions oxygen, when he notices Sennebier's observations; according to which, more carbonic gas is exhaled by plants during the night in close vessels, than there is oxygen disengaged in sunshine.

I beg leave to further observe, that the plant above treated of is, in my opinion, truly a species of Cotyledon, *with which it perfectly agrees, in habit and generic characters; the only difference being in the number of the parts of fructification which in* Cotyledon calycina *are one-fifth less than in the other species of the genus, a difference, however, that according to the principles of the Linnean System, does not form a sufficient grounds for separation.*

I have the honour to be, etc.

20th April, 1813.

Benjamin Heyne

And then there was the eminent French scientist, de Saussure, who once uttered the immortal words *"l'acide carbonique, est elle essentielle pour la vegetation?* " Here was a plant physiologist whose phenomenal contributions have been sadly neglected. His studies of respiratory quotients (mass of carbon dioxide evolved divided by mass of oxygen taken up) are far too many to consider here even though they were immensely informative. (Try asking how a castor bean can respire vigorously for many days without losing weight). In 1804 de Saussure referred briefly to an experiment in which he

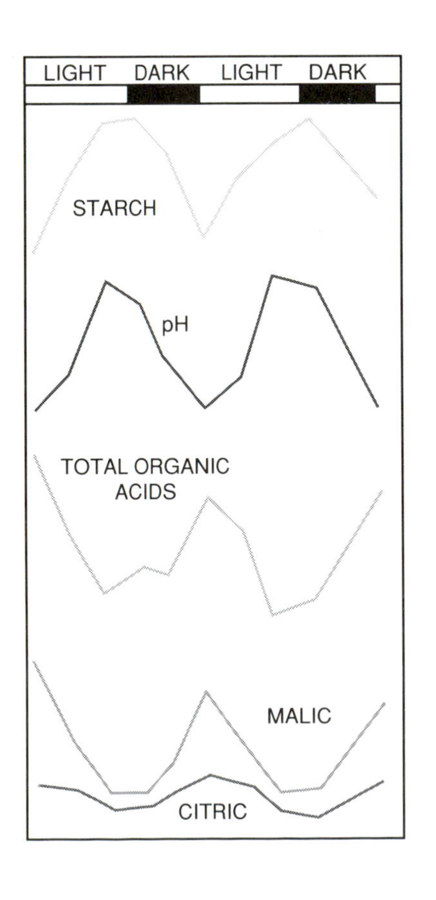

"placed raquettes of the cactus Opuntia *in CO_2 enriched atmospheres and found that CO_2 and oxygen were absorbed simultaneously"*

The implication, with hindsight, being that respiration occurred as usual but that, not only was the CO_2 so produced not released but, together with CO_2 absorbed from the environment, was incorporated into some metabolite (the amounts too large to be accounted for by solution in the abundant sap within the leaves). Imagine Thomas's chagrin in 1947 when he first came across de Saussure's observations in an obscure journal having himself independently re-discovered CAM some 143 years later. It was a measure of his stature that he was quick to acknowledge de Saussure's prior work as soon as he was aware of it. Nevertheless, it was Thomas and his colleagues, not de Saussure who, as I have already noted, firmly established the formation of malic acid from carbon dioxide by night and "de-acidification" by day. In Thomas's own words:-

"The experimental methods were not so refined as those which could be used now; but there was a general agreement in the pattern of the results obtained, by Beevers in the late spring and summer of 1946 for Bryophyllum *and* Kalanchoë *leaves and, later, by Ranson during several CAM seasons for succulent organs of these and other genera of the* Crassulaceae *and of genera, including the cactus* Opuntia, *belonging to other families"*

It was also Thomas who gave me, as a late-comer to his team in the early 1950's, the task of finding, if I could, the enzyme responsible for acid formation.

This, I might add, at a time when Ochoa's malic enzyme (Eqn. 5.30) was the only known enzyme which might conceivably fix carbon dioxide. Moreover it was in a laboratory with no experience of enzymology, no spectrophotometer, no cold room, no refrigerated centrifuge and a pH meter which put pH measurement into the same class as finding water with a forked hazel twig. Here, when I needed NADP (or TPN as it was then called) I could not buy it from SIGMA (a supplier of fine biochemicals of great renown) but went instead to the local abattoir in search of freshly killed pig hearts and embarked on a long and arduous process of extraction; bucket biochemistry at its most basic. Thomas, of course, said that it was most unlikely that I would ever find anything and even told me that he had a more mundane problem in reserve, to give me when I admitted defeat. But luck was on my side. Some time later Axelrod and Bandurski described an enzyme, from wheatgerm, which they called PEP carboxylase. It seemed an ideal candidate for the task of fixing carbon dioxide in CAM and, buoyed up by what I had learned by then at Harry Beever's elbow in far off Indiana, I set about fishing it out of *Kalanchoë*. In the end, I succeeded but that, like *"How the Elephant got his Trunk"*, is quite another story.

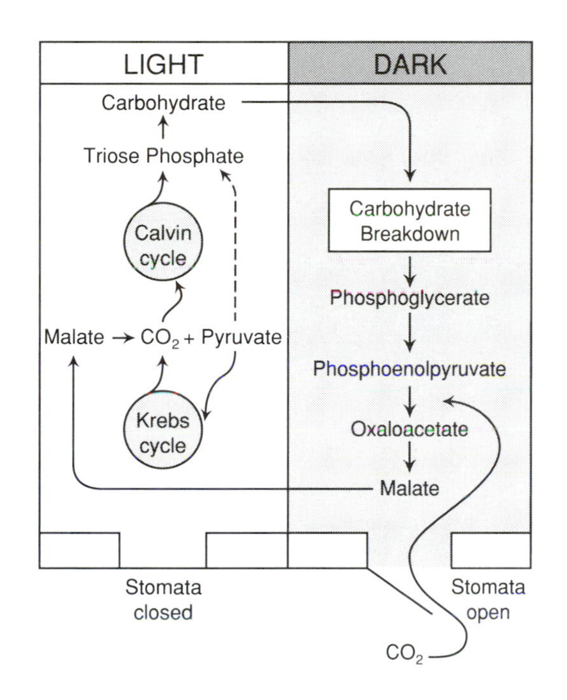

Fig. 5.11 Crassulacean Acid Metabolism (CAM) was a term first employed by the late Meirion Thomas, in whose Newcastle laboratory much of the early work on this type of photosynthesis was done. CAM is typical of (but not restricted to) the family *Crassulaceae* and it involves the accumulation of malic acid - hence the name. Like C4 photosynthesis, it involves preliminary fixation of CO_2 by PEP carboxylase, subsequent decarboxylation of malic acid and refixation of CO_2. Indeed C4 photosynthesis was once flippantly referred to as "CAM mit Kranz" and, however inadequate this description, it does remind us of the fact that in C4 photosynthesis preliminary fixation and refixation occur in separate compartments whereas, in CAM, these processes are separated in time. CAM is basically an adaptation to water shortage. CAM plants conserve H_2O by keeping their stomata closed by day, when high temperatures normally bring about high water loss by transpiration. Behind closed stomata the Calvin Cycle proceeds happily in a CO_2 enriched atmosphere produced by decarboxylation of malic acid (itself resulting from carboxylation of PEP during the preceding night when the stomata were open). Many CAM plants show a degree of succulence, related to the storage of malic acid which is a necessary feature of this process. Opening of stomata by night and closure by day is the converse of the stomatal behaviour of C3 species.

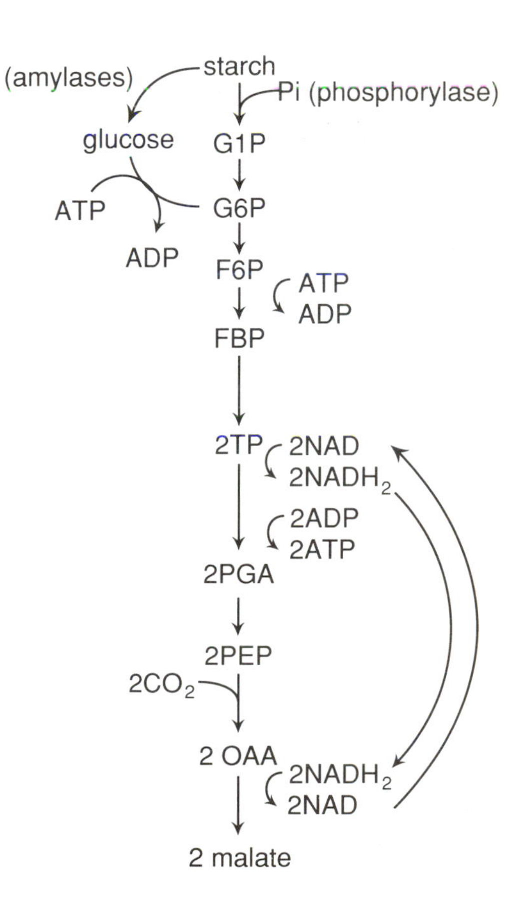

5.14 Dark Acidification

It was once said, half jokingly and half seriously, that C4 photosynthesis was *"CAM mit Kranz"* . Certainly it cannot be disputed that C4 and CAM have some features in common but it should be noted that CAM plants lack the "Kranz anatomy" (Fig. 5.11) which is as much a part of C4 photosynthesis as the metabolic processes that go on within their mesophyll and bundle-sheath compartments. What might be regarded as the closest and most important common feature is that both C4 and CAM plants join carbon dioxide to phosphoenolpyruvate (to form oxaloacetate) prior to decarboxylation and refixation of CO_2 in the Calvin Cycle. But, as we shall see, there are all-important differences in the time and place in which these processes occur. This first carboxylation, or pre-fixation step, is catalysed by PEP carboxylase

$$\text{PEP} + CO_2 + H_2O \rightarrow \text{oxaloacetate} + \text{Pi} \qquad \text{Eqn. 5.26}$$

Phosphoenolpyruvate, like RuBP, is a high-energy substrate in the sense that the addition of carbon dioxide to it, which gives rise to the formation of oxaloacetate and orthophosphate leads to a large decrease in free energy. In round terms, about 6 Kcal (25 Kjoules) is needed to drive the carboxylation but the decrease in free energy associated with the overall reaction is sufficiently large to give a net change of about -7 Kcal. Accordingly the equilibrium position of the reaction is such that it is (like the reaction catalysed by RUBISCO - Section 5.2) virtually irreversible. Once fixed, the carbon dioxide stays fixed.

[It should, perhaps, be noted again here (c.f. Section 1.1) that energy is always required to break bonds and released when bonds are formed. A carboxylation involves joining carbon dioxide to an existing module. This implies bond formation but other bonds have to be broken in this process which requires a net energy input of about 6 or 7 Kcal - roughly equivalent to the free energy of hydrolysis of ATP to ADP. The fact that PEP like RuBP has been previously "wound up" (energetically) into an "uncomfortable" configuration ensures that it welcomes the opportunity to slip into something more comfortable. This sort of metabolic device ensures that reactions of this sort go to completion in the desired direction and that, for substrates to be carboxylated, the low carbon dioxide concentration in the atmosphere offers no impediment. PEP carboxylase also has a very high affinity for carbon dioxide so both the ability of the catalyst to work well in low carbon dioxide concentrations *and* the favourable equilibrium position combine to favour effective carboxylation.]

Oxaloacetate, the end-product of the reaction (Eqn. 5.26) is unstable and readily decarboxylates, given half a chance, to yield pyruvate and carbon dioxide

$$\text{oxaloacetate} \rightarrow \text{pyruvate} + CO_2 \qquad \text{Eqn. 5.27}$$

but it doesn't normally get the chance to hang about. As in C4 photosynthesis, it is either reduced to malate or transaminated to aspartate. Reduction to malate predominates in CAM plants because it is in the form of malate that the initial product of CO_2 fixation is stored (mostly in the vacuoles) throughout the night. The reduction is catalysed by malic dehydrogenase and again the equilibrium is such that the reaction goes largely to completion in the direction indicated (Eqn. 5.29).

$$\text{oxaloacetate} + NADH_2 \rightarrow \text{malate} + \text{NAD} \qquad \text{Eqn. 5.28}$$

DARK ACIDIFICATION

I'm just going to slip into something more comfortable

[Throughout these descriptions "oxaloacetate", "aspartate", "malate", etc., have been preferred to "oxaloacetic acid", etc., (implying a more or less neutral pH). In many circumstances this is not only convenient shorthand but a reasonable reflection of the hydrogen ion concentration (see Section 1.7) of the cytosol in which many of these reactions occur. In CAM plants, however, it needs to be remembered that such large quantities of malic acid (and some other acids) accumulate in the leaves that the process of dark acidification can be followed by taste and measurement of "titratable acidity".]

The formation of aspartate involves an amino donor such as glutamate and an enzyme described as a "transaminase" or an "aminotransferase"

$$\text{oxaloacetate} + \text{glutamate} \rightarrow \text{aspartate} + \alpha\text{-oxoglutarate} \qquad \text{Eqn. 5.29}$$

Although malic acid and aspartic acid are the most important contributors to "dark acidification" and although malic acid predominates and accumulates in large quantities in the vacuoles of CAM plants during the period of darkness, it should not be supposed that these are the only metabolites derived from PEP. All are important compounds in so-called *"intermediary metabolism"* sometimes lying at, or near, metabolic cross-roads which lead in many other directions. Similarly, not all CAM plants are the same and, for example, isocitric acid, almost invariably a product of dark fixation, occurs in relatively large quantities in some species.

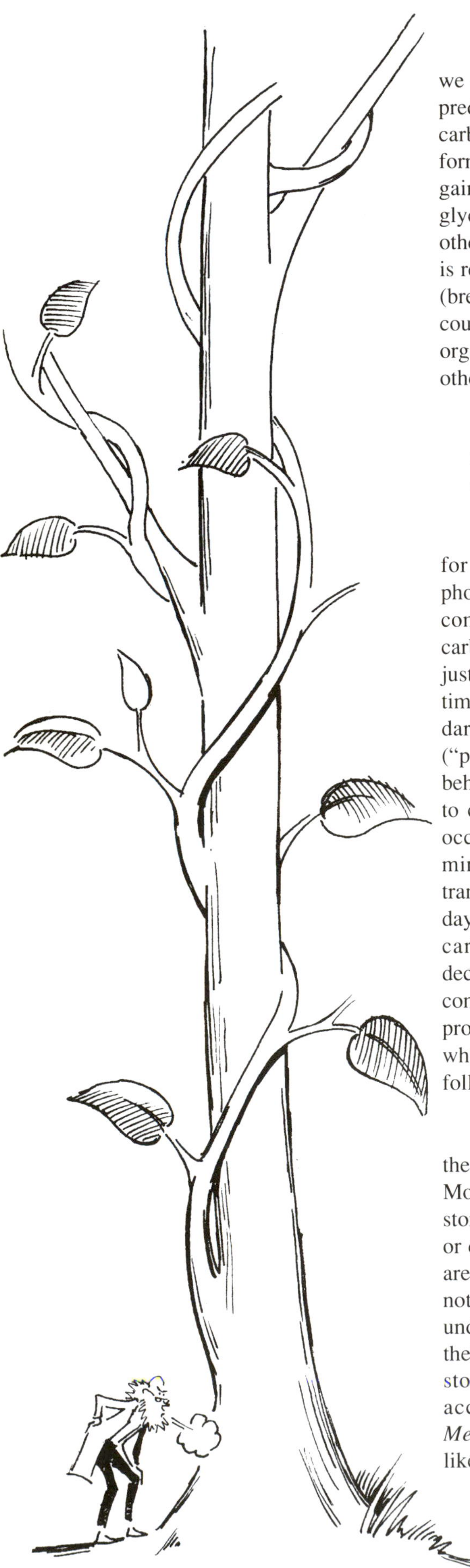

Whatever the precise mixture of acids which results from dark acidification we can, because of easy inter-changeability, regard the accumulated acid as being predominantly malate, (as it so often is) and ascribe to it the function of a temporary carbon dioxide storage product. We must not forget that, as a consequence of forming malate from PEP (Eqn. 5.26 and 5.27) the leaf has not yet made a permanent gain or acquisition. The substrate (PEP) is derived from starch or sucrose by glycolysis (the same sequence of reactions which most living organisms, including other plants and Man, employ in "dark respiration" prior to the "Krebs Cycle" which is responsible for the closing stages). In other words, PEP formation is a "catabolic" (breaking down) rather than an "anabolic" (building up) process. In the normal course of events, most of the PEP which is formed during dark respiration (by organisms other than CAM plants) is consumed, via the Krebs Cycle, giving rise to other building materials and to ATP.

[*Some* PEP carboxylation is a general feature of plant biochemistry. It is needed to replenish key intermediates, such as oxaloacetate, which occupy a central position in intermediary metabolism.]

It is clear then, that in CAM plants as well as C4 plants, the substrate (PEP) for the initial carboxylation is formed at the expense of previously formed photosynthetic products. In C4 metabolism the PEP is carboxylated in an outer compartment of the leaf (the mesophyll); it is spatially separated from the carboxylation catalysed by RUBISCO which is responsible for the net gain in carbon just as it is in C3 and C4 species. In CAM the two carboxylations are separated in time rather than in space as they are in C4 species. PEP is carboxylated during darkness, malate accumulates in the vacuole. During this period the stomata ("pores") which allow carbon dioxide to enter the leaf are open, the converse of the behaviour of C3 stomata which open by day and close by night. CAM plants are able to conserve water in this way because loss of water vapour through open stomata occurs under conditions in which evaporation, and therefore transpiration, is minimal. If C3 stomata behaved in this way they would not only diminish transpiration loss but also succeed in bringing photosynthesis to a halt because of daytime closure. CAM plants, on the other hand, can happily photosynthesise in a carbon dioxide enriched atmosphere behind closed stomata as malate is decarboxylated (Eqn. 5.30) to pyruvate and carbon dioxide. At this stage it is the conventional Calvin Cycle which takes over and it is this "normal" photosynthetic process which allows CAM plants to accumulate organic carbon (including the starch which will give rise to PEP) so that the whole cycle can be repeated during the following night.

CAM plants are not generally noted for high rates of photosynthesis although there are some (e.g. pineapple) which are obviously a commercial proposition. Mostly they occupy arid parts of the environment (such as walls and outcrops of stone in otherwise favourable environments which are untenable for most species) or deserts. Cacti are some of the more spectacular CAM species. Many CAM plants are "succulents" i.e. they have fleshy leaves or stems but there are those which are not succulent and if the degree of succulence can be manipulated by growing plants under different conditions it can also be shown that this does not necessarily affect their CAM function. Succulence, in fact, appears to be more directly related to water storage than to CAM metabolism as such, although it may facilitate malate accumulation. There are also some *"facultative"* CAM species such as *Mesembryanthemum crystallinum* which behaves like a C3 plant if well watered and like a CAM plant if kept dry.

[The archetypal CAM plant is a member of the family *Crassulaceae*, hence the name but there are many CAM species (most notably members of the *Cactaceae*) which belong to a range of families including those which are predominantly comprised of C3 species.]

5.15 Light Deacidification

"In 1804 De Saussure reported that, even in the absence of an external supply of carbon dioxide, illuminated branches of Opuntia *can produce oxygen. He suggested that the plant itself, by consuming its own substance, gives off carbon dioxide, which is then used in carbon assimilation"* ....Thomas (1949)

This really says it all and, in retrospect, it is remarkable that it took so long to arrive at our present understanding of CAM i.e. that daytime photosynthesis, behind closed stomata in a CO_2-enriched atmosphere (which largely precludes photorespiration) supplies the plant with its requirements for organic carbon and leads to starch accumulation. At night, starch breakdown leads to the formation of a CO_2-acceptor (PEP) and carboxylation of this acceptor to the formation of malate, which accumulates. By day, in turn, the malate is broken down and CO_2 released for photosynthesis.

Malate is decarboxylated by "malic enzyme". Before PEP carboxylase was discovered, malic enzyme was once regarded as a serious candidate for carboxylation, although its unfavourable equilibrium position and low affinity for CO_2 did not favour this idea. Working as a decarboxylase these characteristics are not a problem:-

$$\text{Malate} + \text{NADP} \rightarrow \text{pyruvate} + CO_2 + \text{NADPH}_2 \qquad \text{Eqn. 5.30}$$

All three products of this decarboxylation are then available for re-assimilation. The CO_2 of course, is used in photosynthesis but there is still uncertainty about the precise fate, or site of utilisation, of the other two. Some CAM plants use a separate means of decarboxylation. Here the key enzyme is PEP carboxy*kinase* which catalyses the reaction:-

$$\text{oxaloacetate} + \text{ATP} \rightarrow \text{phosphoenolpyruvate} + \text{ADP} \qquad \text{Eqn. 5.31}$$

Such a route offers the advantage that, unlike the pyruvate formed in reaction 5.30, PEP can be readily converted back into starch by reversal of glycolysis. On the other hand reaction 5.31 brings its own problems because the OAA formed from malate by malic dehydrogenase is produced in a reaction with an equilibrium position which greatly favours oxaloacetate reduction rather than malate oxidation:-

$$\text{malate} + \text{NAD} \leftrightarrows \text{oxaloacetate} + \text{NADH}_2 \qquad \text{Eqn. 5.32}$$

and, as in the Krebs cycle, this reaction has to be pulled in the direction written above (Eqn. 5.32) by very efficient removal of its end-products. Moreover, some fairly sophisticated, but yet ill-defined, control mechanisms must exist to prevent futile cycling because, if PEP produced by reaction 5.31 were accessible to PEP carboxylase in an active state, it would be promptly reconverted to oxaloacetate rather than re-assimilated to starch.

5.16 Energy Costs

C3 carbon assimilation, if we regard it as finishing with triosephosphate, requires 2 $NADPH_2$ and 3 ATP per molecule of CO_2 fixed (the utilisation of "assimilatory power" by the Calvin Cycle). Many calculations have been made of the energy costs of C4, CAM and photorespiration and table 5.1 (from Anderson & Beardall) summarises some of the relevant values which can be derived from this sort of consideration. While it cannot be denied that this sort of arithmetic is perfectly valid or that a process of carboxylation (C3) obviously costs less than carboxylation, decarboxylation and recarboxylation (C4) the arithmetic is rather like the distance between Frankfurt and Milan or the population of these two cities - much depends on where you start and stop counting. For example there are grounds for claiming that sucrose can, if anything, be regarded as the main end-product of photosynthesis because this is the material which mainly carries newly fixed carbon to where it is

Table 5.1 Energy costs of C3, C4 and CAM (after Anderson & Beardall).

Energy inputs per CO_2 assimilated		
	ATP	NADPH
C3	3	2
C4	4/5*	2
CAM	5.5/6.5*	2

*depending on the variant in mechanism

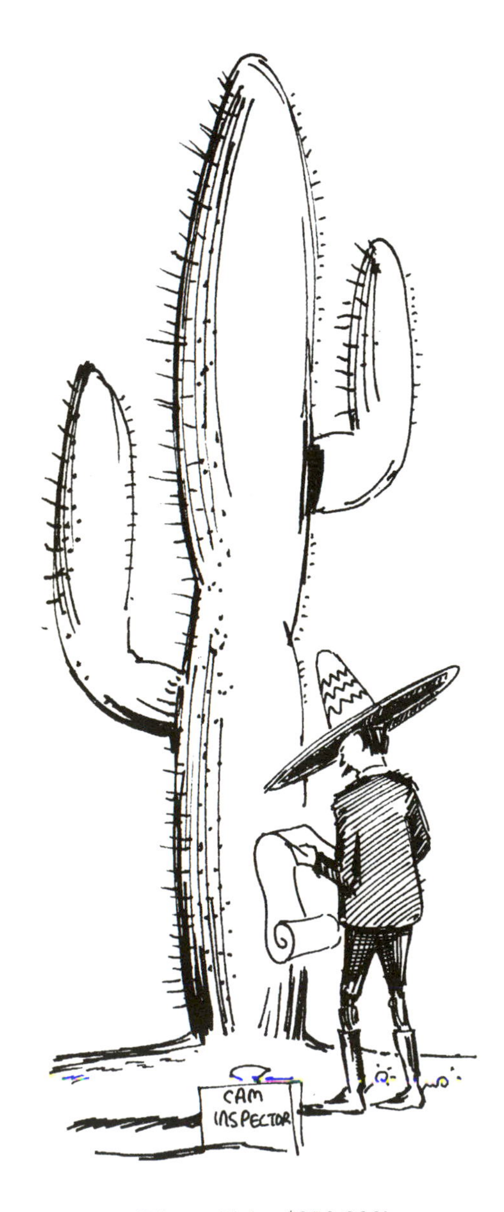

'That will be $350,000'

consumed. If this is accepted it is perhaps as valid to include the costs of sucrose synthesis (or even starch synthesis) to C3 plants as it is to include the costs of reconverting pyruvate to starch in CAM plants. Similarly it may "cost" CAM plants to store malate in their vacuoles and while it costs C4 plants to operate what amounts to a CO_2 pump it also costs C3 species to scavenge the consequences of oxygenation (Section 5.7). Overall there may be little difference in real energy costs between C3, C4 and CAM. Much the same might be said of the supposed advantages afforded by these different types of assimilation. The water economy of C4 (water lost per CO_2 fixed) may be better than C3. CAM may be even better than C4 but, bearing in mind differences in climate, season and age it would be a mistake to suppose that these are anything other than evolutionary answers to environmental problems. As I write, I can glance up at the Cheviot Hills, seven or so miles from the Scottish border. No one thinks about growing maize at this latitude but there is no problem about potatoes which can produce amazingly high yields despite a short growing season. My garden wall provides a happy environment for wall-pepper (*Sedum acre*) and for a species of *Sempervivum* which makes it patently obvious, in flower, why (as well as the mundane "house leek") it also carries the splendid name "welcome home husband however drunk you might be). *Sedum acre* is most certainly CAM. *Sempervivum* is probably CAM but both share their strange environment with a species of *Dianthus* ("pink") which appears to pluck its water from the air as well as the Spanish Moss (CAM) which grows happily on telephone wires in equatorial America. Similarly Death Valley California (notoriously as hot and dry as might be imagined) has, as we would expect, its fair share of C4 and CAM species but it also has specialised C3s which do just as well. Adaptation to an environment is not simply a matter of the underlying biochemistry. There are endless additional factors to be taken into account, anatomical as well as physiological and biochemical. Many of these (including water loss and the ability to resist desiccation) are only marginally related to carbon assimilation. The stinging nettle (*Urtica dioica*), for example, may owe part of its success to its ability to deter predation but it also has a remarkable facility for consuming nitrogen, mobilising leaf nitrogen and storing it during senescence, rather than losing it to the environment by decay.

"notoriously as hot and dry as might be imagined"

5.17 Should I compare thee to a summer's day?

Enzymes are biological catalysts. Catalysts are, as we are all supposed to know, materials which facilitate reactions without themselves being consumed by such reactions. So enzymes make biological reactions go faster. In theory, they do not permit reactions to take place which would not otherwise occur nor alter the equilibrium position of a reaction (i.e. the extent to which a reaction proceeds). Again, in theory, no reaction is irreversible and, if A and B react together to give C and D, a point is reached at which the forward and reverse reactions are going at the same rate. At this time the whole system is in balance; a state of equilibrium has been attained. Enzymes are not, in the general way of things, supposed to affect this equilibrium, only to speed its establishment. In practice, the theory is sometimes hard to accept. This is because some reactions are so slow (at biologically permissible temperatures) that, for all practical effects and purposes, they would not occur at all in the absence of enzymes. Similarly biological reactions often occur in succession and if a second reaction removes the end-product of a previous reaction as soon as it is formed, even an unfavourable equilibrium can be displaced. The reduction of PGA to triosephosphate (Fig. 5.4) is a good example. In photosynthesis this is a key sequence in the Calvin Cycle, one which runs "up hill" by consuming all of the $NADPH_2$ and two thirds of the ATP required by this part of the dark biochemistry. In respiration, on the other hand, it runs "down hill", liberating energy which is conserved as $NADH_2$ and ATP. Whether the sequence goes forward or backwards in this freely reversible situation is determined by the rate at which substrates are supplied and at which end-products are removed. Other reactions are much less susceptible to this push-pull situation. The reaction catalysed by RuBP carboxylase (RUBISCO), for example, cannot be made to run backwards. As we have seen RuBP is a molecule subjected to internal stresses. Once these are relieved, by combination with CO_2, not all the king's horses nor all the king's men can put RuBP back together again (i.e. the equilibrium position, determined by the large decrease in free energy associated with the formation of PGA) is such that for all practical purposes the reaction goes into completion and cannot be reversed.

High affinity and low affinity

So, the extent to which a reaction will go, and the direction in which it will go, are primarily a matter of thermodynamics. A freely reversible reaction is associated with a relatively small, negative or positive change in free energy. One which exhibits a large decrease in free energy will go virtually to completion when circumstances (e.g. the presence of the appropriate enzyme) will permit it to occur. One with a large positive free energy change will not occur at all, unless pushed from behind and pulled from in front in such a way that the equilibrium position is displaced by continuous replenishment of substrates and removal of end-products. Reaction rates, on the other hand, are largely determined by enzymes. Thus if we start with a situation in which a freely reversible reaction has scarcely begun, or a reaction which would go largely to completion because of its very favourable equilibrium position, the rate is determined by the amount of enzyme present and how much it likes its substrate. Accordingly if we plot rate against substrate concentration we get, at the outset, a linear relationship. Twice as much substrate gives twice the rate, so does twice as much enzyme. However, at a low substrate concentration, an enzyme with a high affinity for its substrate will catalyse a reaction faster than one with a low affinity (clearly one with no affinity at all would not accelerate the basic, uncatalysed, rate). Affinities can be measured and expressed in terms of the amount of substrate which permits an enzyme to work at half its maximum rate. Maximum rates are hard to measure because, as the substrate concentration is increased, a law of diminishing returns causes departure from linearity (of the rate v substrate relationship) so that doubling the substrate no longer doubles the rate. Indeed, plots of rate v substrate classically define a rectangular hyperbola in which maximum rate requires infinitely high substrate. Using low mathematical cunning, however, such a hyperbolic relationship can be converted into a straight line by plotting 1/Rate against 1/substrate concentration. At infinitely high substrate concentration (i.e. when 1/substrate concentration equals zero) the intercept on the vertical axis of such a plot is the maximum rate and the (negative) intercept on the horizontal axis is the substrate concentration (Ks) which gives half this rate.

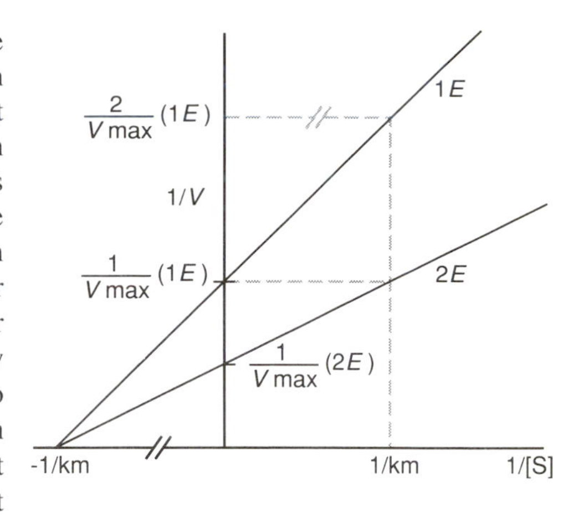

Where does all this take us? At its simplest it gives us a characteristic by which to judge an enzyme. One which requires a small concentration of substrate to work at half its maximum rate is deemed to have a high affinity for its substrate. Herein, lies a story. When RUBISCO was described and finally characterised it apparently required 5% CO_2 to work at half its top speed. In the absence of a CO_2 concentrating mechanism, this was evidently a nonsense because atmospheric CO_2 is about 0.035% and, if it is to diffuse into a chloroplast along a diffusion gradient, the concentration at the enzyme surface would be even less. So there was a great mystery, only to be resolved when fully functional chloroplasts were isolated and the enzyme in them (unlike that somehow changed by purification) was found to have an appropriately high affinity for CO_2. PEP carboxylase, unlike RUBISCO, was never a mystery in this regard although it had such a high affinity for bicarbonate (which it utilises rather than dissolved CO_2) that this was difficult to measure precisely. All of this led to a degree of confusion. Thus C4 species were thought, by some, to have an advantage because CO_2 in their leaves was first fixed by an enzyme (PEP carboxylase) with a higher affinity than RUBISCO. In fact the relative affinities of these two enzymes are so similar as to be unimportant. The advantage of having PEP carboxylase catalyse CO_2 fixation in the cytosol of C4 plants resides in the fact that, unlike RUBISCO, PEP carboxylase is not subjected to competition by oxygen (Section 5.7). Then, as we have seen, decarboxylation of malate within the inner, bundle-sheath compartment, allows RUBISCO to operate in a CO_2 enriched environment which ensures that CO_2 mostly wins the competition with oxygen. Despite the marvellous intricacies of C4 photosynthesis it must be remembered that, in the end, the Calvin cycle fulfils the same vital role that it does in C3 and CAM. Ultimately PEP is derived from the Calvin Cycle and this cycle, and only this cycle, has the autocatalytic ability to function as a breeder reaction producing more potential substrate than it uses. Without this unique ability, plants, and therefore most living organisms, would have no capacity for growth.

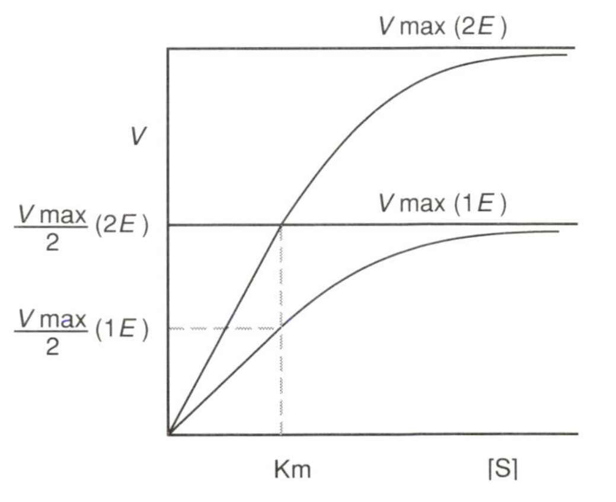

5.18 Summary

Carbon dioxide is assimilated by green plants in a process (the Calvin Cycle) in which assimilatory power (ATP and $NADPH_2$), generated by photosynthetic electron transport, is consumed. Carbon dioxide is joined to a pre-existing acceptor containing 5 atoms of carbon (ribulose bisphosphate) in such a way that two molecules of product (3-phosphoglycerate), each containing 3 atoms of carbon are formed. Other reactions in the Calvin Cycle are concerned with the reduction of 3-phosphoglycerate (PGA) to triose phosphate and the regeneration of three molecules of the acceptor (RuBP) from five molecules of triose phosphate. For every three turns of the Calvin Cycle one molecule of triose phosphate is produced which is not needed for regeneration of the acceptor. This constitutes product and it may be stored within the chloroplast as starch or transported to the cytosol in exchange for inorganic phosphate. In the cytosol it is made into sucrose for transport to other parts of the plant or consumed, directly or indirectly, in other metabolic processes such as respiration.

Carbon assimilation which involves only these processes is described as C3 photosynthesis because the first product of fixation (PGA) contains three carbons. Most temperate species are C3 plants. There are, however, two other groups of plants (and others which can be loosely described as intermediate in behaviour) which have additional mechanisms. The first product of carbon dioxide fixation in C4 species is, as the name C4 implies, an acid, oxaloacetate, which contains four carbons. This preliminary fixation process occurs in an outer (mesophyll) compartment in the leaves of these plants and the oxaloacetate, having been converted to malate and aspartate and transported to an inner (bundle sheath) compartment, is decarboxylated. This serves to increase the carbon dioxide concentration within this inner compartment so that re-fixation by the Calvin Cycle in the bundle sheath is favoured and photorespiration is suppressed. This prevents the decrease in carbon dioxide fixation which occurs in C3 species as a consequence of oxygenation of RuBP (competition between oxygen and carbon dioxide) at the site of carboxylation. This results, when oxygen wins the competition, in the production of one molecule of phosphoglycolate and *one* molecule of PGA rather than *two* molecules of PGA. In C3 species there are mechanisms for returning phosphoglycolate to the Calvin Cycle (as PGA) but this involves the loss of carbon dioxide as two molecules of phosphoglycolate are converted into one molecule of PGA. Ammonia is also lost in photorespiration because the regeneration of phosphoglycolate involves, in effect, the consumption of amino acids.

What amounts to a carbon dioxide-enriched greenhouse within their leaves benefits C4 species in another important way because the preliminary carbon dioxide-fixing enzyme (PEP carboxylase) has a very high affinity for carbon dioxide. Accordingly, for a given stomatal aperture, C4 species can maintain a steeper diffusion gradient for carbon dioxide, between atmosphere and site of fixation, than C3 species. They, therefore, have a higher water use efficiency, losing less water per carbon dioxide fixed than C3 species. CAM plants carry this strategy further by separating the preliminary fixation of carbon dioxide (into oxaloacetate) in time (rather than in space) from subsequent fixation into RuBP. Their stomata open by night and oxaloacetate formed in the dark (by carboxylation of PEP) is stored, as malate, until the next day. Following decarboxylation behind closed stomata, the carbon dioxide is then refixed in the Calvin Cycle; high loss of water by day is eliminated and photorespiration is diminished.

It must be remembered that all plants ultimately depend upon the Calvin Cycle which is the only metabolic sequence which has an in-built autocatalytic function which can convert product (triose phosphate) into new acceptor (RuBP) and therefore permit growth. C4 photosynthesis and CAM are "additions" to the Calvin Cycle, *not* substitutes for it.

Chapter 6

of Plants & Men

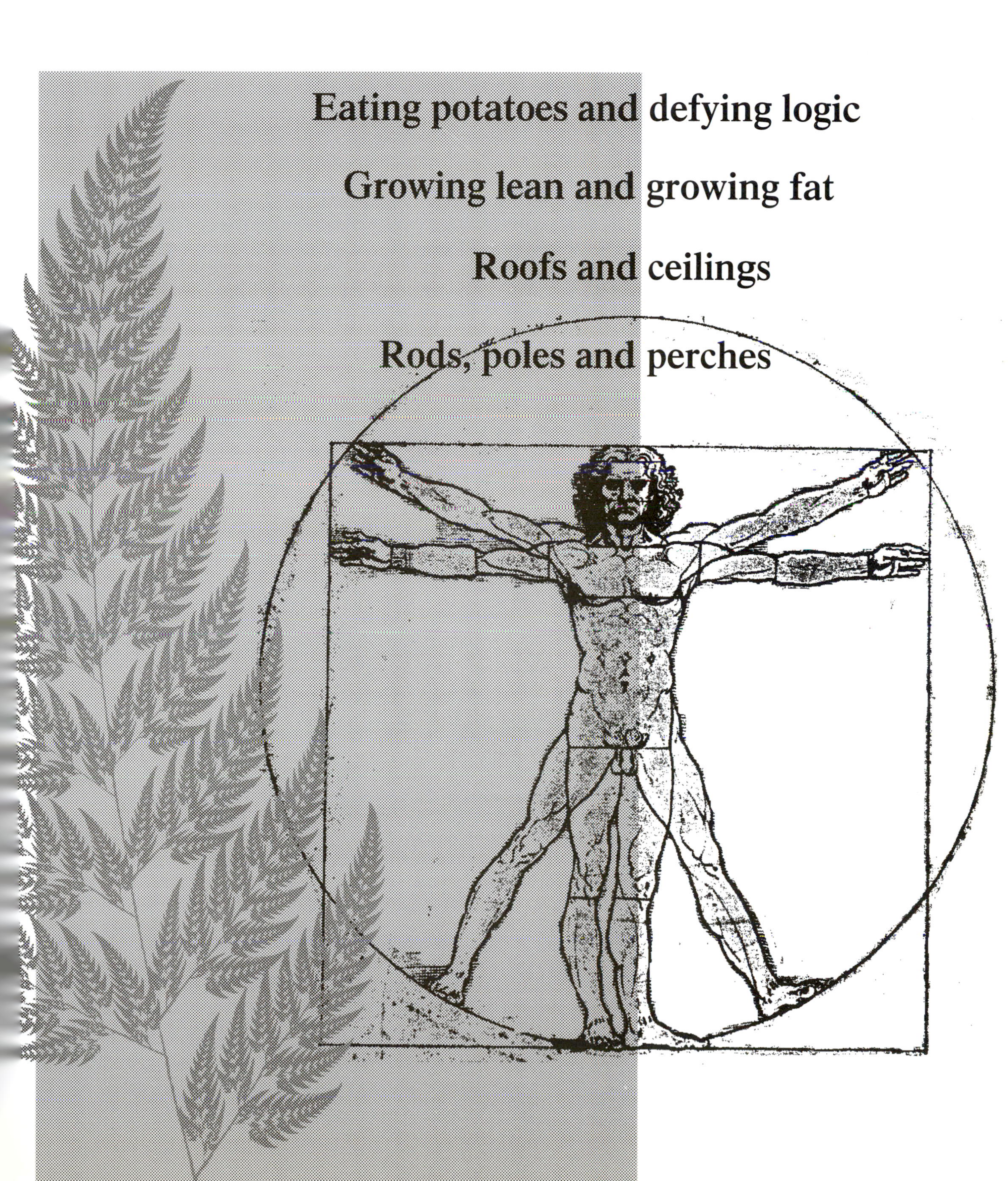

Eating potatoes and defying logic

Growing lean and growing fat

Roofs and ceilings

Rods, poles and perches

6.1 All Flesh is Grass

In the proceeding chapters we have seen that Man's existence derives from nuclear fusion in the sun. Electromagnetic radiation from this source, which arrives at the Earth's surface as visible light, is transduced, by photosynthesis, first into electrical and then into chemical energy. Apart from some bacteria, all other living organisms then utilise plants as their source of energy. Those Eskimos who still live in the old ways are probably the most carnivorous human race and the Gauchos who used to follow vast herds of cattle across the South American Pampas must have been a close second. Even so, the ultimate dependence on plants is obvious. The cattle literally grazed on grass and the Gauchos ate the cattle. The food chain which ends with the Eskimos is larger and more tenuous but it still starts with plants in the shape of marine phytoplankton.

If we wish to pursue this theme of energy, plants and Man we now need to consider some related matters and questions. If we are so dependent on plants, and because there are so many of us and we are increasing our numbers at such an alarming rate (by about a third of a million a day), it would seem sensible to see how many plants we need and how we are changing the environment in which we, and they, must exist.

ALL FLESH IS GRASS

To start with we will attempt to arrive at some idea of how much food we require and then we will try to grapple with the even more difficult problem of how much land we must have in order to maintain existing populations. It will be immediately evident that these are not matters which can be considered in a purely scientific way. Many have little choice about what they eat but, in the West, there is no longer the dependence on staple items of the sort which caused famine in Ireland when the potato crop failed. Moreover, some farmers in the UK are currently being paid to "set-aside" land (i.e. to take it out of crop-production) rather than add to the "food mountains" which accumulate in consequence of the "European Common Agricultural Policy". Years ago, ships sailed into Odessa to load grain. In order to avert famine in 1992, the West is air-lifting food into what was the Soviet Union and what is still, by far, the largest land area (2/5 of the Earth's land surface) in the world; an area still fully capable of growing grain for export. Such famine was once a normal feature of life on the Indian sub-continent. Now regular, famine of an extensive and disastrous nature has been "permanently" averted by investing huge amounts of energy in the synthesis of nitrogenous fertilizers. A desert in Central China, which once advanced, year by year, is now retreating in an equally relentless fashion for the simple reason that the farmers on its fringes have been permitted to sell their crops for personal profit. (Conversely, the same search for profit was a major factor in creating the dust bowl in Oklahoma in the 1930s). The Chinese also soon hope to have many millions of additional refrigerators (for which they have

insufficient electricity) and which, because of CFCs, will add to world-wide damage to the environment (Chapter 7 and 8). At the same time they ask, with some justification, why they should forego modest luxury while the West is so clearly reluctant to give up anything. In short, Man's energy requirements do not stop with food and Man's energy "production" is as much a matter of economics and politics as it is of science.

Where in all of this might we begin? Let us first consider a few basic facts about our metabolic requirements.

6.2 Eating Potatoes

"Man cannot live by bread alone.."

Peanuts are reputedly the cheapest way of fending off starvation in the United States and no one raised on peanut butter and jelly sandwiches - surely the most awful mixture known to man, would find this too great a hardship. The British and Australians, more accustomed to eating autolysed yeast in the form of marmite and vegemite respectively, could only look on and marvel. If they choose to do so, or are obliged to do so, most adults, can survive indefinitely on a totally vegetable diet. In terms of efficiency, such a short food-chain is more effective than a long one. Some of us would feel deprived if we were never to eat another steak but, if we were solely concerned with not going hungry, we would be better advised to sow and eat wheat than to raise cows. In a year, a bull might easily eat as much as 10 people and would not, once finally slaughtered, provide food for 10 people for more than a fraction of a bull's life-span. An Eskimo, eating seals, will be dependent on immensely more photosynthesis than someone eating rice, because, along the way, the marine phytoplankton (on which he ultimately depends) might have had to provide the energy and substance for any number of species, each eaten by the next one in line. At every stage energy would be dissipated in movement etc., and lost in the conversion of food to muscle, bone, or whatever. For these reasons it is very difficult to say how many plants the Eskimo must have working on his behalf (Kelly suggests a conversion efficiency of as little as 0.1%) but let us see what we can do with land-based vegetarians.

How many plants, then, must a person eat in order to maintain himself or herself in good health? Clearly the question can only be answered in general terms because cabbages are less nutritious than peanuts and a large man engaged in heavy manual work in a cold climate will use substantially more energy than a small woman sun-bathing on a beach.

Table 6.1

"A loaf of bread, a jug of wine..."

A sedentary male could derive his daily calorific requirement of approximately 2,400 Kcal (2,400 Calories) as follows:-

As bread	=	880 g	30oz	1.9 lb
or potatoes	=	3400 g	119 oz	7.5 lb
or sugar	=	620 g	21 oz	1.4 lb
or lettuce	=	24000 g	848 oz	53.0 lb
or mutton	=	1000 g	35 oz	2.2 lb
or beef	=	1100 g	38 oz	2.4 lb
or olive oil	=	271 g	9 oz	0.6 lb
or beer	=	6.8 litres	14.4 pints	

(Note that the "Calorie" of diet fame, i.e., Calorie with a capital C, is one thousand calories or one "kilocalorie" (Kcal))

So we shall attempt to arrive at reasonable approximations, expressed in terms of "average" energy requirements of man, just as we shall subsequently arrive at "average" efficiencies of conversion of light energy to chemical energy by crops (Section 6.8). For simplicity of arithmetic we can argue that a sedentary male requires 2,400 Kcals (approximately 10,000 Kjoules) a day or 100 Kcals (418 Kjoules) per hour. The choice of a sedentary male is not sexist. In general, women use rather less energy than men; on average they are smaller, but a modest degree of activity by a women would readily make her metabolically equivalent to a sedentary male. A fit man, in his prime, undertaking heavy manual work might easily require 6,000 Kcal (25,000 Kjoules) a day and, by the same token, the young and active use more than the old and sedentary. Anyone, sitting quietly behind a desk, will use 85% of their metabolic energy to maintain their body temperature but, on a squash court, or at speed on a bicycle, the same individual will expend most energy in mechanical work. At rest, or asleep, there is still a need to maintain bodily functions such as heart beats.

Table 6.1, then, has to do with the man sitting behind the desk or *"sitting on the Clapham omnibus"*. In that sense it is entirely realistic but, in other regards, it is an academic exercise for the obvious reason that, while the Gaucho might have been moderately happy on a couple of pounds of beef a day, few could even contemplate eating 53 pounds of lettuce. Although a diet of gin and lettuce has been suggested as one of the more pleasant ways of losing weight, there are obvious practical difficulties in attempting this, just as there are in foregoing all solid food

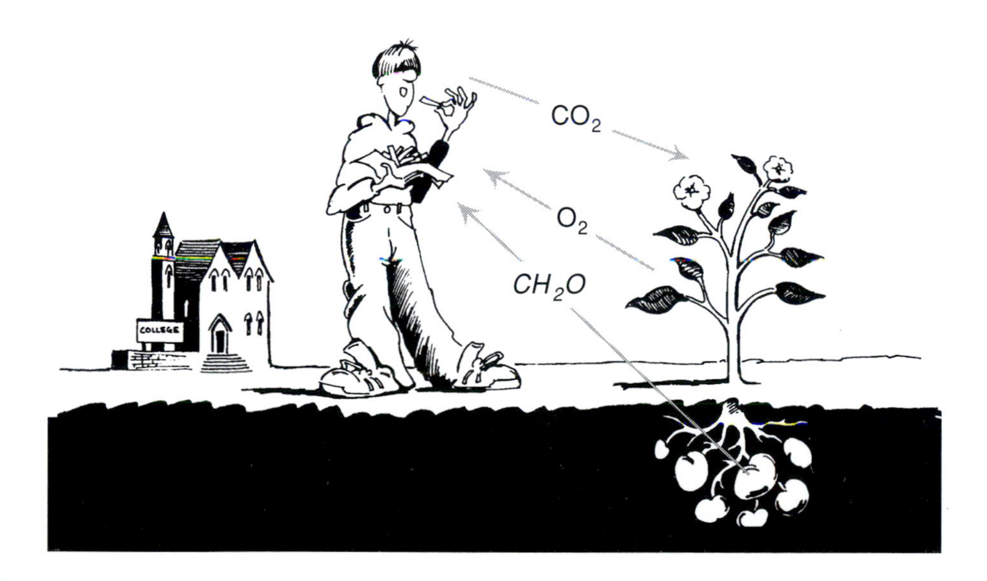

Fig. 6.1 Cycling of metabolites between plants and Man (after Laetsch). Carbon dioxide released by animal (and plant) respiration is reduced to CH_2O in photosynthesising leaves using hydrogens ($H^+ + e^-$) from water. Oxygen released in this process is made available for respiration. CH_2O is transported from potato leaves, as sucrose, to the underground tubers, where it is stored as starch. Digestive breakdown of foodstuffs, such as starch (into simpler molecules) initially releases little energy but in its final stages, recombination of hydrogen (H) and oxygen (O) releases the energy used in movement, to maintain body heat and function and to support the synthesis of new molecules and body tissues.

in favour of beer. The potato, on the other hand, is quite a different matter (Section 6.5) because, some years ago, a Danish keep-fit enthusiast lived for a year on nothing but potatoes. He even drank expressed potato juice. At the end of this period he was neither fatter nor thinner and judged to be as fit as when he started. This is because the potato, totally unadulterated, *really does constitute a complete diet.* In broad terms, the potato might be 75% water and 23% starch, but hidden in these rough proportions there is, as in any living tissue, a significant quantity of protein and amino acids (*"all proteins are but permutations and combinations of some twenty constituent amino acids"*). Indeed, it may come as surprise to some, who regard potatoes as nothing but a mass of starch, to learn that they were once regarded as an important source of protein used in commerce.

> *"The subject is well worth attention because the raw materials would be cheap compared with eggs, blood or fish-roes"* ....R. Haldane, 1885 (in a procedure for extracting albumin, a protein, from potatoes)

The uncooked potato, unlike oatmeal, contains, in addition to other vitamins, enough vitamin C to prevent the scurvy which was common, long ago, in British seafarers before the importance of fresh fruit or vegetables was recognised. (Hence the term "Limey" which was applied to the seamen when limes were introduced into their diet as a prophylactic against scurvy).

> [There is an apocryphal story about a Scottish student who thought to emulate his medieval forbears by surviving one university term on a sack of oatmeal. Wishing to economise on effort as well as on diet, he made a huge pan-full of porridge which he poured into a chest of drawers, breaking off a piece of solidified oat-cake when ever he felt hungry. Soon he became ill and a surprised Scottish doctor diagnosed the first case of scurvy in his country in living memory. Somewhat chastened, the student thought to abandon his experiment but his medical advisor said he would "do fine" so long as he also ate an occasional orange.]

soul food

No one, for choice, is likely to opt for potatoes as sole food but it simplifies the arithmetic enormously, whilst retaining credibility. In past "Imperial" parlance, half a stone (seven pounds or 3 Kg) of potatoes a day is all that it is needed to keep body and soul together, indefinitely. This will allow us to turn to the question of how large an area we need in order to grow a year's supply of potatoes. The short answer is something like 100 square meters but that statement patently requires qualification and it also takes us into all sorts of other matters, including the efficiency of the potato plant (i.e. the efficiency of photosynthesis -Section 6.6). In the end, all of this has to do with energy transduction - light energy into plant material, plant material into metabolic and non-metabolic energy. In order put energy transduction on to a quantitative basis we must needs have a little aside about units (c.f. Section 2.4) but even at this stage we ought to take into account that what Man needs, in the way of his metabolic energy requirements, is usually the least of his problems. There are those of us (like the Eskimos, the Gauchos, the Pigmies of Equatorial Africa) who, certainly once upon a time, had no need, or scarcely opportunity, to burn plant material. If there are still such peoples (who get by on what they eat uncooked) they must be very few and far between. As we progress (geographically) from Africa northwards and westwards the additional energy requirement continues to increase until, in North America, energy is used for almost every conceivable purpose and even for some manipulations which are best overlooked in polite society. For this reason the demand for non-metabolic energy rises from zero, in very primitive societies, to about 10 kilowatts *per capita* in very sophisticated societies. Given "Third World" aspirations, we ought perhaps to talk in terms of a requirement for one to ten kilowatts. Man's metabolic requirements on the other hand are two orders of magnitude lower, at about 100 watts. This follows from the fact that:-

$$2400 \text{ Kcal/day} \cong 100 \text{ Kcal/hour}$$

$$100 \text{ Kcal/hour} \cong 100 \text{ watts}$$

6.3 Defying Thermodynamics

"I know a girl who's mighty sweet,
big blue eyes and tiny feet
her name is Rosabelle McGee
and she tips the scales at 303" .....Hoagy Carmichael

During the last war, obesity was not a problem in Europe just as it is not a problem today in most of the "Third World". By the 50s, however, food that was now more than adequate, together with increasing affluence, led to an increasing pre-occupation with diet. At least in the UK this opened the flood-gates to a never ending spate of pronouncements about "fattening foods". Most of these took great liberties with thermodynamics. Notable at this time was the remarkable notion that obesity derived from eating "starchy" foods. Steaks were in. Potatoes were out. There was also the idea, still favoured to this day, that there are some amongst us who eat almost nothing and become immense and others who eat everything and stay lean.

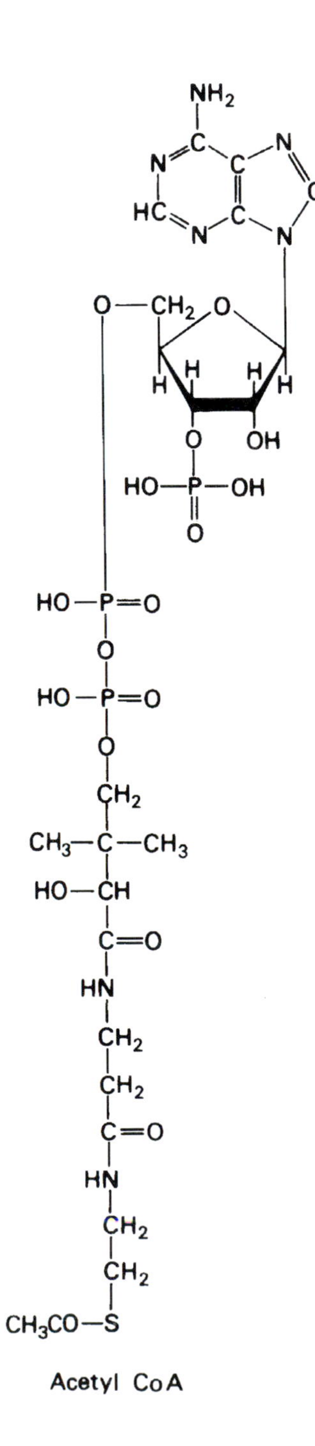

Acetyl CoA

The underlying biochemical fallacy in all of this is that there are a very small number of metabolites, such as acetyl coenzyme A, which lie at the junction between carbon, nitrogen and lipid metabolism. True, there are some things eaten which are excreted but, whether we are talking about lipids (fats), proteins or carbohydrates we are talking, by and large, about strings of carbons with various other bits and pieces attached. Apart from oxygen, these bits and pieces rarely contribute much to the mass or energy content of these compounds. Living organisms contain a lot of water, a lot of carbon, quite a bit of oxygen but not much of anything else; even though the relatively minor constituents such as nitrogen are absolutely essential. So we have substances which are more oxidised and less oxidised (Section 1.2). Those which are less oxidised are relatively energy rich because they release more energy when they are "burned" in the body. Thus olive oil is highly reduced and packs a great many calories per gram (like peanuts). A potato on the other hand contains a lot of water and its dry matter, mostly starch, is much more oxidised than olive oil. Meat is also largely composed of water and the oxidation status of what is left (mostly protein) lies somewhere between starch and oil. This sort of consideration is the basis of why some foods have a high calorific value and some a low value. It is also why we get fat or stay thin. Our bodily functions require energy. If we supply more energy than is needed, the thrifty body stores some substance away for a rainy day and we put on weight. "Ah yes", you might say, "but isn't it easier for the body to make fat from fat?" Broadly speaking the answer is "not really". Food is digested in the alimentary tract. Digestion involves breakdown (of starch, proteins and fats) to small molecules which can be taken up in the blood stream. These small molecules (sugars, amino acids, small-chain fatty acids etc.) are either oxidised to CO_2 and H_2O (with the associated excretion via the kidneys of unusable nitrogen etc.) or resynthesised into body tissues and storage products such as glycogen (animal starch) and fats. Some amino acids derived from digested protein can be used more or less unchanged and proteins are a necessary part of diet because they are the source of nitrogen (and sulphur) used in new protein synthesis but, by and large, the body can convert proteins into fat as readily as it can convert starch into fat or ingested fat into body fat. This was nicely summarised by Hans Krebs who was awarded the Nobel Prize for his work on the "tricarboxylic acid cycle" or "Krebs cycle" as it is so often called.

"The production of energy by the combustion of foodstuffs in living matter may be said to proceed in three major phases. In phase 1 the large molecules of the food are broken down to the small constituent units. Proteins are converted to amino acids, carbohydrates to hexoses and fats to glycerol and fatty acids. The amounts of energy liberated in these reactions are relatively small, of the order of 0.5 per cent of the total free energy that can be released in the breakdown of polysaccharides and proteins and no more than 0.1 per cent in the case of polysaccharides and triglyceride fats. These quantities are not utilised, except for the generation of heat. The reactions of this first phase thus do not yield utilisable energy: they merely prepare foodstuffs for the energy-giving processes proper. They take place in the intestinal tract and also in tissues when reserve material is mobilised for energy

production. In Phase II the diversity of small molecules produced in the first phase -three or more different hexoses, glycerol, about twenty amino acids and a number of fatty acids - are incompletely burned, the end-product being, apart from carbon dioxide, one of three substances: acetic acid in the form of acetyl coenzyme A, a-ketoglutarate, or oxaloacetate. The first of these three constitutes the greater amount: two thirds of the carbon of carbohydrate and of glycerol, all carbon atoms of the common fatty acids and approximately half the carbon skeleton of amino acids yield acetyl coenzyme A. α-ketoglutarate acid arises from glutamic acid, histidine, arginine, citrulline, ornithine and part of the benzene ring of tyrosine and phenylalanine. The details of the pathways cannot be discussed here in full, but one matter of principle should be emphasised: the number of steps which living matter employs in order to reduce a great variety of different substances to three basis units is astonishingly small and could certainly not be equalled with the tools at present available to the organic chemist.

The three end-products of the second phase are metabolically closely inter-related. They take part in the Phase III of energy production: the tricarboxylic acid cycle, the common "terminal" pathway of oxidation of all foodstuffs.

As each step of the intermediary metabolism requires a specific enzyme it is evident that a common pathway of oxidation results in an economy of chemical tools. Surveying the pathway of the degradation of foodstuffs as a whole one cannot but be impressed by the relative simplicity of the arrangement. The total number of steps required to release the available energy from a multitude of different substrates is unexpectedly small. But the economy of stages does not end here. Energy is not released at every intermediary step of metabolism. Most though not all of the available energy contained in the foodstuffs is liberated when the hydrogen atoms removed by dehydrogenations react with molecular oxygen to form water. Such reactions in which 40 - 60 kcal. are liberated per mole occur at Phase II and Phase III. Roughly one-third of the total energy of combustion is set free at Stage II and two-thirds at Phase III." .....Hans Krebs.

Here, indeed, we have the final putting together again (unlike Humpty Dumpty) of the H-O bonds broken by the photolytic process in the early stages of photosynthesis (Chapter 1).

And what about these fat people who eat like birds? To put it starkly, there were no fat prisoners in Belsen. Some people have a propensity to become more fat than others but this, is largely because they have a tendency to eat more and exercise less. If they are not ill, people have a body temperature of 36.9 °C. Hot bodies radiate more heat than cold bodies but the possibility that one sedentary human can loose much more energy, as heat, than another, is obviously limited in normal human society. If we were to walk naked in the cold we could eat more without getting fat but we don't. In all of these matters, as Einstein has told us, energy and mass are interchangeable. The only way that we can dissipate energy, other than by radiation or by work, is by using it in bodily functions. At rest, people of equal size use very similar amounts of energy for this purpose. When not at rest it is a matter of how much mechanical work they are engaged in. Of course, in this sense, work has to be carefully defined. Some can sit with very little movement. Others, lean and active, never seem to be at rest, even when relaxing.

Obviously every movement counts but these inadvertent body movements do not add up to a great deal. A lean and nervous individual may stay as thin as a rake while consuming 6,000 kcal a day but not while sitting all day at an office desk and every evening in front of a television. Such an individual would stay thin only by engaging in very hard labour for long periods, or perhaps in activities which might invite very hard labour for long periods.

> *"Give me men about me that are fat*
> *sleek-headed men that sleep o' nights*
> *yon Cassius has a lean and hungry look*
> *such men are dangerous"* .....Shakespeare (Julius Caesar)

Having now at least a rough idea of how much energy Man needs, and how much he would currently like to have at his disposal for non-metabolic uses, we can go on to ask how much energy there is available (as light) and how effectively plants can transduce light energy into the chemical forms of energy (food) which we can utilise. But before we go any further into this we should define a few more units (please also refer to Sections 2.4 and 2.5)

6.4 Rods, Poles and Perches

In the U.K. at least, school children are no longer expected to be familiar with traditional units of length such as rods, poles, perches, chains etc. Even hands and feet are becoming unfashionable despite the fact that we always carry these about our person. However, in the United States and Britain, and some other English speaking countries, correspondingly archaic units of area are still in every day use (see also Chapter 2). The most persistent of these is the "acre", an area of 4,840 square yards. Anyone who has not even given this a second thought may be startled, on reflection, to find that 4,840 does not have a square root in whole numbers. In other words, we have here a measurement of area which is a rectangle rather than a square. This is because the acre derives from the British feudal system in which ten acre fields were divided into 10 rectangular strips each 22 yards wide and 220 yards long. These relate in turn, of course, to the mile (one eighth and one eightieth of a mile respectively). More than that, the width of the single acre strip is the length of a cricket wicket and, therefore, more or less guaranteed to persist into the foreseeable future. The French, who are demonstrably more practical and less romantic than the English, use the "*are*" rather than the "acre". The "are" is ten times ten metres and the term "hect*are*" (100 ares or ten thousand square metres) is now widely used as a measurement of area in the metric system. One hectare is approximately two and a half (2.471) acres and one acre about two fifths (0.467) of a hectare.

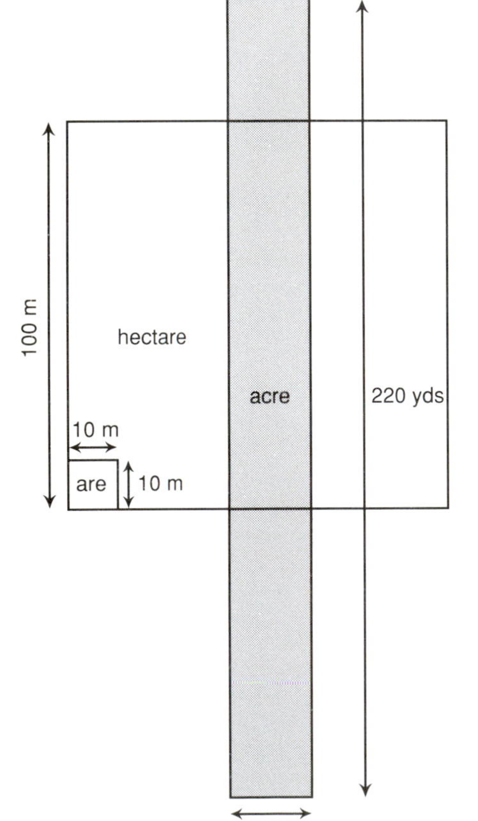

In passing, since we will soon be considering mass as well as area, we ought to take a quick look at the ton. In the British or Imperial system the ton is 2,240 lbs (each lb equalling 453.59 grammes). In the United States, there is also the short ton, equal to 2,000 lbs. Happily the metric ton or "tonne" is not too different from either. The tonne is 1,000 kilograms (equals 1.1023 short, American, tons and 0.9842 long, Imperial, tons). In considering plant yields we can regard all of these as approximately equivalent and acres and tons (or hectares and tonnes) are convenient units when we are considering crops. Light, on the other hand, is best expressed as μmole quanta.$m^{-2}.s^{-1}$ and this is consistent with modern instrumentation. In terms of Man's energy requirements, however, it is easier to think of light-energy delivered in watts per square metre. Once we are aware of how much light energy arrives each day at the surface of the Earth and we know how much of this energy can be converted, by plants into food, we can begin to calculate how many plants each of us needs to have photosynthesising, desperately, on our behalf.

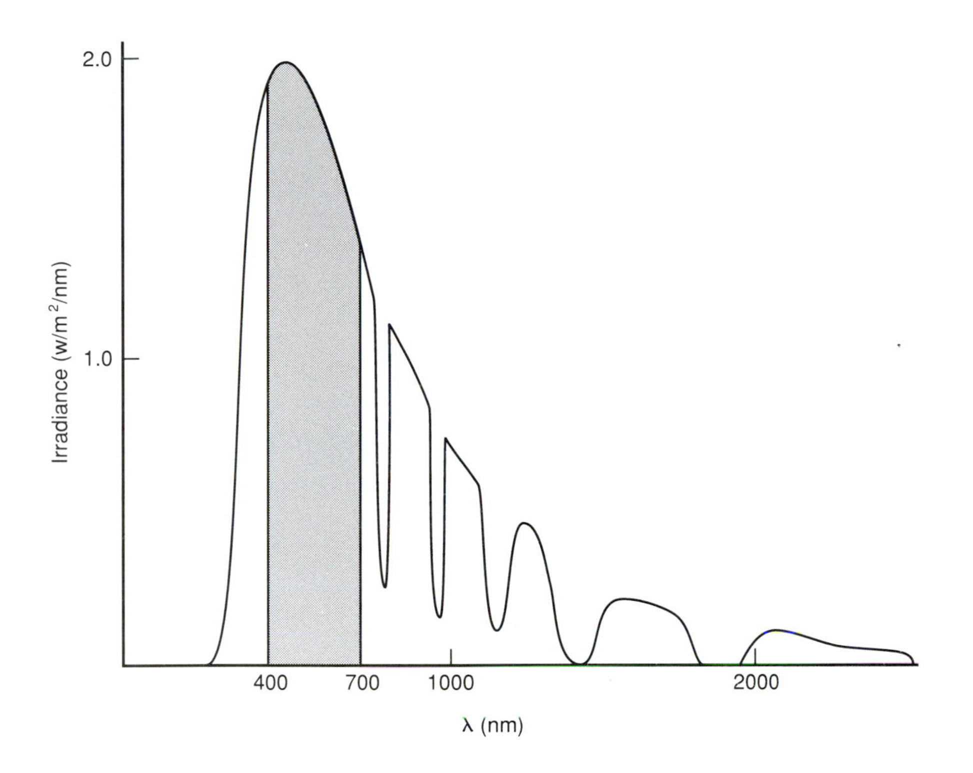

Fig. 6.2 Photosynthetically active radiation (PAR). Shown (shaded) as a fraction (400-700 nm) of total solar irradiance. (Troughs above 700 nm are caused by absorption by CO_2, H_2O vapour etc.)

6.5 Sunbeams into Cucumbers

"lux et veritas"

Until quite recently, light (i.e "light intensity") was measured in foot candles or the metric equivalent (metre candles or "lux"). As the name implies, this is the light given out by one standard candle and perceived at a distance of one foot. A standard candle emits 4 lumens (1 candle power) and, if placed at the centre of a sphere of 1 foot radius, produces a luminous flux at its surface equal to 1 lumen/ sq. ft. If the radius is increased to 1 metre the same light is spread more thinly, the flux decreasing according to the square of the distance. Since there are 3.2808 feet to the metre and since 3.2808 squared = 10.8 it follows that one foot candle (1 lumen/ sq. ft.) = 10.8 lux or metre candles (1 lumen/sq. metre).

The human eye is a very good device for comparing the intensity of weak light sources, but very poor for determining absolute intensity values. This is because the iris diaphragm closes in bright light, so protecting the retina. Accordingly we tend to overestimate the intensity of light if we enter a well-lit room on a dark night. The light intensity in such a room is unlikely to be more than 200 ft. candles, whereas in full sunlight it may be as much as 10,000. For accurate measurements, the human eye was replaced by instruments such as photographic light meters. A contemporary "light meter", however, is much more likely to be a quantum sensor. This, in principle, is like the photosynthetic apparatus, i.e. one quantum ("parcel") of light energy) causes one photochemical event (c.f. sections 2.8 & 9.11). In the instrument this generates an electric current which is made to display the number of quanta. In full sunlight, at noon, at sea level near the equator, the reading might be as high as 4000 μmole quanta.$m^{-2}.s^{-1}$ but only half of this would be visible light or photosynthetically active radiation (the two are more or less equivalent). The remainder (Fig. 6.2) is ultra-violet (below about 400 nm) and infra-red (above about 700 nm). This is fine, so far as it goes, but it does not help our calculations too much if we are thinking in terms of how much solar energy we can convert into cucumbers or potatoes. At the equator the sun only shines half of the day and indeed, averaged over the year, this is true of any part of the globe. Generally speaking early morning light and evening light is less strong. So what we need is the mean annual irradiance i.e. what reaches the surface, averaged over 24 hours a day, 365 days a year. It is also helpful for this purpose to think in terms of watts/square metre. Full sunlight (about 4,000 μmole quanta.$m^{-2}.s^{-1}$) is equivalent to 1 KW/square metre, but in major crop growing areas such as the wheat fields of N. America this falls to 200 watts/ square metre (mean annual irradiance) of which only about half is photosynthetically active. Even so, if photosynthesis were 100% efficient this would be enough to meet the energy requirement of one man. Indeed a fully autotrophic green man would be almost feasible, given this impossibly high efficiency of conversion of light energy into chemical energy. This follows from the fact that the sedentary male can also be rated at 100 W, i.e. such a man consumes, on a continuous (albeit fluctuating) basis, as much energy as a 100 W electric light. Thus:-

$$\begin{aligned}
2400 \text{ Kcal / day} &= 100 \text{ Kcal / hour} \\
100 \text{ Kcal / day} &= \frac{100}{60} \text{ Kcal / min} = \frac{100}{3,600} \text{ Kcal / sec} \\
&= 24 \text{ Kcal / sec} \\
24 \text{ Kcals / sec} &= 24 \text{ x } 4.128 = \text{approx. } 100 \text{ joules / sec} \\
100 \text{ joules / sec} &= 100 \text{ watts.}
\end{aligned}$$

Unfortunately, photosynthesis is not 100% efficient and, as will be seen in Section 6.7, the *maximum* efficiency of a crop is unlikely to exceed 5% and the corresponding *minimum* "land equivalent" would be about 20 square metres in the UK and about 11 in the USA (Fig. 6.3). Under some conditions, nearly 5% conversion of light energy to chemical energy is actually achieved by close stands of vegetation, but the reader would be rightly sceptical if he or she were asked to accept that one man (in the UK) could produce enough food on a piece of ground 10 metres long by 2 metres wide (16.5 feet x 13 feet) to maintain him for one year. What *is* possible can be derived from the experience of our Danish "keep-fit" enthusiast who lived for a year on a diet of potatoes (Section 6.2).

A calorific requirement of 2,400 Kcal/day (100 Kcal/ hr), can be met by about 7 lb of potatoes a day or approximately one ton/year (7 lb x 365 = 2,555 lb). It is not difficult to grow 12 tons of potatoes to the acre in the UK and the highest reported yield (in *exceptional* agricultural practice) was 88 tonnes/hectare (about 36 tons/ acre). So, in the U.K., it has proved possible to grow enough potatoes on 1/36 part of an acre (i.e. on 6 x 22 yds or approximately 112 square metres) to feed one person for one year (Fig. 6.3). Turning to the unexceptional, it was not unusual for an Irish farmer and his wife, to raise a family of five and maintain several milking cows and a couple of horses, on a thirty acre farm as recently as the early 1960s.

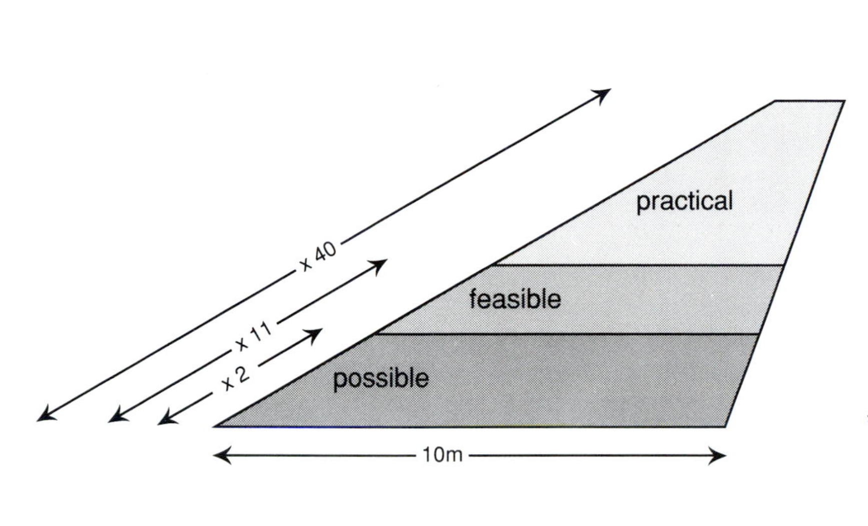

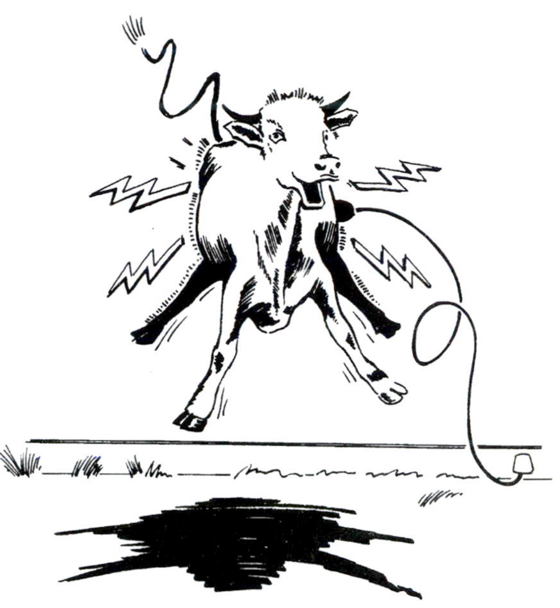

Fig. 6.3 Land areas needed to feed one person. Theoretically 20 sq metres would be sufficient to grow 12 tons of potatoes. Best recorded crop would have called for an area of about 110 sq. metres and 300-400 sq. metres would be required in practice.

There are 14 million acres of arable land in England and Wales alone and even at the very modest yield of 10 tons of potatoes/acre this would support a population of 140 million. (The present population is nearer 60 million). This disregards many more millions of acres which could be used for sheep-rearing etc. The principal reason that the UK is not currently self-sufficient (despite the fact that it already produces enough grain to meet the calorific requirement of its entire population) is that most people prefer to get their calories second-hand. A man is rated at about 100 watts, a cow at about 1 KW. Feeding grain to beef cattle may lead to something more palatable than bread alone but it is hardly the best way to use declining energy resources. Cattle, sheep and pigs require about 7 calories of plant product for every calorie of meat and fat they produce.

Countries like the United States have much higher populations than the U.K. but, proportionately, much more land per head of population. The United States is, therefore, still capable of producing much more grain than it requires for domestic use, although this is largely attributable to mechanisation and the development, in the 1950's of selective herbicides which did so much to free the prairie wheat fields of weeds. The fact that the erstwhile Soviet Union failed to emulate the United States in this regard is more a matter of inadequate harvesting, storing and distribution than a basic inability to grow more. This, in turn, is largely attributable to political interference.

[Following an infamous visit to the United States, Kruschov is reputed to have insisted on major introduction of maize (corn, *Zea mays*) into Soviet agriculture (climatically much better suited, on the whole, to producing wheat) thereby causing wide-spread famine.]

No matter how we look at this subject we come back to two simple and irrefutable facts. The first is that, if we take a small densely populated island in the northern hemisphere, much of it bleak moorland, unsuitable for growing arable crops, and subjected to relatively cold winters, we find that it is still capable of growing enough food to support 2 or 3 times its present population. The second, and more sobering thought, is that this is only made possible by the extensive use of fossil fuels. If the U.K. had to depend for *all* of its energy on renewable biomass, the feasible population would fall, by two orders of magnitude, from round about 150 - 200 million to 2 or 3 million (Section 8.11). This, of course, is in accord with population levels before the industrial revolution and the wide-spread use of fossil fuels.

6.6 Oil into Potatoes

While there is no doubt that even a tiny and highly populated country, like the U.K., could readily maintain itself on potatoes and sheep, it is equally true that much

Table 6.2 Potatoes made of oil

Budget for one acre of maincrop potatoes				
ENERGY INPUTS				
	(From Fossil Fuels)		Kcal/unit	MCal
Fertiliser	N	151 lb	7,350	1,110
	P	151 lb	900	136
	K	210 lb	1,000	210
				1,456
Tractors	25 hrs. (45-50 hp)		42,000/hr.	1,050
Other machinery				1,290
Sprays				331
Sundries				271
Irrigation				500
	TOTAL TO FARM GATE			4,898
	ADD TRANSPORT TO CITY			500
	GIVING TOTAL INPUT OF			5397
	Assume Yield of		13.25 tons	
	Deduct, for seed		1.03 tons	
	Net Yield		12.22 tons	

Assume that daily calorific requirement of 2400 Kcal (for a sedentary male) is met by 7.5 lbs of potatoes, then 12 tons would feed 10 people for 1 year ($7.5 \times 356 \times 10 = 27.375$ lb = 12 tons)

Utilisable calorific content $\therefore$ equals $2400 \times 365 \times 1 = 8760$ Mcal

NET GAIN 8760 - 5379 = 3363 (38%)

This balance sheet shows that, although an acre of potatoes can feed 10 people for one year, most of the energy (about 60%) comes from the sun via fossil fuels rather than by direct utilisation of sunlight.

[One kilocalorie (Kcal) is 1,000 calories or 4,180 joules; one megacalorie (Mcal) is 10^6 calories or 4,180 Kjoules. Note that the Imperial "ton" is so similar to the metric "tonne" that either could be used here without significantly altering the conclusions.]

intensive agricultural practice is little more, energetically speaking, than a somewhat inefficient means of converting fossil fuels into foodstuffs. Table 6.2, which is based on work by Leach but makes slightly different assumptions, shows a net gain of about 40% when the use of fossil fuels is offset against light energy "conserved" as chemical energy in the form of edible carbohydrate. In the western world, the energy subsequently spent on packaging, marketing and cooking, converts this modest gain into a substantial loss. It has been calculated that, on average, approximately 5 calories are expended for each calorie of food made ready for consumption. If fuels are not used to produce fertilisers and herbicides and to drive farm machinery, the crop is likely to be much smaller. But it would be a mistake to suppose that an advantage would necessarily derive from a reversion to past, more primitive, practice. It is true that a horse doesn't run on fossil fuels but neither can it be switched off, like a tractor, when not in use. For this reason its total energy requirement during a year (as much as that of ten people), is likely to exceed that of machines used to perform comparable tasks.

Considerations of this sort highlight the dilemma which man must face as his stocks of fossil fuels decline. It is no longer enough to seek the best possible yield. It is also necessary to attempt to achieve the maximum net energy gain. In many circumstances this may be synonymous with 'best financial return' but normal market forces may be distorted by subsidies and intervention of a political or social nature. Such problems have created the simultaneous existence of European "food mountains" and starving Ethiopians.

6.7 Photosynthetic Efficiency

"In a perfect nature photosynthesis is perfect too" ....Otto Warburg

At this stage we might wish to determine how efficiently plants convert light energy into chemical energy and even ponder the possibility of improvement. If we know how much light energy is put in and how much chemical energy we get out we have a measure of how well a plant can convert one form of energy into another. Of course there are plants and plants and, if we are not careful we could get into some rather uncertain ground. It could be a bit like asking what time a man might take to run a mile. In my youth there were two barrel chested Swedes called Arne Anderson and Gundar Hague who, if my memory is not at fault, took the time for the mile down to about 4 minutes 6 seconds. Four minutes seemed unattainable until Roger Bannister achieved the impossible and, thereafter, sub four minute miles became unremarkable, even commonplace. That is not to say there is no limit to how fast a man might run over a given distance. Some gifted human physiologist might come up, if pressed, with a realistic estimate. We could ourselves, remembering that 10 seconds is a very good time for 100 yds (and that there are 1,760 yds in a mile), declare it unlikely that anyone will ever run the longer distance without flagging and thereby suggest that anything close to 3 minutes is unlikely in the extreme. If we apply the same sort of crude arithmetic to plants we would probably settle for something like 10% (conversion of light energy to chemical energy) as the equivalent of the three minute mile and something closer to 5% for a reasonably safe bet (perhaps equivalent to a slightly lower time for the mile than our best runners can manage now).

Let us have a look at this in more detail and examine some of its many implications. It is after all an important subject - one on which lives may ultimately depend.

In Chapter 2 we invoked a hypothetical carbohydrate with a formula CH_2O. There are a number of real carbohydrates with the formula $C_6H_{12}O_6$ and if these are burned in a calorimeter they release about 672 Kcal (2,822 Kjoules). Clearly we cannot get out more heat than it cost (in terms of light energy) to join together six molecules of carbon dioxide (to form $C_6H_{12}O_6$) in the first place. On this basis, to form CH_2O from one CO_2 must have cost at least 672/6 = 112 Kcal (470 Kjoules). In Section 2.3, moreover, we learned that red light (at 680 nm) had an energy content of about 42 Kcal (175 Kjoules) per quantum mole (Avogadro's number of photons). So, if we only had this crude energy balance sheet to contend with, we could make one mole of CH_2O at the expense of only three quantum moles (3 x 42 Kcal) of light energy. One thing is certain; there is no way that we could get under two quanta of red light per molecule without breaking the laws of thermodynamics and they are notoriously difficult to break. Similarly nothing that we know, at least when we get into the chemical end of things, seems to work without some frictional losses, so four photons (per molecule of CH_2O formed, carbon dioxide fixed, or oxygen evolved) might seem more realistic.

Otto Warburg, a Nobel laureate and German biochemist of great distinction (the man who discovered NADP, pioneered the use of spectrophotometry in biochemistry and gave us "Warburg manometry") thought that photosynthesis was perfect and claimed that his data put the quantum requirement (the number of photons needed per CO_2 fixed, or O_2 evolved) at between 3 and 4. Robert Emerson, a young American who did his Ph.D. in Warburg's laboratory in Berlin took a different view and put the minimal requirement at 8. A great controversy developed. The debate was passionate, even bitter. *Chlorella* was the preferred species. It was necessary, so it seemed, to grow it in special ways. A north-facing window and Berlin spring or well water seemed to confirm some remarkable advantages. Measurement of light was as difficult as ever and the present generation of photon counters were yet to be invented. Warburg manometry was used to monitor gas exchange and refined to the nth degree, with flasks rectangular in cross-section and horizontal microscopes to detect minute changes in the level of manometer fluid. As always, the measurement of photosynthetic oxygen evolution was bedevilled by respiratory oxygen uptake. Whether or not respiration continues unchanged in the light was a question as impossible to answer with certainty at that time as it is now. Warburg travelled to the United States to repeat his experiments in Emerson's laboratory. It took him months to organize apparatus and algae, eleven or twelve hours to make each determination of quantum requirement. He was unable to repeat his Berlin results, at least not in Emerson's presence. Warburg blamed inadequate algae. The following letters, from Otto Warburg and Robert Emerson to F.C. Steward (reproduced here, verbatim, by kind permission of Geoffrey Hind), gives some inkling of the underlying currents at the time. Emerson, soon to die tragically in an air disaster, lamented the years wasted in futile argument, ironically attributing his differences with Warburg to a failure by the great man to recognise an inadequacy (a lag) in manometry. It can be scarcely doubted, given Emerson's many other important research contributions, that his preoccupation with this controversy cost photosynthesis dear.

Urbana/Ill. South Coler Avenue 801 *January 2nd, 1949*
Dear Professor Steward !

Some months ago Dr. Robert Emerson asked me if I possibly could lecture in your institute on the yield in photosynthesis. Thus I suppose that you are interested in this problem.

As you know I found a requirement of 4 quanta in 1923 and the same requirement in 1945 when I repeated my experiments considering all the objections of Dr. Emerson. The university of Illinois invited me to decide here in Urbana which value is the right one, 4 or 12 quanta.

The first thing I did here was a to simplify the procedure. I devised a chemical radiometer for the visible part of the spectrum so that it is now possible to measure quanta intensities manometrically (photo-oxidation of thiourea sensitized by chlorophyll or porphyrin, dissolved in pyridin. This quantum yield is under suitable conditions one).

The determination of the yield in photosynthesis is now extremely simple. First you give in two vessels the same amounts of cells, but different volumes of acid culture medium and measure the pressure changes in both vessels, in dark and light. Thus you get for every period the oxygen development and CO_2 absorption or the reverse.

Then you put in the same light beam in the same position the radiometer vessel (of the same dimensions), and measure the oxygen consumption in light for the same time.

A few days ago I got cells to measure the yield in this way and I found, in the presence of two impartial observers, 4 quanta per molecule oxygen. Proceeding in exact [sic] *the same way the next day two assistants of Dr. Emerson found for a different culture about 20 quanta per molecule of oxygen. So I came to the conclusion that the Chlorella cultures of Dr. Emerson are not suitable or at least unequal and I declared here that I could continue the experiments only if I had the control over the cultures of the chlorella.*

This is a long story. If here in the US you wish to decide the problem finally it seems to me necessary that I show in an impartial institute how to get four quanta, being able of course to cultivate the chlorella myself. If on the other hand I go now back to my institute in Dahlem and repeat it there I have no opportunity to show the yield to other people and the situation remains as it is.

Is there a possibility to do these experiments in your institute or in the institute of the Eastman Company? I have here a lecture at January 19 and would then be free for about two months. The trouble is that I must be paid because it is not possible to change german money in dollars.

It would not be advisable to ask Dr. Emerson about this plan, But Dr. Tippo, the head of the department of Botany in Urbana is acquainted with the problem and the situation.

I have manometers and vessels and a prisma to produce red light. The required intensities are 0,3 to 3,0 micromoles of quanta per 10 minutes, (630 to 660 nm) or better if possible monochromatic red light.

Do not answer this letter if you see no possibilities.

I am yours sincerely

Otto Warburg.

Professor F.C. Steward
Botany Department
University of Rochester
Rochester 3, New York.

January 28th, 1950

Dear Mr Steward,

I appreciated receiving your account of Burk's performance in New York. One of our graduate students was there, too, and gave us his impressions, but the field is so new to him that he couldn't give us as full an account as you do. We are amused that Burk had no time for discussion of results with scientific colleagues, but had plenty of time to spill a big story for newspapers reporters.

As I may have suggested in my last letter to you, we are now pretty sure that the major cause of error in the Burk-Warburg work is their assumption (stated in Science Sept. 2nd and reiterated in the Meyerhof Festschrift of Biochemica et Biophysica Acta) that there was no "Physical lag" in their manometric system. On the basis of this unsupported assertion (which we find to be strictly contrary to fact), they based most of their photosynthesis measurements on 10-minute exposures to light, alternated with 10-minute intervals of either darkness or unmeasured light. This does not lead to correct values of photosynthesis in either phosphate or carbonate buffers. However, there are additional factors in their work which are still obscure, and we shall not be satisfied until we can give a quantitative explanation of all the major inconsistencies. We do make progress toward this objective, but it's disappointingly slow.
Yes, Burk gives one this impression that he is making an intentional effort to confuse issues, rather than to clarify them. I'm inclined to agree that an ethical problem is involved, as well as a question of scientific fact. I'll appreciate advice on how to deal with the ethical issue, but I'm inclined to let it go until we have settled the facts.

With best wishes,

Sincerely

Robert Emerson.

So why did it take so long to distinguish between 4 and 8?

[The Z-scheme (Section 4.3), of course, demands 8 photons but, at that time, it was still to be proposed and authenticated].

The answer is really quite simple. Given the techniques available it was extraordinarily difficult to measure either the light absorbed or the oxygen evolved. It should also be borne in mind that the highest yields only occur in low light, i.e., when the changes to be measured were at their smallest. Moreover, photosynthetic organisms respire like any others. How to make an appropriate allowance for dark respiration, relatively large in relation to photosynthesis in these circumstances, could pose problems. In the long-run everything gradually happened together. Techniques improved, the Z-scheme was proposed and relentlessly authenticated, four photons (per carbon dioxide fixed) became untenable. These days the actual quantum requirement for leaves, following work by Demmig and Bjorkman, is put at about 9 and many species, provided that they are in good conditions and not stressed give values close to this figure. Why 9 and not 8? A quantum requirement of 8 (i.e. a quantum yield of $1 \div 8 = 0.125$ molecules of carbon dioxide or oxygen per photon) is probably the real underlying value because that is what the Z-scheme demands and the Z-scheme become increasingly unassailable. The rest is down to measurement again. These are still very difficult to get exactly right and are biased to more (rather than to less) because it is not yet possible, in a leaf, to measure photosynthesis *per se* without inputs from other associated processes (e.g. sucrose synthesis). Also many sorts of stress will increase the apparent quantum requirement a little. Real stress such as photoinhibition (Sections 5.9 and 6.9) can increase it a lot. In short, hydrogen transport from water to NADP (Chapter 4) cannot be easily measured in leaves without inadvertently measuring a few energy requiring reactions which push up the apparent quantum requirement. The consistency of the requirement for about 9 photons is also very striking. It is as if a wide range of species have gone as far as thermodynamics will permit. If this is so, there is little or no possibility of increasing quantum efficiency (the reciprocal of quantum requirement). In this regard, at least, green plants may be as efficient as it is possible to be. (Section 6.4)

Where does this leave us in the present context? If we wish to measure photosynthesis, as a function of light intensity, in a sensible fashion we should do it at near-optimal temperatures in saturating CO_2 i.e. under conditions in which light, and only light, is thought to be limiting. In such circumstances the relationship is that shown in Fig. 6.4. At low photon flux densities the initial slope is linear, the graph only departing from linearity at PFD's above about 75 -100 μmole quanta.$m^{-2}.s^{-1}$. Given Bjorkman and Demmig's consistent values of near 9 for quantum yield this means that at (for example) a photon flux density of 80 μmole quanta.$m^{-2}.s^{-1}$ most, if not all green plants, will give rates of about 0.76 μmoles $m^{-2}.s^{-1}$ (O_2 evolved or CO_2 fixed). In other words, in the absence of stress or other complicating factors, *all plants photosynthesise at the same rate in low light.* (Section 6.8).

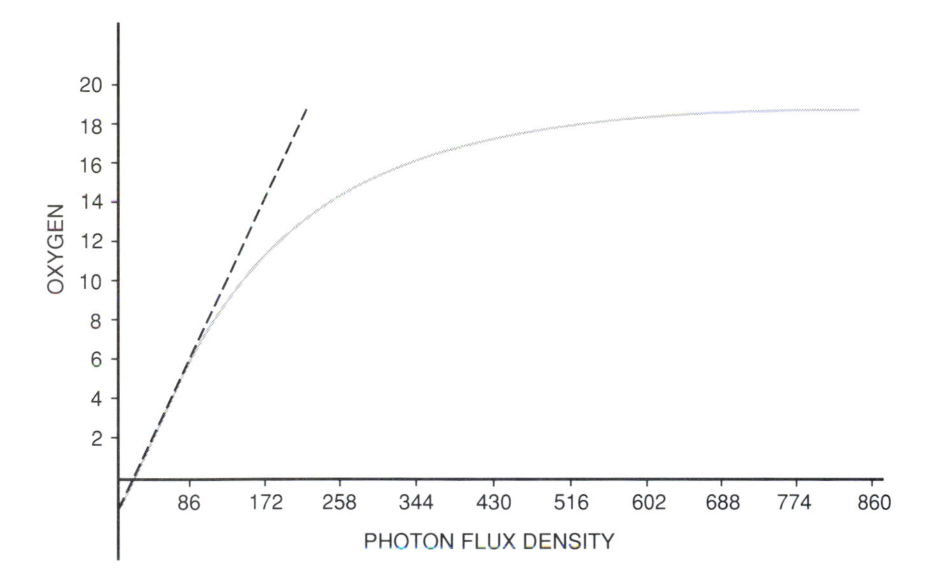

Fig. 6.4 Rate of photosynthesis as a function of light intensity (PFD). At first the rate increases, with increasing light (PAR), in a linear fashion. At relatively low PFDs (about one twentieth of full sunlight - see Table 2.2) the relationship departs from linearity because of metabolic constraints. The maximum (quantum) efficiency is calculated from the initial slope. In this light-limited region the theoretical maximum is 0.125 molecules of O_2 evolved (or CO_2 fixed) per photon (a quantum requirement of 8). At a PFD of 400 μmole quanta.$m^{-2}.s^{-1}$ (PAR), equivalent to one fifth (800 μmole quanta.$m^{-2}.s^{-1}$ full sunlight the rate of photosynthesis is already close to its maximum. This particular figure was derived from computerised measurements undertaken with a Hansatech "leaf-disc electrode"

[In the rate v PFD plot the initial slope is a measure of quantum requirement. Thus for a PFD of 80 and a rate of 0.76 the quantum yield would be 80/0.76 = 9.3]

Let us look again at the overall efficiency of energy transduction with this in mind. For a value of 112 Kcal (for CH_2O) we would have, for electron transport from water, through the Z-scheme, to CO_2:-

$$\frac{112}{8 \times 42} \times 100 = 33\%$$

i.e. 8 photons, each worth 42 Kcal (175 Kjoules) would be consumed and a product worth 112 Kcal (468 Kjoules), when burnt in the calorimeter, would be produced. Here, you might suppose we are being generous in the extreme so let us see if, and how, we might come nearer to reality. First of all, even very modest photoinhibitory stress (Fig. 6.5) may decrease the initial slope of the rate v PFD relationship to an extent equivalent to a quantum requirement of 12. These values relate to red light and saturating CO_2 and if we put a value of 112 Kcals/mole on CH_2O and 42 Kcals/mole photons on red light at 680 nm, the efficiency of conversion (light-energy into chemical energy) becomes:-

$$\frac{112}{12 \times 42} \times 100\% = 22\%$$

But only half of sun-light is photosynthetically active radiation (PAR) and 50 Kcal per quantum mole a more realistic value than 42, given the spectral composition (Fig. 6.2) of sun-light. (On sunny days we have blue skies rather than red, and blue light carries more energy per photon than red). On this basis the above arithmetic would become:-

$$\frac{112}{12 \times 50} \times \frac{1}{2} \times 100\% = 9\%$$

Then we have photorespiration (Section 5.7) and dark respiration. Quantum requirements of 9 come from measurements in saturating CO_2, in which photorespiration is believed to be totally suppressed. In air, C3 plants photorespire. C4 plants do not photorespire but then they have greater quantum requirements

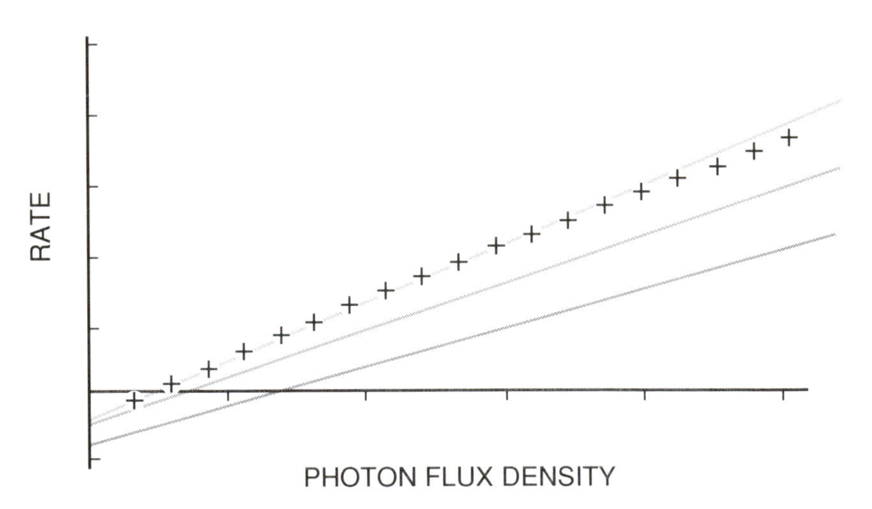

Fig. 6.5 A degree of photoinhibition. Rate v PFD showing slight photoinhibition. Measurements were made in the light-limited, (linear) range, before (A) and after (B) 5 minutes exposure to high light in CO_2 -free air. This treatment decreased the initial slope (i.e. it decreased the quantum yield and therefore increased the apparent quantum requirement) indicating a degree of photoinhibitory damage to the photochemical apparatus. A further 15 minutes exposure (C), to high-light in CO_2 free air, decreased the initial slope still further and also increased the rate of dark respiration (the negative intercept on the vertical axis). This type of photoinhibition is fully reversible and may, in itself, afford a degree of safe dissipation (Section 6.10).

anyway because of the energy costs of their CO_2 pumps (Table 5.1). To what extent photorespiration decreases net carbon assimilation is arguable but few would question an additional correction factor of 0.6 to take care of photorespiratory and respiratory losses. In addition, leaves reflect and transmit light (mostly green), which is therefore not used in photosynthesis, and absorb light in ways which also make no contribution to carbon assimilation. A further correction of at least 0.85 must be invoked to cover these losses. These two additional corrections take us down to a final round figure of 4.5%.

This estimate (4.5%) *is absolute tops.* The correction factors put into the arithmetic are, to be frank, educated guesses at best. Moreover, in the field, maximal quantum yields, are rarely approached. At high light intensities (high PFDs) the quantum efficiency (Section 6.6) falls, particularly in C3 crops (such as wheat, barley, rice, sugar-beet and potatoes) rather than in C4 crops (such as sugar-cane, maize, sorghum etc). Nevertheless there are rare, highly favourable, circumstances in which stands of natural vegetation (Section 6.8) come close to our calculated maximum, allowing us to conclude that 4.5% is not a complete nonsense. For example, Cooper notes that a closed canopy of *Pennisetum typhoides* grown in Katherine, W. Australia, with a daily solar energy input of 5,100 Kcal/sq.metre (PAR), showed a daily dry weight increase of 54g/sq.metre (some 4.5% efficiency). However, it should not be forgotten that *Pennisetum typhoides* is a C4 species and, as such, much less readily light-saturated than most C3 species.

6.8 Efficiency of the standing crop

Although the various losses listed above diminish the return from a theoretical 33% to a practical 4.5% the farmer would be delighted if he could approach the maximum of 4.5% even on rare occasions. This is because the crop in the field is much less efficient than the plant in the laboratory and the estimates based on high quantum yields only apply to light-limiting conditions. The average C3 crop species will photosynthesise at near maximum rates at about one fifth of full sunlight (Fig. 6.4). Thus, at PFDs above about 800 μmole quanta.$m^{-2}.s^{-1}$, most of the incident light has to be dissipated, as heat, (Section 6.9) rather than utilised in carbon assimilation. In most types of agricultural practice, the crop does not cover the available ground throughout the growing season and obviously irradiation of bare earth adds little to the balance sheet, apart from soil warming. Conversely, as the season advances efficiency will be diminished as the upper leaves begin to form an unbroken canopy which shades those leaves below it. In areas removed from the equator or at high altitudes, there will be periods when the temperature is too low to support appreciable photosynthesis, and at temperatures above 30°C, photorespiratory losses will become increasingly important. Water shortage will limit growth as will deficiencies in essential and trace elements. Indeed, availability of water is often the most important limitation in agriculture. Carbon dioxide, at about 0.03%, i.e. 300 parts per million (ppm), will frequently be a limiting factor (particularly in C3 species which may photosynthesise two or three times more rapidly in augmented CO_2) and this limitation may be aggravated locally, by active photosynthesis in still atmospheres. Photorespiration will take its toll. To all of these must be added the usual agricultural hazards such as disease, damage and loss. These will frequently combine to decrease yields to much less than 1%.

The maximal dry weight yield can be calculated for any country at any level of efficiency. For example, the UK receives sunlight at 100 joules/sq. metre/sec. on a continuous basis. At 4.2 joules/cal this equals about 24 cals/sq. metre/sec.

$$= \frac{24 \times 3600 \text{ Kcals / sq. metre / hr}}{1000}$$

$$= 86 \text{ Kcal / sq. metre / hr.}$$

Five per cent of this value is 4.3 Kcal and, at 4.25 Kcal/g dry weight, it may be calculated that:-

1.0 g/sq. metre/hr. = 1.0 x 365 x 24 g/sq. metre/yr.
= 8.8 kg/sq. metre/yr.
= 88 tonnes/hectare/yr.
= 36 tons/acre/yr.

This allows some very rough comparisons to be made. It is difficult to obtain accurate figures of total yield (as opposed to crop yield) but the best ever reported potato crop in the UK was about 36 tons (*fresh weight*, tuber) per acre (88 tonnes/ hectare). Very similar yields have been reported from the United States.

Dry weight measurement of a total potato crop (including leaves and roots) has been only rarely attempted. However, values obtained by Lorenz put the total *dry* weight at a little less than one quarter of the *fresh* weight of the tubers and this would suggest a maximum dry weight yield of approximately 9 tons/acre (22 tonnes/ hectare). This is equivalent to 25% of the theoretical maximum, or 1.25% conversion of light energy to chemical energy. The average potato crop is perhaps one third of the maximum. These values would have to be recalculated for each crop to be considered but the values cited in table 6.3 could be taken as a first approximation.

Table 6.3

Percentage Conversion of Light Energy to Chemical Energy
(i.e Dry weight yield of potatoes, given different efficiencies of photosynthesis at the crop level)

	YIELD		
	% conversion of light to chemical energy	Total dry weight in tons/acre or (tonnes/ hectare	Fresh weight of potatoes in tons/acre or (tonnes/ hectare)
Possible Max.	5.0	36 (88)	144 (352)
Feasible	1.25	9 (22)	36 (88)
Average	0.4	3 (7)	12 (29)

It will be seen that in this Table there is a four fold difference between "possible" and "feasible" whereas in Fig. 6.3 the difference is 5.5. It should be noted that this discrepancy relates only to the nature of the crop and to the assumptions made about its calorific content and fresh weight/dry weight ratio.

6.9 Roofs and Ceilings

As light increases in brightness, i.e. as the rate at which photons strike a leaf increases, the rate of photosynthesis increases. Accordingly, if we plot rates of photosynthesis against light intensity (or more properly, photon flux density or PFD) we get straight lines at low intensities (e.g. Fig. 6.5) which reflects this direct relationship between (quantum yield) the arrival of photons at the leaf surface and the photosynthetic output. At higher PFDs however, this linear relationship is lost (Fig. 6.7). This initial slope (the roof) gives way to a curve which gradually approaches a new, biological, limitation (the ceiling). This is analogous to a man frying eggs. At first the rate at which he can fry eggs will be limited, in part, by the rate at which he is handed eggs to fry but, very soon, the biological limitations of *his* handling capacity will dictate how many he can fry. He will soon have reached his own biological ceiling.

As well as horizontal ceilings, most houses have pitched roofs. The ceiling is biological and variable and has to do with the ability of the leaf to utilise incoming photons. The maximum slope or pitch of the roof (quantum yield) is a thermodynamic constraint, governed by the light energy required to evolve one molecule of oxygen or reduce one molecule of carbon dioxide. Each photon entering each photosystem (Fig. 6.6) raises one electron to a higher energy level. Since there are two photosystems, and since four electrons must be transferred in order to satisfy the equation:-

$$2H_2O + CO_2 \rightarrow CH_2O + O_2 \qquad \text{Eqn. 6.1}$$

this means that eight photons are required per molecule of oxygen evolved or carbon dioxide fixed.

We can measure the efficiency of a motor vehicle in terms of fuel consumed per mile travelled or from the initial slope of a graph in which the rate of fuel

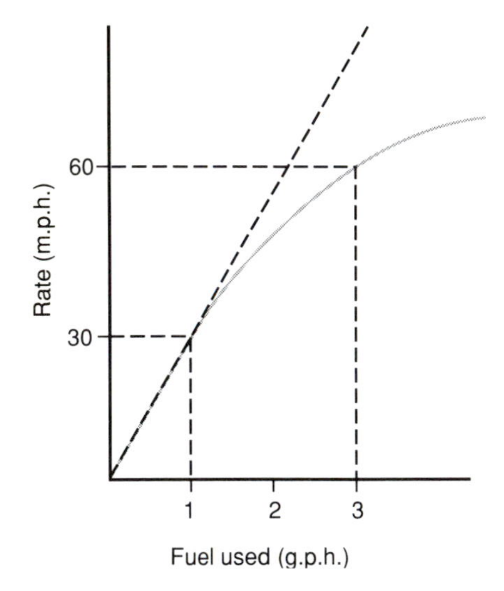

consumption is plotted against the speed of the vehicle (left). The initial, linear, part of the rate of photosynthetic oxygen evolution v PFD plot allows us to derive photosynthetic efficiency in a similar way but, in terms of µmoles of oxygen evolved per µmole of photons used rather than miles per gallon or kilometres per litre. Surprisingly, many diverse species display very similar photon requirements if they are allowed to photosynthesise under good conditions. The implication, yet to be proved, is that many, or possibly all, species have acquired the ability to utilise light as well as is thermodynamically possible. If this is correct, molecular biologists are unlikely to be able to improve the maximal photosynthetic efficiency of plant species (the roof) not least because they will have no genetic variability to draw upon. The roof (the initial slope in Figs. 6.4 - 6.8) is already pitched as steep as the Z-scheme permits and, if not stressed, most plants may operate close to this limit. Raising the ceiling is quite another matter because here the constraint is biological rather than thermodynamic and there is wide variation amongst species and varieties. Even the same species grown in a different environment exhibit something like a tenfold difference in maximal photosynthetic rate. Even so, this ceiling may not be of over-riding importance. If a driver is required to observe a rigorously applied speed limit of 70 m.p.h. it will not be to his advantage to have a vehicle capable of 100 m.p.h. unless it also possesses other attributes such as better acceleration and stability. Moreover, a rally driver will consistently drive the same vehicle faster than a university professor because he or she will have driving skills (i.e. skills in regulation) which the professor is unlikely to possess. Similarly, plants are always beset by environmental constraints and may differ widely in *metabolic* regulation. For example, take two varieties of plant with identical roofs and ceilings and suppose

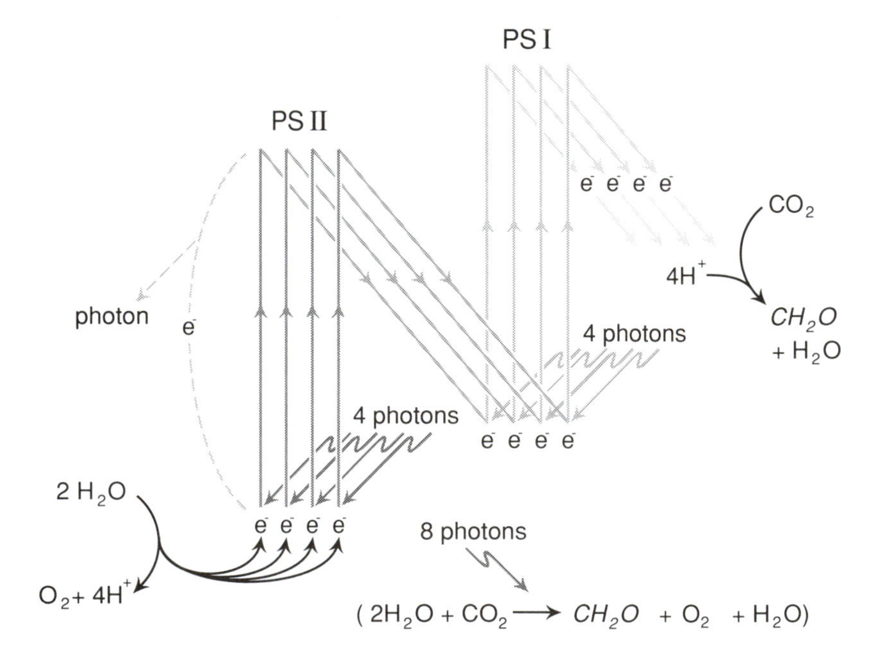

Fig. 6.6. The Z-Scheme. Yet another version of Hill & Bendall's classic representation of electron transport (c.f. Fig. 4.2). This time stripped of detail to emphasise the manner in which 8 photons combine to transport 4 electrons from H_2O to CH_2O. Protons (H^+) separated from hydrogens ($H = H^+ + e^-$) at the beginning of this process, as water is split (releasing O_2) in photosystem II, are restored to CH_2O (via NADP) after electrons have passed through PSI. In leaves, a small fraction of excitation energy is dissipated as fluorescence emission from PSII (left). As the light intensity increases, an increasing amount of excitation energy is dissipated as heat (Section 6.2) and the quantum yield (0.125 molecules of O_2 per photon at maximum) falls accordingly.

that one can convert 10% more of its photosynthetic product into new leaves than the other. The new leaves would also photosynthesise and as arithmetic will show, within 8 weeks, the plant with the leaning to investment would have grown to twice the size of the other. In this instance the ability to convert more photosynthetic product into new photosynthetic machinery could be much more important than modest differences in photosynthetic rate. Thus increased productivity might be found more readily, in improved regulation of metabolism and development, than in faster photosynthesis. This is not to say that there is no room for improvement. Many years ago, a famous Cambridge biologist called Blackman took the view that the roof gave way to the ceiling abruptly rather than in a continuous curve (Fig. 6.7) as limitation by light gave way to limitation by the ability of the leaf to utilise light effectively. Contemporary measurement (particularly when it is based on illumination from above) does not often show such an abrupt change, not least because there are usually several layers of photosynthetic tissue within a leaf, each layer with somewhat different characteristics and each lower layer inevitably receiving less light than the leaf-surface. Nevertheless, the extent to which the rate

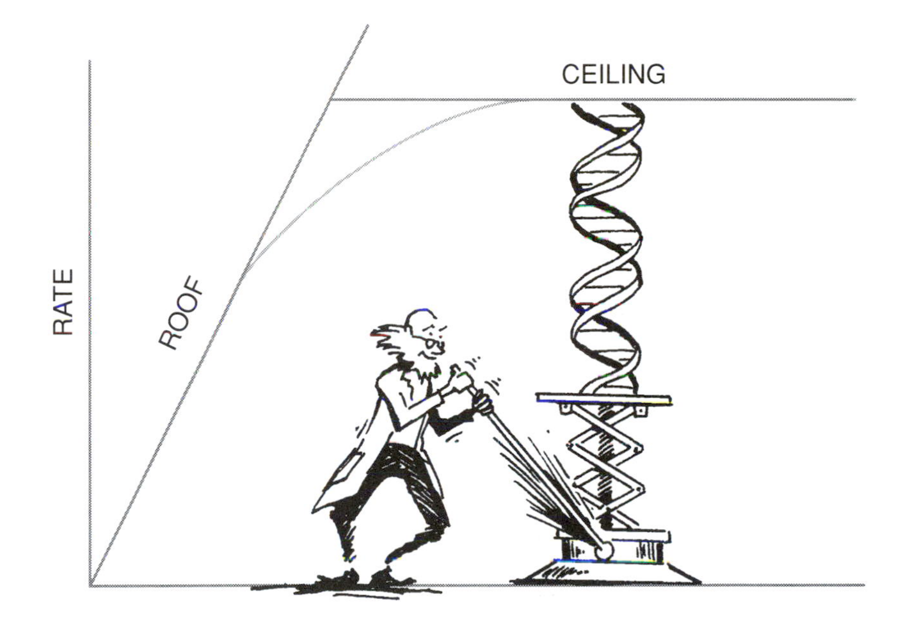

Fig. 6.7 Roofs and Ceilings. The initial slope of the rate v. PFD relationship (i.e. the quantum yield) is a thermodynamic constraint unlikely to be steepened. The ceiling is determined by enzymic complement and regulation and might well be raised by genetic manipulation.

v PFD relationship (the continuous curve in Fig. 6.7) can *approach* the thermodynamic and biological constraint of the perfect Blackman relationship (roof plus ceiling) may well prove to be an important indicator in the assessment of plant performance. Molecular biologists have acquired a magic wand in the sense that, at least in principle, they are now able to create entirely new plants; to move parts of metabolic machinery from one plant to another or to introduce new completely novel features. Their problem is that they need guidance about what to do. It may prove less advantageous to raise the maximum rate of photosynthesis than to seek to ensure that they function as close to existing roofs and ceilings for as much of the time as possible. Resistance to environmental stress and speed of recovery from stresses such as photoinhibition could be crucial in this regard. These are features of a plant which an experienced plant breeder now recognises intuitively when he or she talks about "vigour". It is these that give some species, such as barley, the versatility to grow in widely different environments. As the breeders traditional skills are replaced by the new wonders of genetic engineering, "vigour" and "versatility" will need to be quantified and assessed. Such assessment will undoubtedly involve the application of new techniques to the evaluation of the old Blackman concept of roofs and ceilings and the extent to which crop species will be able to approach these limits.

SAFELY DISSIPATED

6.10 The need for dissipation

Plants are rarely able to move out of the sun and yet are quite unable to use all of the light-energy that may arrive at their leaves. This is particularly true of circumstances in which carbon assimilation is constrained by other factors, such as low temperatures or drought which may induce stromal closure and consequential

"Where we might soon have fainted in that Enchanted Ground but now and then a cluster of pleasant grapes we found"

CO_2 deficiency. There is, therefore, a need for safe dissipation of excess excitation energy. To an extent, modest reversible photoinhibition i.e. damage caused by light to the photochemical apparatus (Fig. 6.5), may constitute a safety mechanism but clearly irreversible photoinhibition would serve no useful purpose. Recent research indicates that there are additional safety valves that dissipate excess excitation energy as heat (rather than fluorescence or carbon assimilation). In some circumstances reduction of O_2 (the Mehler reaction - See Fig. 4.3) may be involved.

If not stressed, most species photosynthesize in low light, at close to the same rate but the linear relationship between PFD and rate is lost; progressively more quickly by leaves grown in shade than by leaves grown in full sunlight (Fig. 6.8).

At first sight, this may seem to be more of a problem for shade leaves than for sun leaves in the sense that they can usually (but not invariably) utilize only a much smaller fraction of full sunlight than their sun counterparts. In Fig. 6.8 the percentages of incident red light used by shade and sun *Helliconia* (in the range of 0-800 μmole quanta.$m^{-2}.s^{-1}$, and assuming a quantum efficiency of 9) were about 20% and 40% respectively. In full sunlight (2,000 μmole quanta.$m^{-2}.s^{-1}$ in the 400-700 nm PAR range capable of being absorbed by chlorophyll) these values for

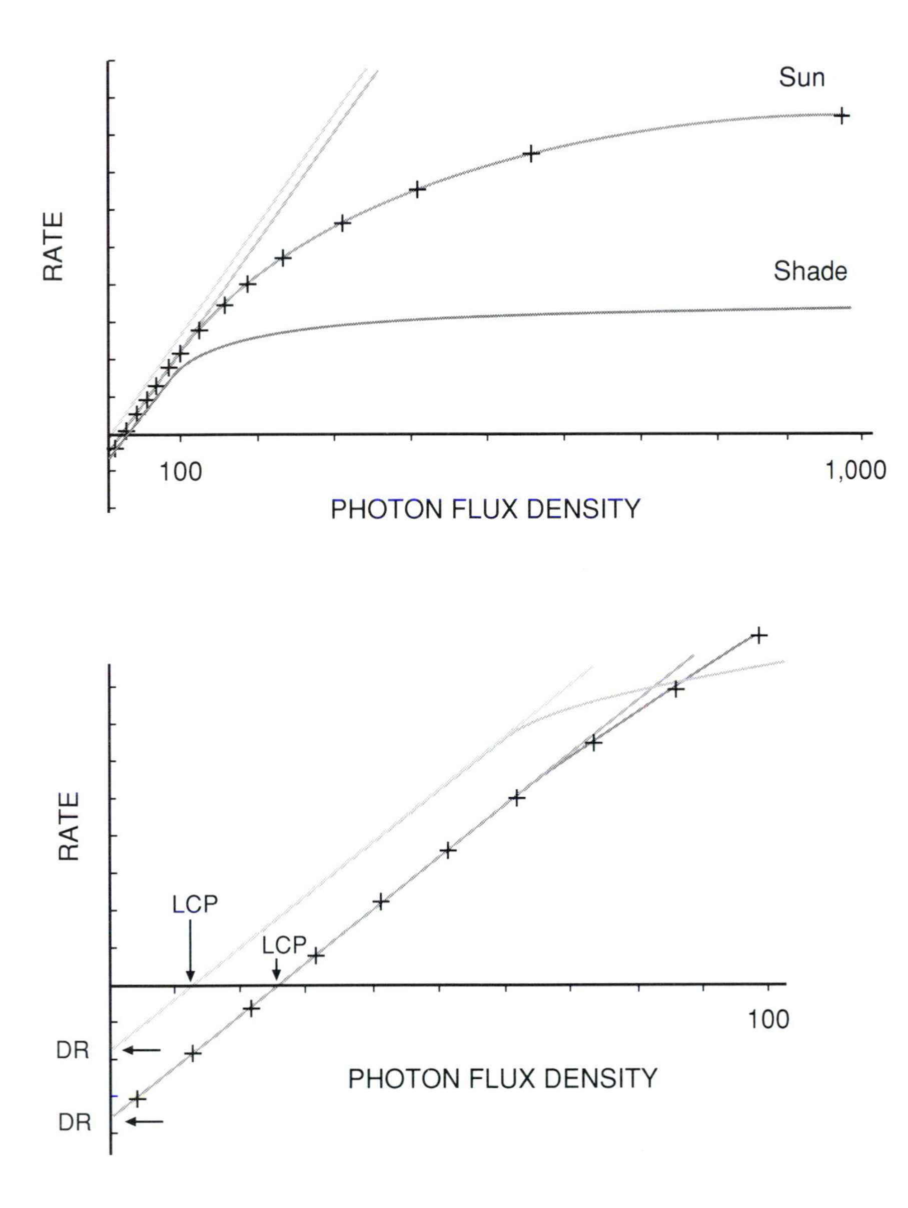

Fig. 6.8 Rate v PFD in sun and shade leaves. Above: Comparison of the rates of photosynthesis displayed by sun and shade leaves of *Helliconia* over a photon flux density range of 0 to 1000 μmoles quanta/per sq. metre/sec. Below: Detail of above emphasising the fact that the initial slope (and therefore the quantum yield and requirement) are the same in a sun and shade leaf. The fact that the latter had a lower light compensation point (LCP), i.e. the light intensity at which photosynthesis, oxygen evolution and dark respiratory oxygen are in balance, results from the lower rate of dark respiration (DR) shown by the shade leaf. [Data from 10th UNEP/UWI Training Course on *"Photosynthesis & Productivity in a Changing Environment"* (Barbados, 1991).]

utilization could be diminished by almost two thirds. This constitutes a problem in safe dissipation. Some plants, such as species of *Oxalis,* can alter the inclination of their leaves in order to minimise light interception. Most species have to rely on biochemistry rather than mechanics. They are not always successful. It is certainly the case that shade leaves are more likely to be damaged by unwonted exposure to high light than sun leaves but, by the same token, in the normal course of events, they are less likely to experience bright light. Damage by excess light, once described as "solarization" and now termed "photoinhibition" seems to be more common than was once thought and there is evidence, for example, that crop species, like wheat and barley, may suffer from the combination of cold and high light which they almost invariably experience during some periods of their life cycle in temperate climates. Similarly, horticulturists, conscious of the costs of heating glasshouses, have more to fear from cold bright days than cold overcast days. Photoinhibition (Section 5.9) is too large a subject to be considered here in detail but is may be noted that much evidence suggest that the photochemical apparatus (in particular the D1 protein in PSII) sustains more or less continuous damage and undergoes more or less continuous repair, the more effective in sun leaves. However, plants have also learned that prevention is better than cure. Botanists have conjectured for many years about the possible protective role of carotenoids. The fact that pigmentation of leaves is often an indicator of stress (e.g. drought or cold) is something familiar to gardeners. More recently this has been put on a more substantive basis and there seems little doubt that the "zeaxanthin or xanthophyll cycle" constitutes one mechanism based on carotenoids which contributes to the dissipation of excess excitation energy in PSII. Both the formation of zeaxanthin (which can accumulate in high light) and its proposed mode of action (zeaxanthin appears to amplify dissipation rather than to initiate it) are favoured by high ΔpH. The full picture, however, is extremely complex and much is still to be clarified. Hydrogen ion concentration, or more specifically, the transthylakoid hydrogen ion gradient ΔpH), undoubtedly plays a major role.

6.11 Acid bath technology

The transthylakoid proton gradient (ΔpH) results from electron transport (Chapter 4) involving all or part of the Z-scheme. Some of the electron carriers involved can only accept electrons, others require hydrogens (i.e. electrons plus protons). Thus electron transport involves taking up and discarding protons and the distribution and behaviour of carriers within the thylakoid membranes of the chloroplast is such that protons are taken up from the outside of the membrane and put down on the inside. Accordingly the stroma of an illuminated chloroplast becomes more alkaline, the thylakoid compartment more acid. Protons discharging through the ATPsynthase in the thylakoid membrane bring about ATP formation by condensation of ADP and inorganic phosphate, according to Mitchell's chemiosmotic concept. Isolated thylakoids, "coupled" to ATP formation, transport electrons more rapidly than they do in the absence of ADP and phosphate. Similarly "uncouplers" (agents which, for example, effectively puncture the membrane) also speed electron transport because they (like ADP + Pi) facilitate the discharge of the proton gradient which otherwise creates a "back-pressure". (Simplistically, it is energetically more difficult to discharge a proton into a space already full of protons than into a space containing few protons and, since electron transport involves vectorial movement of protons from the outside of the thylakoid to the inside (Fig. 4.7), this also becomes more difficult). Thus the role of ΔpH in regulation of electron transport at the site of the cyt b/*f* complex has long been recognised. More recently, however, it has become increasingly clear that ΔpH plays more than one role. Thus it not only affects electron transport between the two photosystems but within PSII and PSI and beyond PSI. Moreover, all of these sites of pH-related regulation tend to interact with one another to a small or larger extent. For example the relatively alkaline environment of the stroma (pH 8.5 or thereabouts) favours carbon assimilation and therefore ATP consumption. ATP consumption regenerates ADP which, in the presence of Pi, discharges the proton gradient through the ATPsynthase. For this reason, removal of CO_2 from the atmosphere surrounding a leaf causes an almost immediate increase in ΔpH In turn, increased ΔpH regulates electron transport. As just noted, one site of regulation lies between PSII and the cyt *b/f* complex but, within PSII, there are all manner of additional possibilities, with ΔpH seen as the common feature.

6.12 The Governor

Plants cannot take shelter from too much sun by seeking the shelter of a nearby tree. Only a few have the capability of rapid leaf movement. At a maximal quantum efficiency of 0.125, temperate C3 species may utilize more than 80% of the red light that they are offered in the 0-800 μmole quanta.$m^{-2}.s^{-1}$ range. The fact that this value may fall below 20% in shade species or shaded leaves is a matter of photosynthetic capacity (enzyme complement) not efficiency. Whether shade or sun, there is a clear need to dissipate the excess excitation energy safely. Safe dissipation can be visualised as analogous to the mechanical "governor" beloved of Victorian engineers. Here a revolving shaft developed centrifugal force in two masses of steel

or iron spinning about a centre, opposite to one another. The greater the energy input, the further the two masses would swing apart and, in so doing, would increasingly disengage the driving force from the machinery being driven. The objective was protection. The machinery had a capacity for work which could not be exceeded without damage. The governor ensured that this safe limit could not be exceeded. In photosynthesis, electron transport creates a ΔpH which discharges through an ATPsynthase to form ATP from ADP and Pi. If these components are not recycled rapidly enough, because carbon assimilation is constrained by inadequate capacity or unfavourable environmental factors, ΔpH (the governor), disengages the driving force. Horton *et al.* propose that this disengagement results from an aggregation of the major light harvesting complexes, induced by protonation of the surface of the thylakoid lumen and promoted by the presence of zeaxanthin. While there is no life for a leaf without excitement, too much excitement can lead to decreased efficiency and photoinhibitory damage. Acting as a safeguard, against all but the worst excesses, the governor exerts a calming influence.

6.13 Summary

All metabolic energy and much of Man's fuels are derived from photosynthesis. Photosynthesis involves the utilisation of light energy in breaking H-O bonds; energy is derived from the products of photosynthesis when these H-O bonds are re-established by respiration or burning. Within chloroplasts the transduction of light-energy into electrical energy and then into chemical energy can proceed at the remarkably high efficiency of 33% but the optimal conditions necessary for maximal efficiency do not obtain in nature and a variety of factors (including photorespiration, dark respiration, incomplete interception of light and decreased efficiency at high light intensities) combine to decrease the *maximum* efficiency of conversion to about 5% in leaves. Even for standing crops, or swards of natural vegetation, 4.5% is only rarely approached. In general agricultural practice the figure is well below 1%. Even so, large human populations could be maintained by modern agriculture - populations much larger than currently exist. Such intensive agriculture is, however, only made possible by investing very large amounts of fossil fuels into fertilisers, herbicides, machinery, transport etc.) so that, in the end, modern agricultural practice constitutes a very inefficient way of converting the products of primaeval photosynthesis into contemporary photosynthetic products. Primitive agricultural practice is not necessarily less energy-intensive and is certainly much less productive.

While plant breeding may enhance plant productivity by improving the versatility of crops it seems unlikely that the fundamental efficiency will be enhanced in this way. Most C3 species seem to operate, under optimal light-limited conditions, close to the maximal quantum yield of 0.125 permitted by the Z-scheme. At low light intensities there is no difference in photosynthetic efficiency between "sun" and "shade" plants. In all leaves the photochemical apparatus accommodates mechanisms which allow safe dissipation of excitation energy in excess of that capable of being used in carbon assimilation.

In the present context, the most important "take home lesson" from all of this is that the "civilised" world we see about us has been made possible by long past photosynthesis. In the first instance this gave us the air we breathe and the food we eat but, since the Industrial Revolution and the advent of modern agricultural practice, it is the extensive use of fossil fuels that has permitted the relentless and continuing expansion of world population. In the remaining chapters we will examine the consequences of this in more detail. For the moment we can simply conclude that present trends cannot be sustained indefinitely on the present basis. Increasingly, fossil fuels will become too scarce and therefore too expensive to use widely in the manner that we use them now and sole dependence on contemporary photosynthesis would demand ten-fold reduction in population. In the short term our insistence on returning carbon dioxide to the atmosphere in quantities which it took millennia of photosynthesis to remove will, in itself, have major impact on our environment.

Chapter 7

Doom & Gloom

Generating hot air and discharging methane

Man's greatest experiment

Life in a warming climate

Rising seas and shrinking opportunities

PREDICTION, SCEPTICISM, OPTIMISM AND PESSIMISM

7.1 Noah and the Ark

"The optimist proclaims that we live in the best of all possible worlds the pessimist fears this is true" ....James Branch Cabell

The story of Noah and the Ark exemplifies some aspects of human thought and emotion which must always have been with us. There are those who tend, like the grasshopper, to live for the day. There are others, perhaps more pessimistic, or better informed or (like Noah) with inside information, These have always viewed the future with a certain amount of gloom. Given the fact that, during this century, there have been two world wars, major attempts at genocide and many "minor" wars (including the most awful, between Iran and Iraq, in which casualties have still been counted in millions rather than in thousands) it is difficult to deny that human pessimism has been entirely misplaced. Until the annexation of Kuwait by Iraq the prospect of future world war and nuclear holocaust seemed mercifully more unlikely than at any time in the recent past. Even so, if we were now to suppose that someone could wave a magic wand, that Arab would embrace Jew, that the Ayatollahs would be kind to Salman Rushdie, that India would warm to Pakistan and that the Irish Republican Army and the Basque Separatists would forgive and forget, it would still be difficult to find grounds for endless optimism about the future of the human condition. Yet even the most cautious and guarded expression of concern is still likely, in some quarters, to be dismissed in the same way as Noah's forebodings about the forthcoming flood. Scepticism, of course, is at the heart of good science and there are endless examples of well based predictions which have been made to seem almost ludicrous by subsequent events. In many Western countries, for example, population doubling once seemed inevitable by the end of this century. Now better contraception, economic pressures, and changed social attitudes (all closely interrelated) have combined to keep young women at work longer before they start families. This, in turn, has decreased the number of children per family and there is now even concern, in countries like Germany, about a future in which there might be too few young people to sustain the older part of the population in comfortable retirement. In the Third World of course, nothing is changed. World population, so we are told, is currently increasing by the easily remembered figure of 365,000 per day (Section 7.18).

So prediction is uncertain. There is even some encouragement to be derived from the fact that world "experts" are agreed that it is difficult, if not impossible, to make realistic predictions about the consumption of fossil fuels into the next century or the fate of the arctic tundra. Some things, on the other hand, are reasonably certain. Let us see what these are.

THE GREENHOUSE EFFECT

7.2 What is it?

"Ask Daddy, he won't know" .....Ogden Nash

The "greenhouse effect" (or more properly the *"enhanced* greenhouse effect") is central to contemporary relationships between plants, Man's energy requirements and his environment. It is also central to the whole concept, now slowly but surely

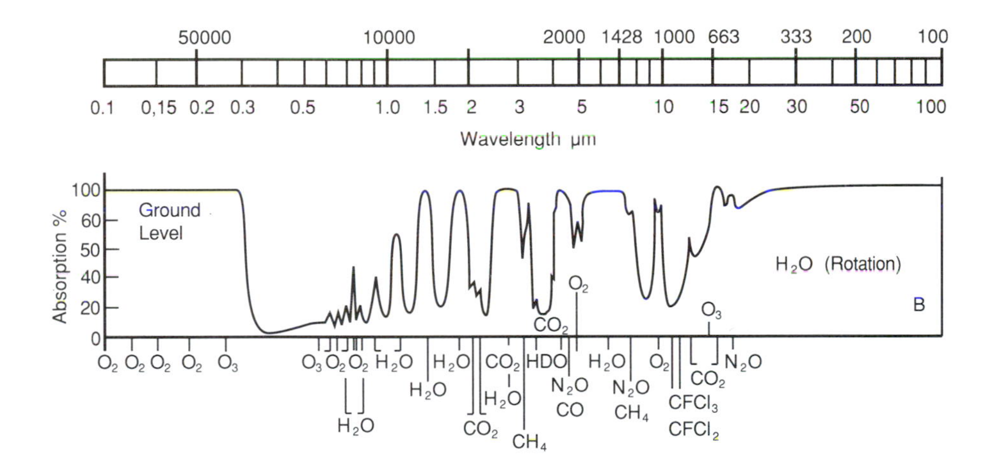

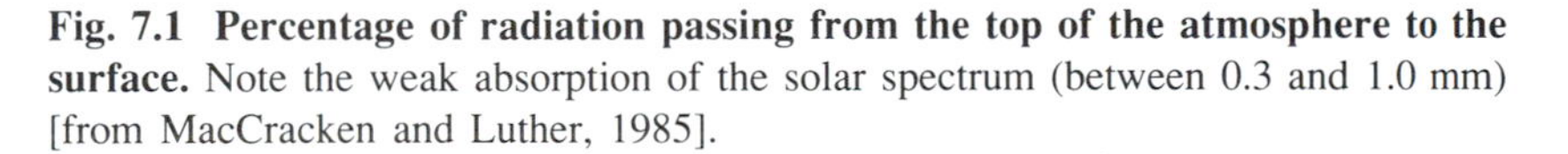

Fig. 7.1 Percentage of radiation passing from the top of the atmosphere to the surface. Note the weak absorption of the solar spectrum (between 0.3 and 1.0 mm) [from MacCracken and Luther, 1985].

being rejected, that Man could do what he wished with "his" planet without eventual and possibly painful consequences.

> [The term "greenhouse effect", as described below, is applied loosely, in accord with present practice, to a situation which existed long before Man and one which allowed the advent of life on Earth. What is currently a matter for concern is the *enhancement* (of this effect) which has occurred, and is continuing to occur, since the Industrial Revolution.]

At least, in the northern hemisphere, *"greenhouses"* are a common feature of horticulture. Much of Europe's tomato and cucumber crop is grown under glass in order to raise the temperature to one which favours these plants. Immense amounts of fossil fuels are burned, in winter and spring, to warm these greenhouses but, even if they are not heated artificially, they are warmer than their immediate environment.

This is primarily because the glass permits sunlight to enter reasonably well but at the same time diminishes the heat loss which would otherwise occur by free circulation of air and by convection. The analogy between such greenhouses and the *natural* "greenhouse effect" (which makes life on Earth possible to Man, by raising the *mean* global temperature from -18°C to +15°C) also relates to warming by the sun and prevention of cooling but there the similarity ends. Instead of glass we have gases. These gases do not interfere with cooling air movements, like the glass in a man-made greenhouse, but they do decrease the rate at which heat is lost by radiation until the temperature increases (on average by 33°C). Clearly, a new balance between heat gain and heat loss will have to be struck because, if it were not, the world would get hotter and hotter. Similarly, it is evident that, although the greenhouse effect currently results in a mean global temperature of 15°C, it is very variable in its effect. These variations contribute to the complexities of climate.

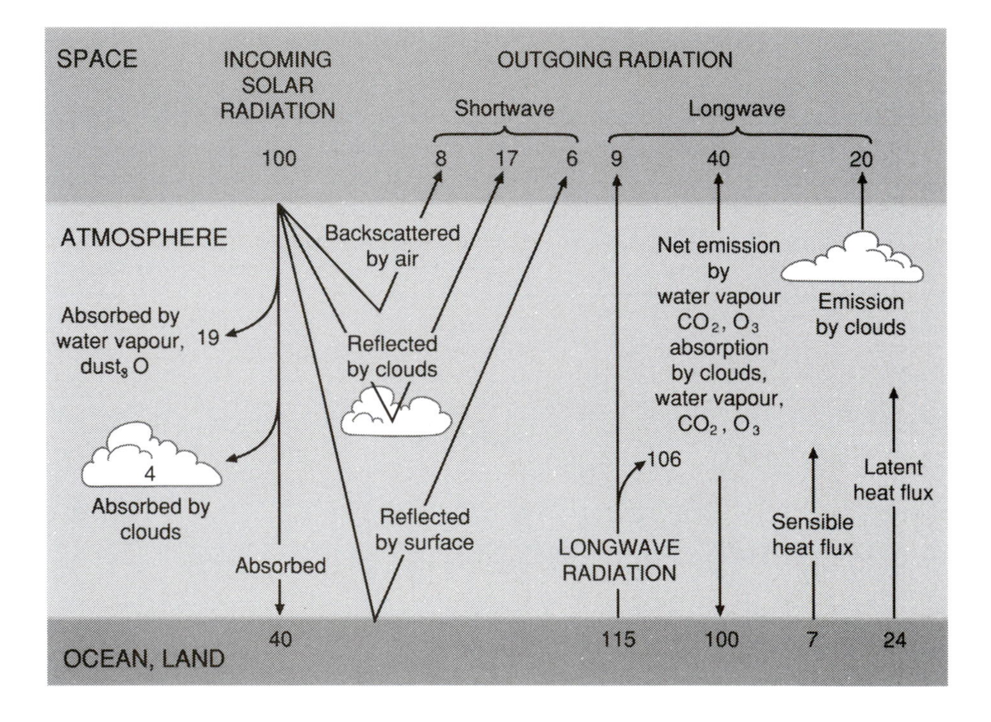

Fig. 7.2 The atmospheric heat balance. The units are percentages of incoming solar radiation. The solar fluxes are shown on the left, and the longwave (thermal infra-red) fluxes are on the right (after MacCracken and Luther, 1985).

Those of us who live in temperate climates are very conscious of winter frosts and their association with clear nights. The famous Scottish phrase *"Its a braw, bricht, moonlit nicht the nicht"* usually presages frost. There are such things as freezing fogs but, by and large, we know that if the stars sparkle in the winter night it is likely to get a great deal colder than if there is thick cloud. In such circumstances, we go to bed beneath some thick cover or blanket which ensures that we do not lose body heat to our immediate environment. Clouds of water vapour in our skies serve much the same purpose. Indeed, most of the natural greenhouse effect is attributable to water vapour and, in northern climates, we are only too well aware of the immediate chill that we can experience when the sun goes behind a cloud. In such circumstances it is primarily incoming solar radiation which is being excluded but water vapour is even more effective in diminishing the loss of heat to space. Even deserts, which become exceptionally hot as the sun beats down from a cloudless sky, can become uncomfortably cold by night without a blanket of water vapour to slow re-radiation of heat. Although water vapour is immensely important in this regard it does not act unaided. As we shall see in Section 7.5 - 7.8 its work is supported by other greenhouse gases of which the most important, and contentious, is carbon dioxide.

The analogy with the "greenhouse" or "glasshouse" now becomes clear. Unless such structures are ventilated, or air-conditioned in some way, they soon become much warmer than their immediate environment because solar heating is no longer as effectively balanced by heat losses of any magnitude. The Earth, of course, is well "ventilated" but the greenhouse gases still combine to slow the loss of heat into space.

7.3 How Does It Work?

Like most aspects of science the greenhouse effect is a great deal more complex in detail than it is in principle but, at its simplest, the story goes like this. Electromagnetic radiation from the sun (Section 2.2) penetrates the Earth's atmosphere. Some is reflected back into space from clouds of water vapour and some from the ice caps and other snow-covered and reflecting areas. Water vapour and other greenhouse gases are not, however, very effective absorbers of *solar* radiation

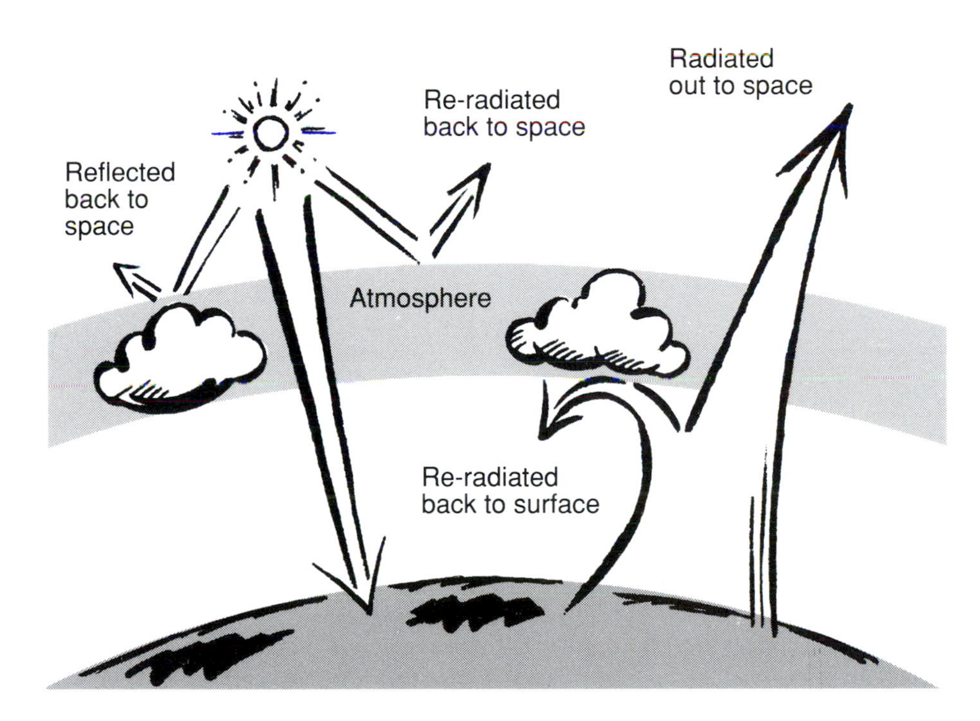

Fig. 7.3 The greenhouse effect. Simplified version of Figure 7.2. Some incoming solar radiation is reflected back into space by clouds, some is absorbed by the atmosphere, and some (mostly visible light) reaches the Earth's surface. In turn, the Earth radiates heat as infra-red. Some of this is again absorbed by the atmosphere which then radiates part of it back to Earth and part of it back into space. This raises the current global mean surface temperature from about -18°C (the estimated steady-state temperature at which heat input and heat lost would come into balance in the absence of an atmosphere) to 15°C.

and accordingly much of the sun's radiation penetrates to the Earth's surface. This is particularly true of electromagnetic radiation in the visible and photosynthetically active band (400-700 nm). Radiation which is absorbed by the planet and its atmosphere is also continuously re-radiated, by night and by day, in the long wave (infra-red) part of the spectrum. If we heat an iron in a fire it becomes "red hot", i.e. it emits red light in the 650-700 nm part of the spectrum. At the same time it emits infra-red radiation (at wavelengths longer than 700 nm which we cannot see) and any object, at any normal temperature, can lose heat by radiation in this way. Part of this long-wave radiation is intercepted by water vapour, carbon dioxide and other greenhouse gases. These re-radiate energy in *all* directions thereby deflecting some heat which would otherwise be lost, back towards the Earth's surface. Water, and water vapour, absorb radiation very effectively throughout much of the long-wave spectrum (the simplest way of cooling light from a tungsten source such as a conventional electric light is to pass it through a reasonably deep layer of water). There is, however, a major "window" between 8,000 to 12,000 nm (12-18 μm) in which both water and carbon dioxide absorb poorly and in which some other gases like methane (Section 7.6) and ozone (Section 7.9) absorb strongly.

In the presence of an atmosphere (Fig. 7.3), much of the loss of long wave radiation from a planet occurs above its surface. A planet, like Mars, with a "thin" atmosphere radiates most of its heat from near its surface and one, like Venus, with a dense carbon dioxide atmosphere and a strong greenhouse effect, radiates most infra-red from well above its surface. The mean radiating height for Earth is currently 5.5 km and if greenhouse gases continue to accumulate this height will increase. The effective radiating temperature of the planet will remain unchanged at -18°C (255K) but the temperature of the troposphere (Fig. 7.14) and the temperature of the surface (which are tightly coupled) will rise.

7.4 Generating Hot Air

"Does a Christ have to die in torment in each generation
to save those who have no imagination?" ....George Bernard Shaw

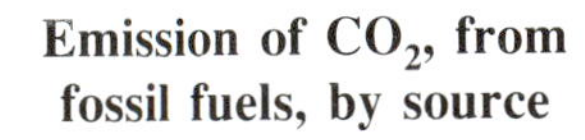

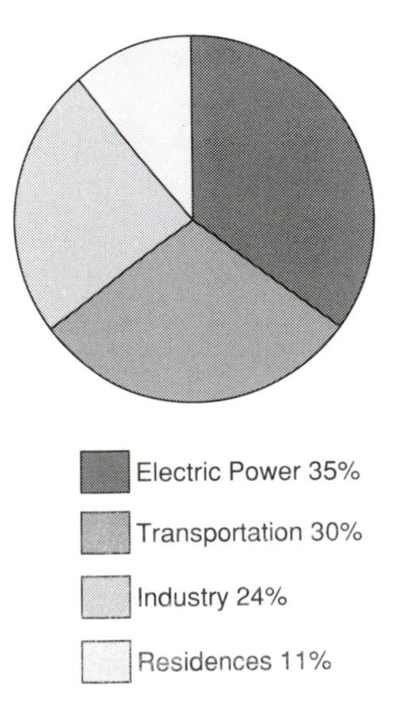

Seen from the twentieth century, the heresies of St. Joan pale into insignificance compared with those of Galileo but nothing so offends the current dogma, be it secular, ecclesiastical, spiritual, ethical, or simply a matter of common sense, than the concept that the universe does not revolve about the world. As always it is desperately difficult for the layman, or the scientist outside of his own immediate discipline to make objective judgements. A scientist is supposed, by vocation and training, to be properly sceptical to be open-minded and, while working within the confines of current dogma, to be able to step outside of existing "truths" once these can be demonstrated to be less than true or even false. These constraints are, by definition, in conflict with one another. Moreover, scientists are human, they can be mistaken and misled. They can be bought. They often have enough intelligence to see where their best interests lie. These human failings, together with very real uncertainties have been joined together by some journalists, television producers and politicians to create a "Green House Conspiracy". This is seen as a self-interested attempt by scientists, anxious for funding, to further their careers by persuading a gullible public that there is real cause for concern about the future when they know full well that such dangers are grossly exaggerated. Given some of the conspiracies in politics and commerce which seem to surface on an almost daily basis it is, perhaps, not unwise to discard the possibility of conspiracies in science but the evidence is hard to find. So far as funding is concerned it is certainly no easier to obtain support for work on global warming, despite its topicality, than for any

comparable branch of science and a proposal which sought to question any of the current dogma would undoubtedly have as much chance as one which embraced it without reservation. So far as science itself is concerned, there is no doubt that there have been scientific hoaxes and that falsification of results is not totally unknown but it would be very difficult to argue that these have ever significantly changed our perception of things. Scientists as individuals are presumably neither better nor worse than business men, politicians, poets or market gardeners. Moreover, cheating at science is a particularly pointless thing to do - almost on a par with cheating at patience. In the long term, truth will out, exaggerated claims will be questioned, charlatans will be discredited. It is very hard, therefore, to give credence to this particular conspiracy theory; almost impossible to believe that there is a dedicated band of greenhouse freaks, indifferent to scientific reality or moral considerations. This is not to say that there is no room for real doubt, uncertainty, or a need to question simple lack of understanding. It cannot be said too often, or too clearly, that the only present certainty is that CO_2 has increased and will continue to increase in the immediate future. What impact that will have, immediately passes into the realm of well-informed guesswork or mere speculation. It has already been noted that the whole concept of global warming, consequent upon Man's activities, is based on the *enhanced* greenhouse effect and that if it were not for the underlying or natural greenhouse effect the mean global temperature would be -18°C rather than +15°C. However, Richard Lindzen, writing with all the authority of a professor of meteorology at M.I.T., points out that this concept is *"seriously incomplete"* and that

"if it were a true representation of the actual greenhouse effect, the Earth would be 95°C rather than the 33°C warmer than it would be in the absence of greenhouse gases. The reason why 77% of the actual greenhouse effect is absent is simply that the Earth's surface does not cool primarily by radiation: it cools mainly by evaporation. Heat in the form of either sensible heat of latent heat (available for

release when water vapour condenses into rain) is bodily carried away from the surface by air motions associated with turbulence, cumulonimbus towers, and larger-scale meteorological systems. It is deposited at higher latitudes and altitudes where there is much less greenhouse gas to inhibit cooling to space."

Lindzen also points out that it could easily be inferred, from some contemporary accounts, that current atmospheric water vapour provides 60-70% of greenhouse warming and carbon dioxide 25%. In fact, he says, over 97% of the greenhouse effect on Earth is due to water vapour and stratiform clouds so that all other greenhouse gases play a *"truly minor role"*. Remarks such as these are particularly telling and carry much more weight than thoughts of greenhouse conspiracies. They should remind us all to approach this subject in the same way as did Robert Boyle when he wrote "The Sceptical Chemist". We must respect Lindzen's caustic candour when he says (of "Climate Change: The PIPCC Scientific Assessment")

"we really have two distinct documents: a Policymakers Summary which suggests that we are certain that serious climatic consequences will result from increasing levels of so-called greenhouse gases, and a lengthy report which finds no evidence for this certainty."

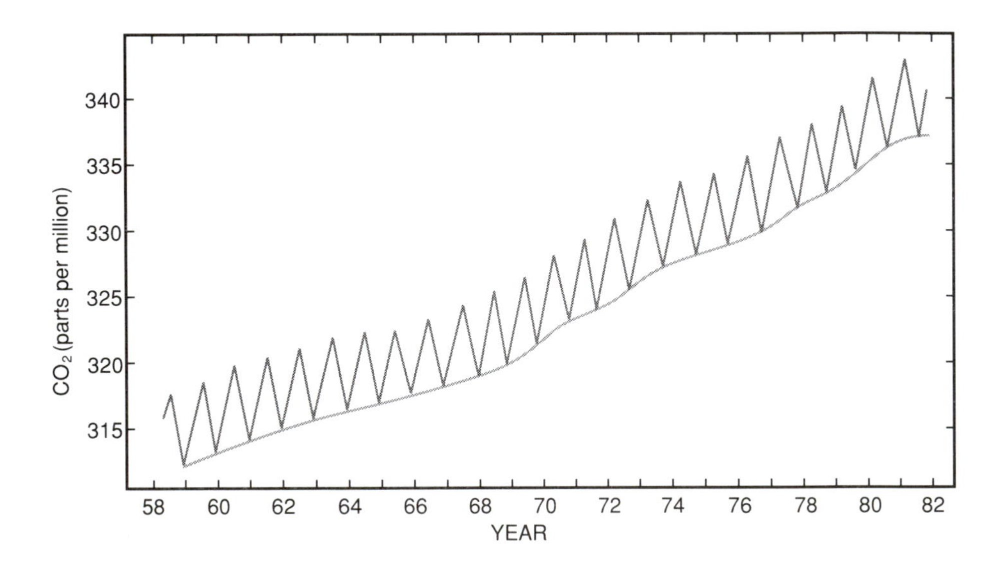

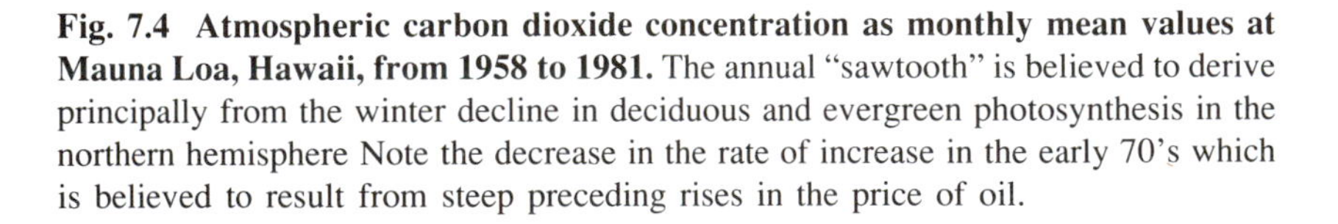

Fig. 7.4 Atmospheric carbon dioxide concentration as monthly mean values at Mauna Loa, Hawaii, from 1958 to 1981. The annual "sawtooth" is believed to derive principally from the winter decline in deciduous and evergreen photosynthesis in the northern hemisphere Note the decrease in the rate of increase in the early 70's which is believed to result from steep preceding rises in the price of oil.

But, if the doubts and uncertainties are so manifestly many, why should we be concerned at all? One obvious, and by no means trivial reason, is that it always helps to be as well-informed as possible about any matter which has become the subject of public debate and which we ourselves might be able to influence for better or worse. Another is more fundamental. We know that, whatever else happens, CO_2 and methane will continue to rise because of Man's activities. We know that these increases will change our environment just as destroying rainforests, killing whales, over-fishing or merely creating rubbish and sewage will change our environment. We might then reasonably ask if we ought to contemplate continuing doing these things at present rates without at least examining the *possible* consequences. Biologists, and most of the public at large, almost invariably find merit in conservation. Conservation is not compatible with carbon dioxide at twice its present level even if this were to have absolutely no effect on global temperatures. The mere fact that CO_2 might double simply underlines the fact that we ought not to gratuitously change the world in such a cavalier and careless fashion. What is more, there are good economic arguments why we should not do so (Chapter 9).

THE GREENHOUSE GASES

7.5 Carbon Dioxide

"Mans' greatest geophysical experiment" ... Rogers Revelle

Carbon dioxide, like water vapour, transmits electromagnetic radiation in the visible range but absorbs it, very effectively in the infra-red, particularly between 12,000 and 18,000 nm (12-18 µm). An invisible and tenuous blanket of water vapour and atmospheric carbon dioxide continuously diminishes the extent to which our planet loses heat to space by infra-red re-radiation. Carbon dioxide and greenhouse gases other than water (methane, N_2O, CFCs, etc) currently absorb about 50% more

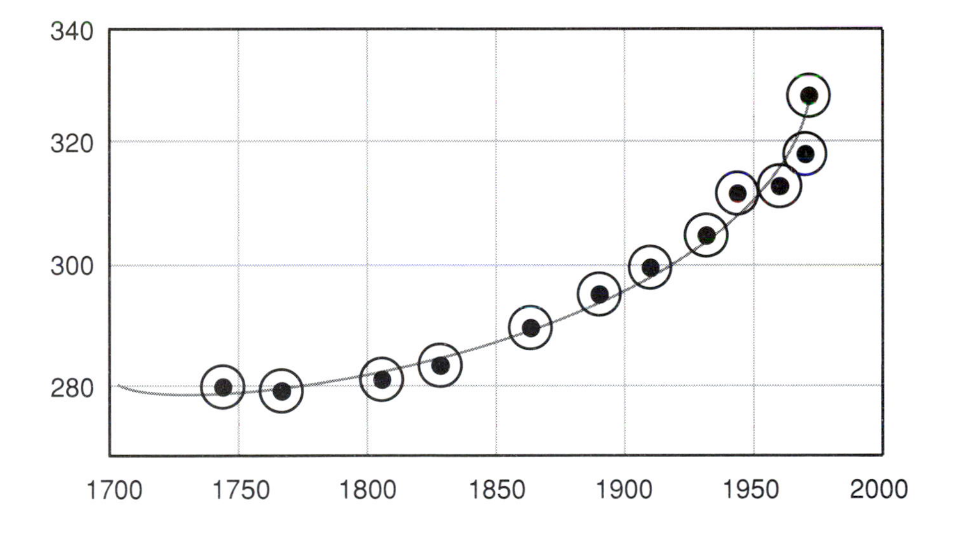

Fig. 7.5 Carbon Dioxide in Glacier Bubbles.

infra-red radiation from the earth's surface than did carbon dioxide alone, in pre-industrial days. As a result of Man's activities, carbon dioxide is produced on a massive scale (5.6 *billion* tonnes, in 1987) mostly by burning fossil fuels, of which, in the immediate future, natural gas will continue to be the most important.

In Section 4.1 we have seen that plants use the energy from the sun to incorporate carbon dioxide into carbohydrates (CH_2O) and that both plants and animals consume CH_2O (and derivatives of CH_2O) in respiration, a process which leads to the release of energy and the return of carbon dioxide to the atmosphere (Section 1.3). Fossil fuels can be regarded as photosynthetic products which have been stored for millions of years and which have been modified, as a result of pressure etc., during storage, into oil, natural gas and coal. Moreover, when fossil fuels are burned in fires, furnaces, electricity-generating plants or internal combustion engines, the process is always roughly analogous to respiration in the sense that a reduced compound (Section 1.2) is oxidised in a more or less controlled

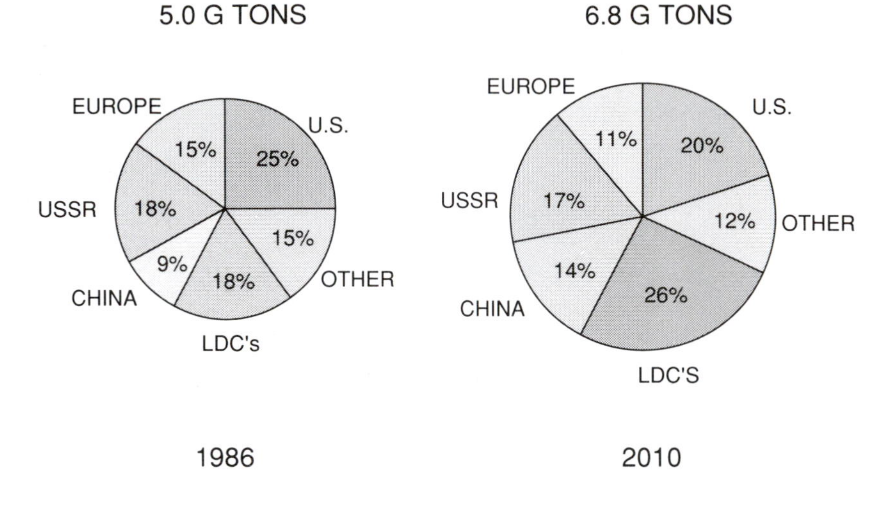

Fig. 7.6 World carbon emissions from fossil fuels.

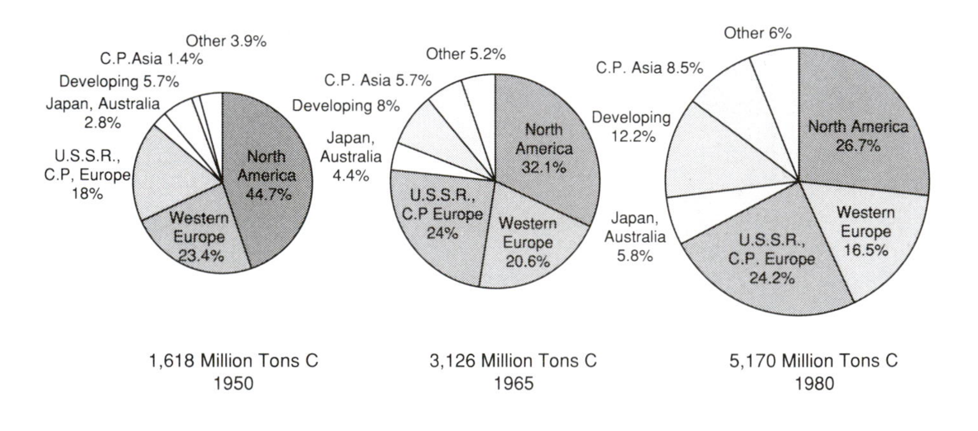

Fig. 7.7 Historical carbon dioxide emission rates, by region.

fashion, and the formation of new C-O and H-O bonds leads to the release of carbon dioxide and H_2O. Respiration itself contributes to the return of carbon dioxide to the atmosphere but respiration is of no real consequence in this regard because it only returns to the atmosphere, carbon dioxide newly removed by photosynthesis (Section 1.3). Plants, animals, and "primitive" Man lived (and still live) in balance with their environment because they cannot continue to release more carbon dioxide than photosynthesis has recently fixed. Even when stretches of forest were cleared, by burning, in the dawn of agriculture, the scale was small and the rate of progress so

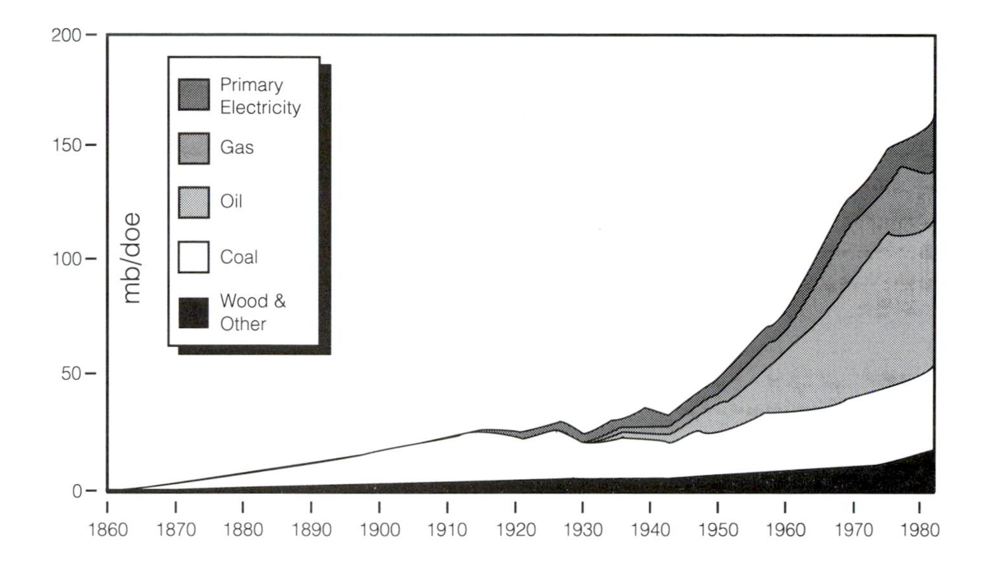

Fig. 7.8 World primary energy consumption 1860 - 1985.

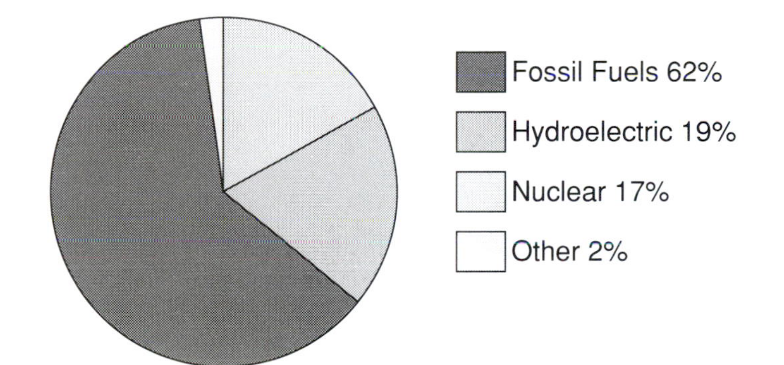

Fig. 7.9 Sources of electrical energy.

slow that the impact on the atmosphere was trivial. Currently, burning of tropical rain forests is making a major contribution to atmospheric carbon dioxide because the scale is so immense and it is all being done so quickly.

[Tropical rain forests (Section 9.13) do not themselves abstract as much carbon dioxide from the atmosphere as is sometimes supposed because, despite their massive photosynthesis, mature rain forests together with their respiring and decaying plant, animal, insect and bacterial populations also return equally massive amounts of carbon dioxide to the atmosphere. On balance, rain forests are more beneficial in all manner of ways than the temporary cattle ranching which so often replaces them. However, it is the rate and extent of current burning which so adversely affects the atmosphere rather than the continuing absence of forests once they are destroyed.]

In the same way, if contemporary Man could rely entirely on biomass and renewable resources for both his metabolic and his other energy requirements, he would constitute no threat to the atmosphere. This is seen very easily if we bear in mind that it makes little difference to *metabolic* energy whether a man is rich or poor, fat or thin, greedy or abstemious. He can not increase his own personal energy rating much above 100W unless he does a great deal of physical work. (Even then we are talking about, say, 300W whereas the rest of us, in relative sedentary occupations, will use only about as much as a 100W electric lamp, burning continuously.) Conversely, if we are relatively rich or profligate or simply live in a Western country with a high standard of living and have been brought up to take cheap energy for granted, our expenditure on food, transport and industry and what we use to warm, cool and light our houses, may add up to 10 kW or more per head. This could be a hundred times as much as we "spend" metabolically. Food, (Chapter 6) figures large in these considerations because *it is quite impossible, at present, to sustain present populations without recourse to fossil fuels.* We depend as much on past

Table 7.1

Per Capita Carbon Dioxide Emissions (Tonnes)*		
	1950	1988
Canada	3.1	4.6
France	1.3	1.6
Italy	0.2	1.7
Japan	0.3	2.2
United Kingdom	2.7	2.7
United States	4.6	5.3
West Germany	1.9	3.0
Czechoslovakia	1.7	4.1
German Dem. Repub.	2.4	5.4
Poland	1.2	3.3
Romania	0.3	2.6
Soviet Union	1.0	3.8
Brazil	0.1	0.4
China	0.04	0.56
India	0.1	0.2
Korea	"zero"	1.3
Mexico	0.3	1.0

* from fossil fuel consumption, cement manufacture and gas flaring (the two latter collectively contributed less than 4% of the total in 1988) - Source Carbon Dioxide Information Analysis Centre, Oak Ridge National Laboratory, USA.

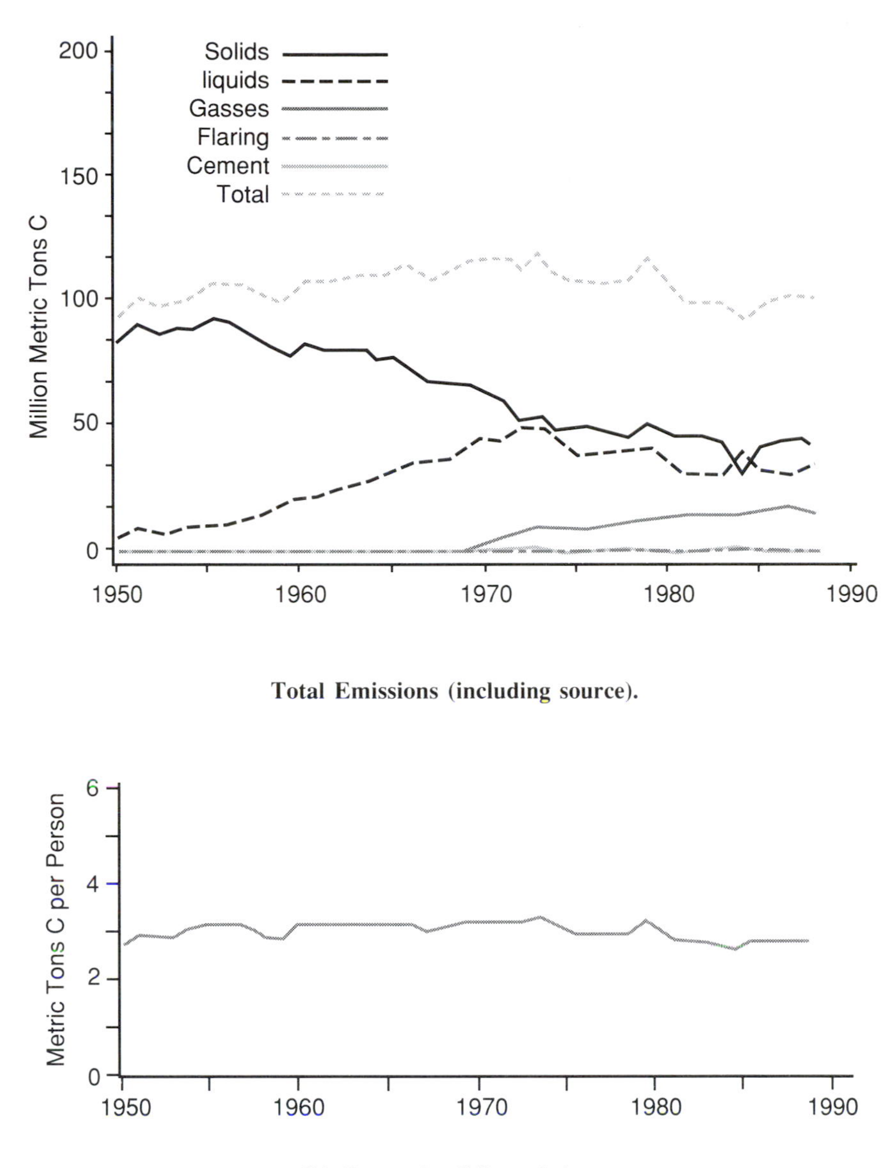

Total Emissions (including source).

(b) ***Per capita*** **CO_2 emissions.**

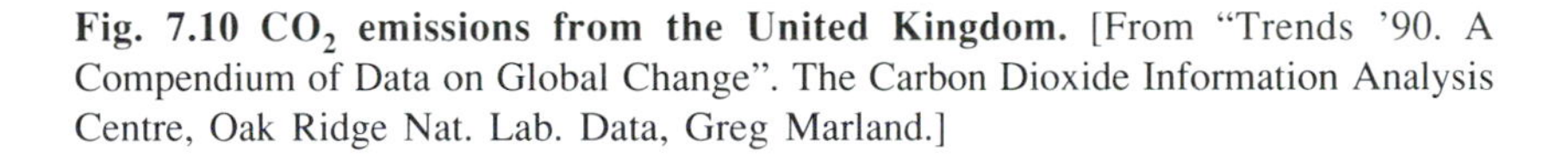

Fig. 7.10 CO_2 emissions from the United Kingdom. [From "Trends '90. A Compendium of Data on Global Change". The Carbon Dioxide Information Analysis Centre, Oak Ridge Nat. Lab. Data, Greg Marland.]

photosynthesis as on present photosynthesis for what we eat. When it comes to fossil fuels there is a measure of choice. Coal is bad because it is *primarily* carbon and goes like this:-

$$C + O_2 \rightarrow CO_2 \qquad \text{Eqn. 7.1}$$

and, if it contains appreciable amounts of sulphur, as it often does, then the combustion of the sulphur (while adding marginally to the calorific value without affecting the amount of carbon dioxide released) gives rise to sulphur dioxide which, as acid-rain etc., damages health, buildings and forests alike.

$$S + O_2 \rightarrow SO_2 \qquad \text{Eqn. 7.2}$$

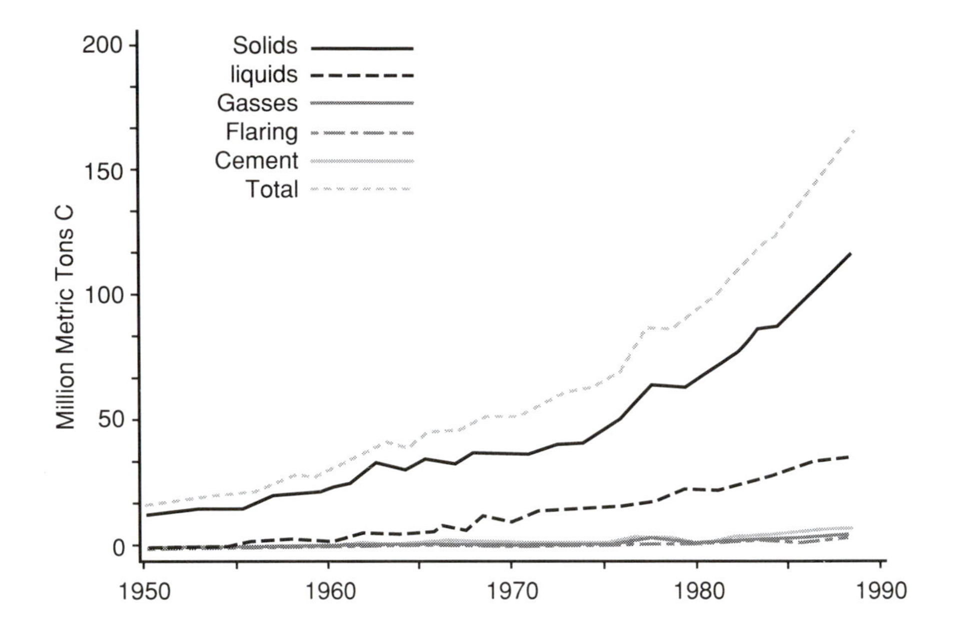

Total Emissions (including source).

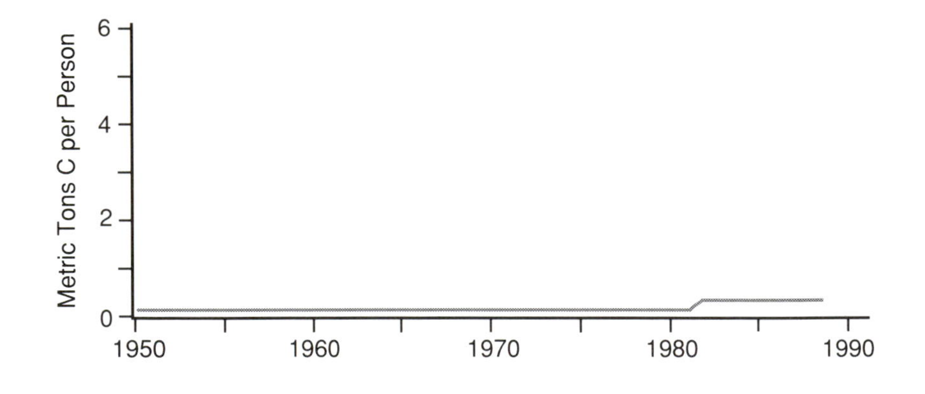

(b) *Per Capita* CO_2 emission.

Fig. 7.11 CO_2 emissions from India [From "Trends '90. A Compendium of Data on Global Change". The Carbon Dioxide Information Analysis Centre, Oak Ridge Nat Lab. Data, Greg Marland.]

Methane (i.e. "natural" gas - Section 7.6) is better in some regards because, as equation 7.3 shows,

$$CH_4 + 3O_2 \rightarrow CO_2 + 2H_2O \qquad \text{Eqn. 7.3}$$

its combustion involves the formation of twice as many H-O bonds as C-O bonds.

The former, while contributing energy (Table 7.3) only add water vapour (rather than carbon dioxide) to the atmosphere. (Given the amount of water vapour already in the troposphere (Fig. 7.15) and the factors which affect its equilibrium with land and oceans, this may seem to be of no consequence, but, in fact, methane oxidation can adversely effect *stratospheric* water vapour thereby reinforcing direct radiative effects (Section 7.6).

Although a shift from coal and oil to natural gas is not, therefore, an entirely unmixed blessing the relevant records give some indication of what might be achieved by intelligent application of appropriate energy policies (Chapter 9) Global emission of carbon dioxide from fossil fuels, cement production, gas-flaring etc. were 5.65 billion (5.65 x 109) tonnes in 1987, just over 1.1 tonnes of carbon for every person on Earth. Citizens of the United States were the most profligate in this regard, accounting for 22% of the total or *about 5 tonnes of carbon per person.* Inhabitants of the United Kingdom are not a lot better but, strangely enough, they have not become worse since 1950 (Table 7.1). During the last forty years or so *per capita* emissions from the U.K. have remained more or less constant. This is not the result of any conscious effort. On the contrary, past indifference and lack of recognition

Table 7.2
Heating Caused by Greenhouses Gases

Gas	Concentration ppm	Principal absorption bands. Position cm^{-1}	Strength $arm^{-1}\ cm^{-1}$ STP	Greenhouse heating Wm^{-2}
Water vapour	~ 3000			~100
Carbon dioxide	345	667	(many bands)	~ 50
Methane	1.7	1306	185	1.7
Nitrous oxide	0.30	1285	235	1.3
Ozone	$10\text{-}100\times10^{-3}$	1041	376	1.3
CFC11	0.22×10^{-3}	846	1965	0.06
		1085	736	
CFC12	0.38×10^{-3}	915	1568	0.12
		1095	1239	
		1152	836	

The main absorption bands and band strengths (a measure of the probability of a molecule absorbing a photon at the band wavelength) are shown for the less abundant gases. (see also section 7.7)

of the consequences of carbon dioxide emissions has given way even the last decade to the philosophy of "market forces" and a government with a desperately poor record in regard to promotion of energy conservation. Why then have outputs of carbon dioxide per person remained constant and total emissions only 10.5% higher in 1988, *"by far the smallest increase among the top twenty nations"*. Britain, once third among the major, carbon dioxide emitting, nations has slipped to seventh place without even trying. It is perhaps not entirely coincidental that the largest sustained decline of 19% occurred between 1979 and 1984, the first five years of office of the present Conservative government. "Market Forces" undoubtedly played a part because, as oil became more expensive, natural gas (which now accounts for 20% of the carbon dioxide emitted from the U.K.) became more attractive. At the same time, the massive decline in heavy manufacturing industry (particularly steel and shipbuilding which uses large quantities of steel) must, to a large extent, have offset the seemingly relentless rise in road-transport. Moreover, mindful of the miner's strike which ended a previous Conservative government, Margaret Thatcher and her colleagues were determined to ensure that the U.K. became less dependent on coal. It is obviously difficult to assess which of these many factors was most telling. Different political persuasions would undoubtedly influence interpretations. Nevertheless the record is indisputable. Whether or not the economy of the U.K. is materially better or worse than it was twenty or forty years ago is a matter for argument but there is no doubt that Britain's carbon dioxide emissions have remained relatively stable during a period (Fig. 7.10) in which global emissions and emission from some broadly comparable countries has increased substantially (Table 7.1). In experimental terms it seems clear that national policies have had a material influence on the extent of carbon dioxide emissions and that wholehearted adoption of better policies (Chapter 9) and in particular policies of conservation could lead to substantial *decreases* in emissions, despite increases in population. Similar trends are evident in Table 7.1. The growth in emissions from third world countries (see also Fig. 7.11) is inevitable, at least in the short term, but there are remarkable differences between western and eastern block countries (e.g. East and West Germany, Romania and Spain) and European countries (e.g. France, Germany and the U.K.) with broadly similar populations and standards of living.

7.6 Methane

The present concentration of methane in the atmosphere (Table 7.2) is extremely small. We are already accustomed to thinking of carbon dioxide as a minor

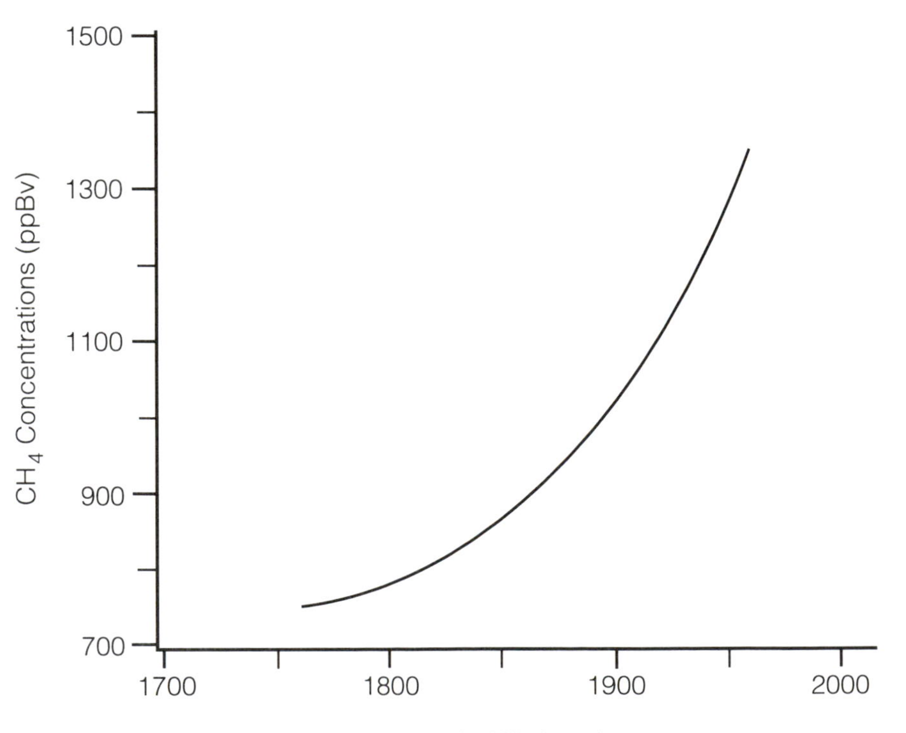

Fig. 7.12 Concentrations of atmospheric CH_4 from ice cores

constituent of the atmosphere, despite its importance. Certainly if we were to represent each molecule of gas in the atmosphere by one inhabitant of the city of London we would need to ask two or three thousand of its inhabitants to pose as molecules of carbon dioxide (to represent a concentration of 345-350 parts per million). On the same basis, however, we would be only looking for ten or fifteen London citizens to play the role of molecules of methane. Why then is methane now regarded as an important greenhouse gas? There are two reasons. One is that, although the concentration of methane is only about 1/200th of that of carbon dioxide, it is 20-25 times more damaging on a molecule to molecule basis. This is because its chemical structure makes it more effective than carbon dioxide in absorbing infra-red and also because it is drawing down a blind across a window in the infra-red absorption spectrum which is still largely unscreened and relatively unaffected by carbon dioxide. For these reasons, even though it is present in such small amounts, methane accounts for about 18% of the *increases* in the greenhouse effect (Fig. 7.15). Another cause for concern is that its concentration in the atmosphere is rising at a rate considerably faster than that of carbon dioxide. It is only in the last decade that it has been possible to make accurate direct measurements of global methane and these show that it is increasing at about 1% per annum. It is also possible to look at the geological record of methane (Section 7.6) in the same way as it is now possible to say with considerable accuracy, how much carbon dioxide was present in the atmosphere in the relatively recent past. In both cases this is done by sampling the contents of the bubbles trapped in arctic ice of known ages. The records show that methane remained remarkably constant for about 30,000 years and started to rise only towards the end of the 17th century. There was then a marked

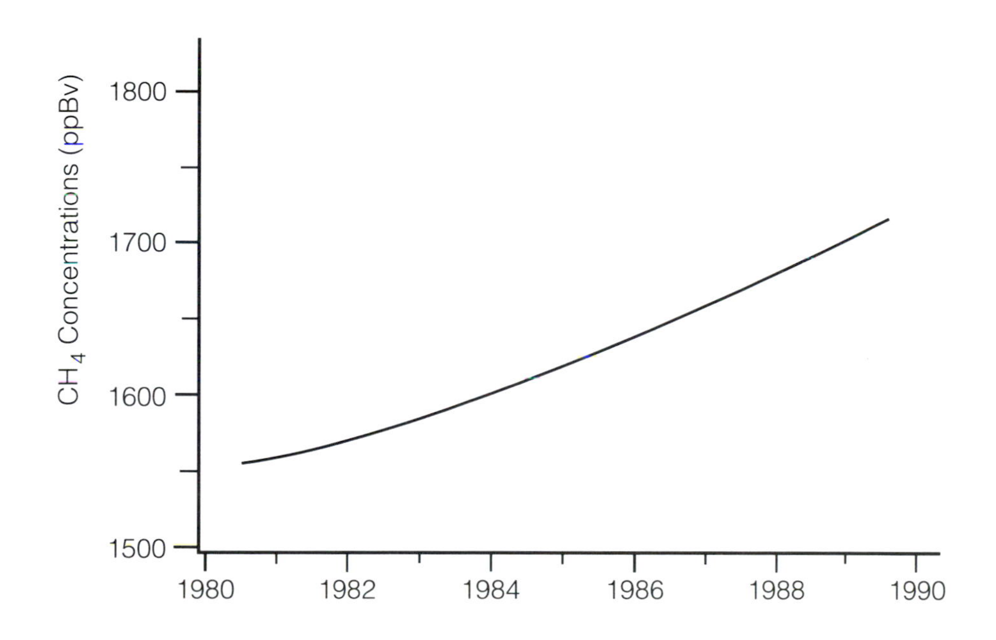

Fig. 7.13 Monthly concentrations of atmospheric CH_4

acceleration at the turn of the present century such that atmospheric methane has approximately doubled over the last 200 years. This is very strong circumstantial evidence that this part of the increase in methane concentration is *"anthropogenic"* (i.e. that it is a consequence of Man's activities). This is based on the striking correlation between the growth in methane concentrations and the growth in human population. Only a little methane is generated by geological activity:-

> *"Methane is bound into lattice-like structures, clathrates, which are present in vast quantities (exceeding all known coal reserves in the ocean sediments), leakage of methane from the ocean floor in fact provides the main source of carbon for some marine populations. But there is no reason to suppose that methane production from these submarine reservoirs has increased in recent times".*

So where is the methane coming from? Possibly as much as 30% derives from incomplete combustion of fossil fuels. Decaying plants and animals make a contribution. The release of methane from garbage in "land-fills" can constitute a real fire risk. Methane is "marsh gas". The old legend of will-o'-the-wisp came from the spontaneous ignition of methane, flickering over bogs and marshes. It has even been

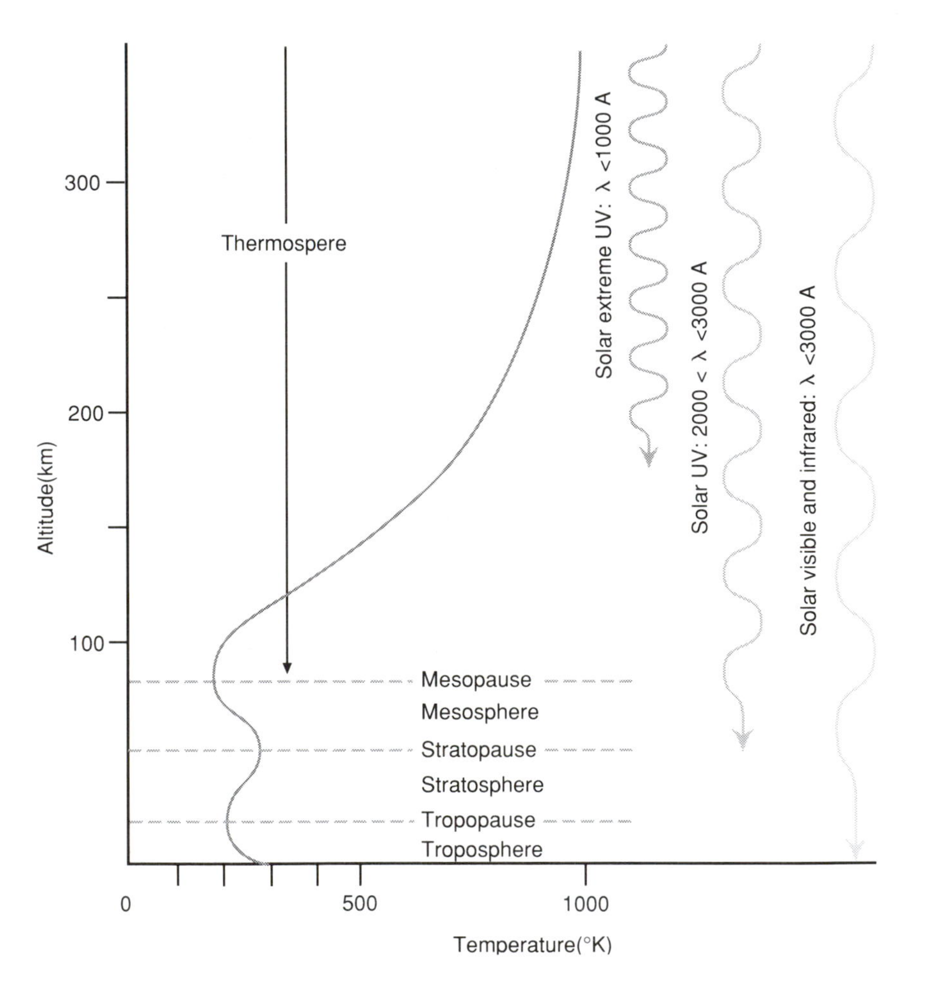

Fig. 7.14 Divisions of the Atmosphere. The atmosphere can be divided into troposphere, stratosphere, mesosphere, and thermosphere, separated by tropopause, stratopause and mesopause respectively. The high temperature regions of the atmosphere result from the absorption of solar radiation in certain wavelength ranges, as shown on the right. (from Goody & Walker, 1972).

known for British students of botany, engaged in peat borings in the middle of raised bogs in the pouring rain, to be cautioned against smoking for fear of igniting some vast conflagration. This, of course, was always as unlikely as the possibility that these wetlands, now rapidly diminishing as the result of draining, might be largely responsible for *increasing* the methane in the atmosphere. There is, however, one such source of methane which will, quite clearly, have increased in parallel with increases in world population. Rice production in the paddy fields of South East Asia has undoubtedly contributed to increasing methane production.

Coal mines and oil wells have made their own contribution, both in regard to methane inadvertently released as a result of these operations, or as a consequence of incomplete combustion of fossil fuels. It is possible to derive some estimate of the amount of methane in the atmosphere produced in this way. These estimates are based on C^{14} measurements (C^{14} is a radioactive isotope of carbon). Oil and coal reserves contain very little C^{14} compared with methane released in other ways. Such estimates put the current atmospheric methane derived from fossil fuels at about 30%. In the U.K. 29% of the methane released has been attributed to deep-mining of coal but the uncertainties in this sort of attribution are underlined by the fact that a very indefinite 10% is related to "gas leaks" (i.e. leaks from pipes which supply "natural gas" for domestic and industrial use).

[Leaks have to be taken into account in considering the relative environmental impact of burning methane, which is often said to be preferable to coal and oil because it yields less carbon dioxide per unit of energy released (Table 7.4). It can readily be seen that this advantage would be largely lost if an appreciable percentage of methane escaped prior to burning and made its own direct contribution as a very effective greenhouse gas. It has been claimed, by Wallis, that leakage from North Sea Gas wells together with land-based leakage could increase the emission of greenhouse gases by a factor of 0.5-2.0 when it is preferred to coal in UK electricity generators.]

Table 7.3

Past and Projected Greenhouse Gas Concentrations and Associated Changes (Δ) in Greenhouse Heating (Q)

Gas	Assumed 1860 Concentration ppm	Q 1860-1985 Wm^{-2}	Estimated 2035 Concentration ppm	Estimated ΔQ 1985 - 2035 Wm^{-2}
Carbon dioxide	275	1.30	475	1.80
Methane	1.1	0.40	2.8	0.50
Nitrous oxide	0.28	0.05	0.38	0.15
CFC11	0	0.06	1.6×10^{-3}	0.35
CFC12	0	0.12	2.8×10^{-3}	0.69
Total		1.9		3.49

Methane is also produced in large quantities in the rumen of cattle. Although cows are clearly herbivores in the normally accepted sense of the word they are carnivores in another. Unable themselves to digest plant constituents, such as cellulose, they rely on bacterial digestion within their rumens. Rotifers etc., which in turn feed on these bacteria, are washed over a "weir", together with other products of digestion, into parts of the cow's gut which can digest this remarkable brew. Methane produced in the rumen is released from both ends of the digestive tract.

[Cattle can become bloated if they are allowed to eat some plants such as Lucerne, (Alfalfa, *Medicago sativa*). This is because chloroplast lipids in these plants contribute to the formation of a stable foam which prevents the release of methane to the atmosphere.]

Table 7.4

Carbon dioxide emissions from fossil fuels*	
Coal	24
Oil	20
Natural gas	14

* Approximate values in Kg carbon released per GigaJoule of energy generated.

In the U.K., 23% of the methane released is attributable to belching and farting cattle and 9% to sheep. Pigs, poultry, horses and people combine to add another 1%. You and I, it seems, contribute between 300 ml and 2 litres a day in this way.

"Some of us fart a lot and some of us fart a little but, whatever anyone tells you, there is no-one, not even the Queen, who doesn't fart at all."

Fortunately (or perhaps unfortunately in some respects) such exhaust gases are not all pure methane. One saving grace about methane is that it does not hang around indefinitely and it is continuously undergoing photochemical oxidation by the hydroxyl radical (OH) in the troposphere (thereby making a modest addition to less effective carbon dioxide)

$$CH_4 + 8\,OH \rightarrow CO_2 + 6H_2O \qquad \text{Eqn. 7.4}$$

Nevertheless

"If the current rate of increase is maintained, concentrations will reach 2.8 ppm in the next 50 years contributing an additional 0.5 Wm^{-2} in radiative heating".

Moreover, (Section 7.5) oxidation of methane may also contribute to stratospheric water vapour, thereby reinforcing direct radiative effects.

7.7 Chlorofluoromethanes

Methane (Section 7.6) is bad enough but methane in disguise is even worse. Methane is a simple molecule, one carbon joined to four hydrogens and, since each of these can be substituted by other atoms, it is possible to come up with all manner of related molecules like chloroform ($CHCl_3$) and carbon tetrachloride (CCl_4), once used widely for removing stains from clothes and doing nasty things to people's livers if used unwisely. *"Chlorinated fluorocarbons"* (CFCs) belong to this family. Chemically, CFC 11 is $CFCl_3$ and CFC 12 is CF_2Cl_2.

> *"The contribution of a gas to the greenhouse effect depends on the wavelength at which the gas absorbs radiation, the concentration of the gas, the strength of absorption per molecule (line strength), and whether or not other gases absorb strongly at the same wavelengths. As a consequence, one molecule of dichlorodifluoromethane (CF_2Cl_2) is about 10^4 times more effective in "trapping" long-wave radiation than one molecule of carbon dioxide in the present atmosphere."*

CFCs are not only much nastier than methane itself because they absorb infra-red in parts of the spectrum that other gases don't reach but also because of their notorious facility for destroying atmospheric ozone (Section 7.8), which would otherwise prevent ultraviolet radiation (from the sun) reaching the Earth's surface. CFCs are released in relatively minute amounts from aerosols, air-conditioners and refrigerators and in somewhat larger amounts from commercial installations producing frozen foods and plastic foams. CFCs are very stable, inert molecules which makes them ideal for this purpose but it also endows them with an unparalleled facility for conveying chlorine from the Earth's surface to the stratosphere. Many other organic molecules are susceptible to the near ultraviolet radiation which reaches the troposphere (Fig. 7.14) but CFCs are virtually untouched. Indeed, when James Lovelock first sampled atmospheric CFCs in 1971, he found that a substantial fraction of the CFCs manufactured at that time had accumulated in the *troposphere*, impervious to breakdown by near ultraviolet light. In the *stratosphere* on the other hand, far ultraviolet light degrades CFCs, releasing chlorine which can react with ozone to give the free radical, chloric oxide.

$$Cl + O_3 \rightarrow ClO + O_2 \qquad \text{Eqn. 7.5}$$

Moreover chloric oxide can regenerate chlorine

$$2\,ClO \rightarrow Cl_2 + O_2 \qquad \text{Eqn. 7.6}$$

so that the whole process can continue

"Rowland and Molina estimated that CFCs would have lifetimes of between 40 and 150 years before being decomposed in the stratosphere. Once that happens, each atom of chlorine could destroy, through its chain reactions, as many as a hundred thousand molecules of ozone. They estimated that, if the current rate of production of CFCs continued, half a million tons of chlorine would be released annually into the stratosphere and this would double the rate of ozone decomposition that occurs naturally. The net result was likely to be a depletion of ozone by between 7 and 13%.

"no imminent danger"

Chlorine might eventually take over the chemical control of the stratosphere with very serious consequences on animal and plant life. Finally if this were not enough to be worrying about, the effects were already ordained in some degree by the long lifetime of the CFCs already released" ...George Porter.

Once it became clear how dangerous CFCs *could* be, there followed a huge argument in the United States involving the CFC industry, the National Academy, a Congressional enquiry and the Environmental Protection Agency, (which put a price tag of $1.5 billion and 9000 jobs on banning CFCs). When the National Academy concluded that the threat was real but distant, "Yale Scientific" said

"The EPA now has the proof. There is no imminent danger to the ozone layer by CFCs. These losses will never be recouped", and went on *"We must distinguish between scientists who present science as science and those scientists who promote science policy decisions as scientific gospel"*

In 1985 the whole picture changed dramatically. Farman, of the British Antarctic Survey, reported a seasonal loss of ozone over Holley Bay in the Antarctic.

"By 1987 comprehensive measurements showed that the springtime depletion of ozone over Antarctica was the most severe yet seen. It was quite spectacular and, occurring elsewhere, could have been catastrophic since 95% of the ozone between altitudes of 14 and 23 km was destroyed in less than two months. The total ozone column was reduced to 40% of its pre-1979 thickness. This was a total surprise and

a grave warning, through it was still unproven that the culprits were CFCS. But that evidence was soon to come. The British findings were confirmed in August of that year by reassessment of data from NASA's Nimbus 7 satellite. This had been taking *measurements of ozone over Antarctica since 1978 - so why hadn't they reported the same low values? Because the computer had been told to reject exceptionally low ozone values which were not expected. When the data were reprocessed with a less intelligent computer they confirmed the measurements of the British team. In August and September 1987 a large airborne American experimental campaign overflew Antarctica and measured the concentrations of several trace gases up to altitudes of 18 km. The results were as convincing as we are ever likely to be blessed with. The free radical, ClO, appeared rapidly and abundantly in the late August and this was followed by a dramatic fall of ozone in mid-September, which showed an inverse correlation even in the smallest details. All other gases such as the nitrogen oxides and even water, H_2O, decreased with ozone; only ClO increased. The 'smoking gun' had at last been found and all the fingerprints were in place"* ....George Porter

There are obvious lessons to be learned from the CFC story in regard to the Greenhouse effect and global warming. Science itself, however, has to do with "truth" and its elucidation. Of course, there are scientists who, like cabinet officers, have been *"economical with the truth"* Some are given to exaggeration, others blatantly dishonest. On the whole, however, a scientist embarked on a dishonest approach to his subject would be like a tennis player who hit every ball out of court. The whole exercise would be self-defeating and, in the end, such a scientist would inevitably be discredited. Although this is manifestly self-evident, scientists who warn about the dangers of radioactivity, or of pollution, or of inadequate hygiene, etc., etc., are frequently attacked and denigrated by those who earn money or political advantage by exploitation of the environment. Scientists are not infallible but "Yale Scientific" (above) ought to have paid more heed to scientists engaged in pure research than to the manufacturers of CFCs or aerosol valves. Attitudes similar to those of "Yale Scientific" still abound (Section 7.4). Time, of course, will tell who is right but, in the meantime, it might be wiser to emulate Noah than to heed conspiracy theories put about by those with vested interests intent on becoming rich or famous at the expense of the environment.

7.8 New CFCs for Old

The world's chemical companies may have done much to improve our lives but they have not be renowned for being friendly to the environment. Much attention has been given recently to the impact of antiquated chemical plant on what, until recently, was East Germany. Refurbishing that out-moded economy has already cost many million DM and will certainly cost many millions more. Much of this will be spent on cleaning up an environmental mess of a sort which would only be tolerated by a totalitarian regime, No doubt the record is better in the West but not so much better that it could be regarded as a course for rejoicing. Similarly there has been a sorry record (of drugs, pesticides and herbicides which have killed and deformed) to put along side those which have done what was intended of them without damage. If we were to take a totally cynical point of view we could say that industry can no longer, in a environmentally aware world, get away with so much nor afford to totally disregard public opinion in its pursuit of profit. If we take a more generous view, we should applaud the present attempts to present a kinder face. In this context attempts to produce alternatives, less damaging to the environment than CFCs, should be applauded even if the applause is muted by the likelihood that the alternatives might not have all the virtues which are claimed for them.

CFCs are nasty because they contain halogens (chlorine fluorine or bromine). For example CFC 12, widely used in refrigeration, releases chlorine in sunlight at an altitude of 20-40 km and this promptly set about tearing holes in the ozone layer.

Moreover CFCs are stable in the lower atmosphere and are eminently suitable for carrying the halogens to the stratosphere where they do so much harm. Currently

Table 7.5

Atmospheric Lifetimes and Ozone Depletion potential

Average values

Compound	Lifetime (Years)	ODP
CFC 11	75	1
CFC 12	120	1
CFC 113	100	0.8
CFC 114	270	0.8
CFC 115	550	0.4
HFA 22	20	0.05
HFA 123	2	0.02
HFA 124	7	0.02
HFA 125	22	0
HFA 134a	19	0
HFA 141b	10	0.1
HFA 142b	27	0.06
HFA 143a	50	0
HFA 152a	2	0
CCl_4	62	1.2
CH_3CCl_3	6.5	0.15

Estimates of lifetimes of HFAs and CFCs. This table also shows ozone depletion potentials, defined as the potential ozone depletion arising from one tonne of the compound compared to that of one tonne of CFC 11, over the whole of their atmospheric lifetimes. Compounds of the methane (CH_4), ethane (CH_3-CH_3) series are called "alkanes" or "saturated hydrocarbons" (because all of the carbon valencies are saturated with hydrogen). Hence the term HFAs includes both HFCs and HCFCs (fluorocarbons containing no chlorine or chlorine respectively). For example HFA 134a is CH_3CH_2F and HFA 123 (a possible substitute for CFC 11) is CH_3CHClF.

there are compounds of a somewhat similar chemical nature such as methylene chloride, trichlorethylene etc which are much less stable and would not act as shuttles, transporting halogens into the ozone layer. Yet others contain no chlorine and are therefore even more desirable. The chemical industry prompted by the Montreal Protocol, has become suddenly interested in an "*Alternative Fluorocarbon Environment Acceptability Study*". This has focused on hydrofluorocarbons (HFCs) which comprise two carbons, fluorine and hydrogen and HCFcs which are similar but still contain some chlorine. For example it is proposed that CFC 12 (CF_2Cl_2), one of the most widely used CFCs, might be replaced by tetrafluorethane (HCF_2HCF_2) in refrigeration and air-conditioners. Such compounds (hydrofluoroalkanes or HFAs) containing hydrogen, are able to react with natural hydroxyl radicals present in the lower atmosphere so that even those containing chlorine (like HFA 123) have a low potential for "delivering" chlorine through the lower atmosphere into the ozone layer.

7.9 Ozone

Ozone is extremely important to life on Earth and should not be regarded as simply one more greenhouse gas. As we can see from Fig. 7.15, apart from water vapour (not shown) the overall picture is dominated by carbon dioxide, methane and CFCs and oxides of nitrogen (7.10). Nevertheless, *stratospheric ozone* owes its additional importance to the fact that it filters out incoming *short-wave* radiation (ie ultraviolet) rather than infra-red. Ultraviolet can cause skin-cancers and damage to the photosynthetic apparatus of plants. In sufficiently large amounts, at particular wavelengths, it can even cause genetic damage when absorbed by DNA.

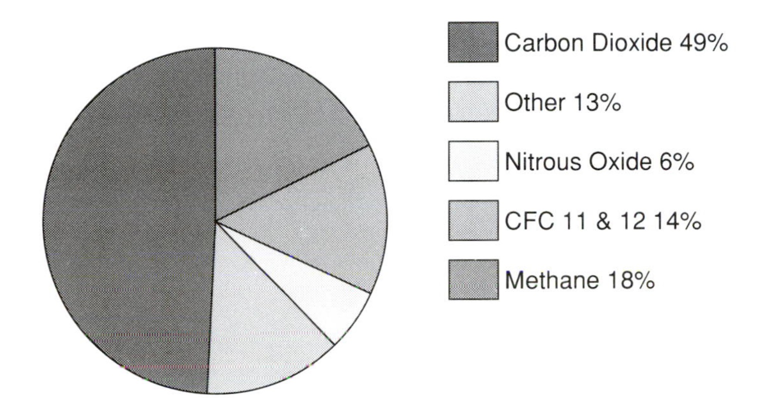

Fig. 7.15 Present Contributions to Increasing Greenhouse Warming.

[Ultraviolet radiation can be classified as UV-A (above 320 nm), UV-B (between 320-280 nm) and UV-C (below 280 nm). Most solar radiation in the 200-290 nm region is currently absorbed by stratospheric ozone but it should be noted that artificial radiation with a maximum at 254 nm readily damages the water-splitting facility of PSII and plastoquinone (Section 3.9).]

Anderson has recently shown that pea plants (*Pisium sativum)*, grown in augmented UV-B, can be damaged, photosynthetically, in several ways including adverse effects on ATPsynthase and RUBISCO activity (Sections 4.7 & 5.2). Plants acclimatised to high altitude growth appear to be afforded added protection by increased concentration of phenolic compounds called flavenoids which, even in plants not adapted to high UV-B, absorb all but a few per cent of incident ultraviolet at these wavelengths. Cadwell, however, has drawn attention to the fact that whereas a 16% decrease in stratospheric ozone would only increase the total UV at the Earth's surface by 1%, the damage to DNA would increase by nearly 50% and lower the viability of pollen, less protected against UV than other plant cells. Kelly, who has addressed some of these problems, points out that the known susceptibility of marine phytoplankton to UV may be less of a problem than this might imply, because lethal amounts of UV already penetrate the first 5 metres of ocean and a 20% increase in UV would increase this depth from 5 to 6 metres without major attenuation of photosynthetically active light.

OZONE
...sought after the world over

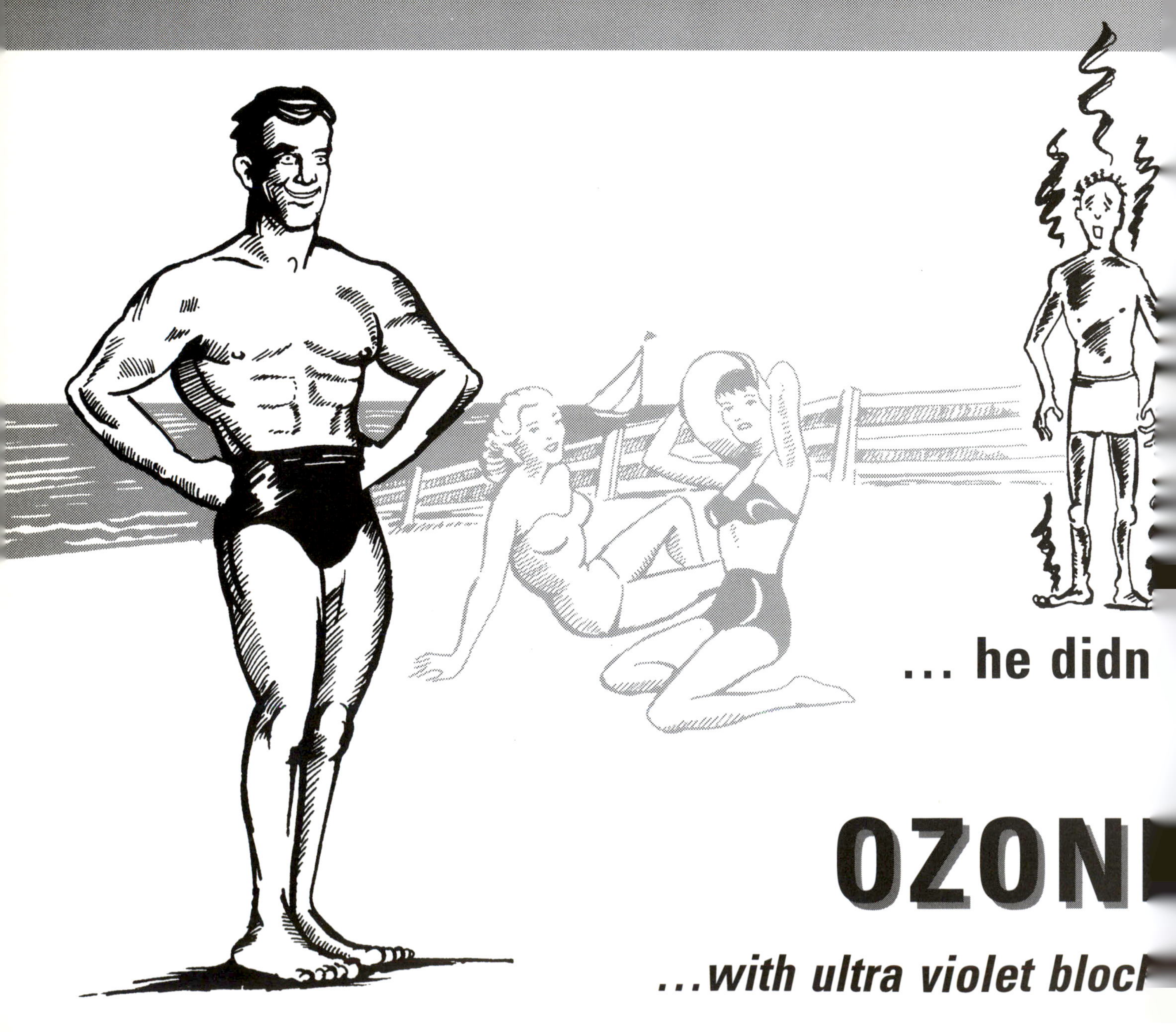

A somewhat more pessimistic note is sounded by Kirk,

" A further, perhaps even more serious, consequence arises from the possible link between oceanic photosynthesis and world climate. The phytoplankton of the ocean fix about 30 billion tonnes of carbon per year. This takes place in the upper, well-illuminated, mixed layer, but about 5 billion tonnes of this carbon, in the form of zooplankton faecal pellets and aggregations of senescent cells, sinks down through the thermocline to the deep sea. This loss, and the accompanying loss due to precipitation of calcium carbonate derived ultimately from weathering of silicate rocks, is made good by outgassing of CO_2 from the Earth in volcanoes and soda springs, and exchange of this atmospheric CO_2 with the surface layer of the sea. Normally a dynamic equilibrium between these processes exists, maintaining an approximately constant steady-state level of atmospheric CO_2. If, however, by destroying the ozone layer and increasing UV-B flux into the ocean, we substantially decrease oceanic photosynthesis, then we shall diminish one of the major sinks for atmospheric CO_2, and so in the long run may exacerbate atmospheric carbon dioxide increase and its associated global climatic effects".

In a way, ozone behaves a bit like clouds in the sky which reflect incoming solar radiation into space. In addition, (Table 7.2) acting in the same way as water vapour, *tropospheric* ozone (about 10% of the total ozone) absorbs long-wave emissions from Earth into space and re-emits part of what it absorbs back towards Earth. The lifetime of tropospheric ozone is relatively short but a 25% increase of ozone throughout the troposphere would contribute about 0.2 Wm^{-2} to greenhouse warming (c.f. Table 7.2).

[It is important to remember that heat and temperature are not the same. An ocean of water at 12°C contains a great deal more heat than a cupful at 90°C. Greenhouse warming tendencies (Table 7.2) can be calculated with some certainty but it does not necessarily follow (Section 7.9) that increases in greenhouse *warming* will give rise to an increase in surface *temperatures*. Changes in amounts of specific greenhouse gases may enhance or ameliorate positive and negative feedback mechanisms (for example, by affecting clouds and precipitation). The thermal inertia of the oceans also slows down the rate at which the temperature of the land surface and the atmosphere might otherwise rise.]

7.10 Oxides of Nitrogen

"Laughing gas from leaky pipes" ....Oliver C. Zafiriou

Stratospheric ozone is formed (equations 7.5 and 7.6) by ultraviolet irradiation of oxygen (the characteristic odour of ozone can be readily detected in the vicinity of U.V lamps). Most ozone at the Earth's surface is formed by industrial activity and motor vehicles (which combine, in some climatic circumstances to produce smog). Nitrous oxide (N_2O) is responsible for about 8% of the greenhouse effect. It also, like CFCs, contributes to the destruction of stratospheric ozone (accounting for 10-20% of ozone depletion). On a molecule for molecule basis it is 200 times more effective as a greenhouse gas than CO_2. It is presently increasing at 0.2% a year (about half as fast as CO_2) and resides in the atmosphere for nearly a century. Much of this increase derives from the extensive use of nitrogenous fertilizers, from burning and refining fossil fuels and from forest clearing. Although the use of nitrogenous fertilization does not constitute a major contribution to global warming it exemplifies the dilemma facing the Third World. In the European Community the continuing, largely unrestricted, use of nitrogenous fertilizers does little more than to increase agricultural surpluses, profits, pollution and political arguments. In the Indian subcontinent it has averted or diminished the inevitability of famine but, at the same time, it contributes to greenhouse warming many times over because of the utilisation of fossil fuels in its manufacture, the release of N_2O in its use, and the inevitable increases in anthropogenic related greenhouse gases associated with increases in population (from about one third of a billion in the last century to about a billion in this).

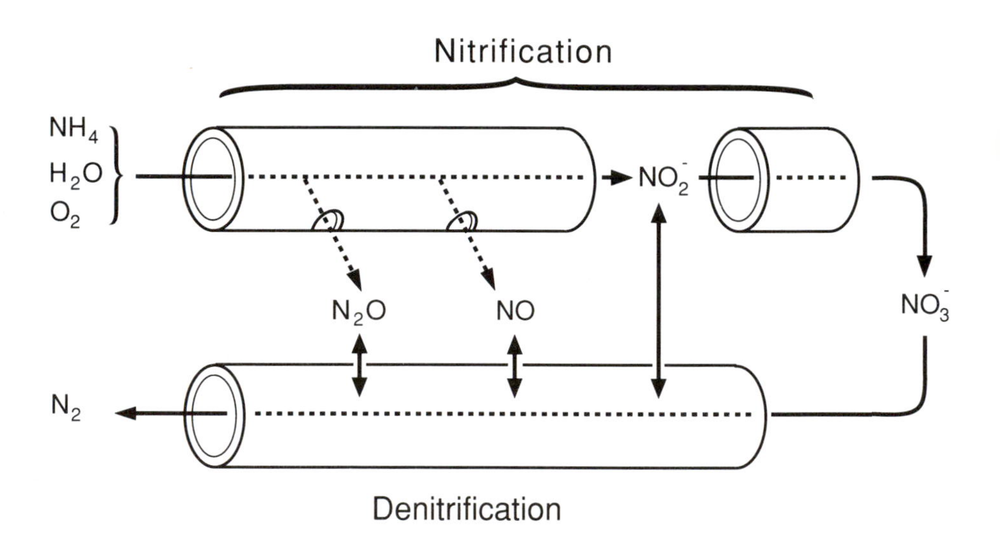

Fig. 7.16 Laughing gas from leaky pipes. This is a concept of Mary K. Firestone elaborated by Oliver C. Zafiriou. Three metabolically discrete types of bacteria are involved in nitrification and denitrification. Nitrification is a process in which ammonia (NH_4^+) is oxidised to nitrite (NO_2^-) and nitrate (NO_3^-). Denitrification consumes nitrate and releases N_2 back to the atmosphere. Here some of the processes involved are represented as pipes which permit multiple pathways through which nitrous oxide (N_2O, laughing gas) and other oxides of nitrogen can be taken up and released.

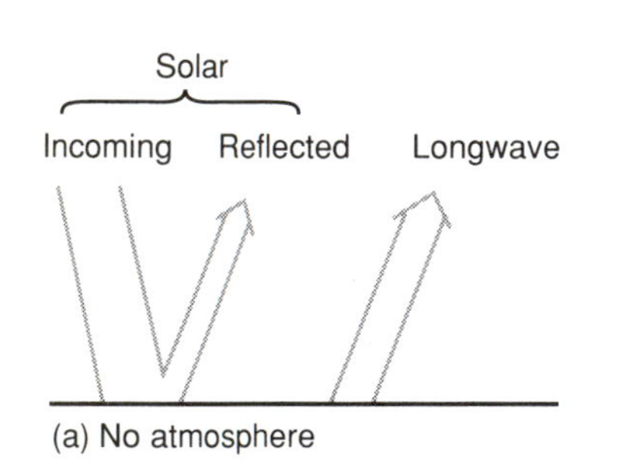

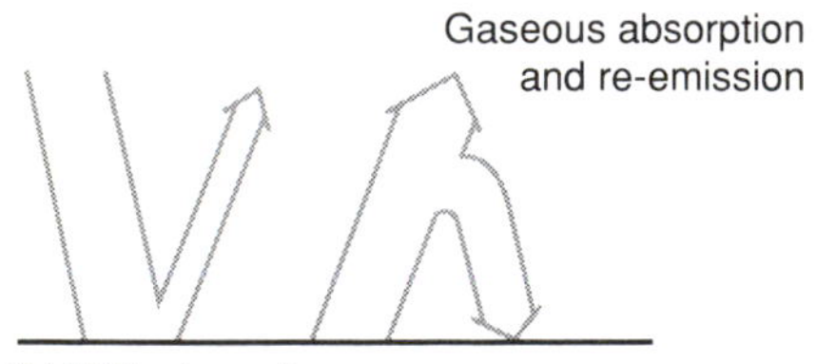

7.11 Summary

Greenhouse gases constitute a blanket which keeps us warm. Without them life on Earth, as we know it, would be impossible. The most important greenhouse gas and the one which Man influences least is water vapour but there are others such as carbon dioxide and methane (which are increasing steadily) and a layer of ozone (currently being torn apart) which are very dependent on human activity. By affecting the concentrations of these gases Man is now having an immense impact on his environment, some of the possible consequences of which are considered in Chapter 8.

Chapter 8

Consequences & Uncertainties

Through a glass darkly

People and rubbish

Malthus revisited

The missing sink

Catastrophe contemplated

CONSEQUENCES AND UNCERTAINTIES.

"One thing is certain, the rest is lies
The flower that once has blown, forever dies" ....Omar Khayam

8.1 What Might Happen

The consequences of unchecked emission of greenhouse gases can not be defined with precision. Nevertheless, despite the many uncertainties about the magnitude of the effect and the time-scale *"the direction of future change is clear and unambiguous"*. It is believed (says Irving Mintzer, of the World Resources Institute in Washington, D.C.) that, even if the lowest estimates are taken,

> *"today's activities are propelling the planet over a period of a few decades towards a climate that will be warmer than any experienced since the last glaciation. Unless actions are initiated soon, to slow the rate of greenhouse gas build-up, national economies and natural ecosystems are likely to suffer severe consequences from this rapid alteration of the global climate"*.

One of the few things that we can be reasonably certain about is that there will be no significant decrease in emissions of greenhouse gases in the immediate future. It is conceivable, but most unlikely, that current rates of emission might be markedly diminished by immediate and concerted political action like the Montreal Protocol, which seeks to put an end to the use of CFCs. On the other hand, as we shall see, there is massive uncertainty about what might happen (and when) if we continue to introduce greenhouse gases endlessly into the atmosphere at the going rate. Moreover, there is no firm evidence that the increases in carbon dioxide, which have undoubtedly occurred in the last century, have led to a *detectable* increase in global temperatures. There *seems* to be an increase in air and sea temperatures (Fig. 8.1) but the maximum predicted increases are small, past measurements often inaccurate, records poor, and fluctuations, brought about by other factors, very large.

WHAT MIGHT HAPPEN

It is, perhaps, unlikely that we shall see oxen roasted whole over bonfires on the frozen Thames next winter but there are many still living who can remember the

misery of the bitter British winter of 1947. Climatologists attach no particular significance to that winter or to the remarkably warm summer that residents of the United Kingdom experienced in 1989. In scientific parlance, the noise to signal ratio in the temperature record for the last 100 years is too large to permit unequivocal interpretation. Whether average temperatures increase or not, there will still be exceptionally hot summers and remarkably cold winters in the next century, as there were in the present and the last. So we are, in the absence of proof, back to prediction and at least, in regard to global warming, this can be done with reasonable confidence. *If* greenhouse gases continue to be emitted more or less as they are now, so that carbon dioxide (for example), doubles in the next 50 to 100 years, *average* temperatures may rise by between 2°C and 5°C. This is largely a matter of arithmetic based on the known physical properties of the gases concerned. The lack of precision about the actual temperature changes comes from making educated guesses about positive and negative feedback mechanisms, i.e. whether or not some possible changes will enhance or diminish the underlying trend. For example, higher temperatures can give rise to more evaporation and more clouds. Clouds intercept incoming solar radiation (causing cooling) and out-going long-wave radiation (causing warming). Precipitation, as rain, also releases latent heat of condensation. It is difficult to predict whether these conflicting feed-back mechanisms will reinforce or diminish greenhouse warming but current models seem to favour reinforcement. The uncertainty about the timing relates to the equally difficult task of relating emission to accumulation and to the almost total impossibility of arriving at *precise* estimates of fossil fuel utilisation, forest burning etc., in the next century.

If we accept the arithmetic, we are offered "good" scenarios (two degrees Celsius increases) and bad (five degrees Celsius increases) but if we attempt to look beyond increases in temperature to the consequences of such increases we may, if we are not careful, slide into the realms of science fiction. The truth of the matter is that ecosystems are even more complex than economic systems and it has so far been beyond the wit of governments to do a great deal about their national economies. In addition, we shall not simply see events unfold without political response. At present, global warming is regarded as matter for concern rather than of impending disaster. If perceptions swing towards the latter then there would be

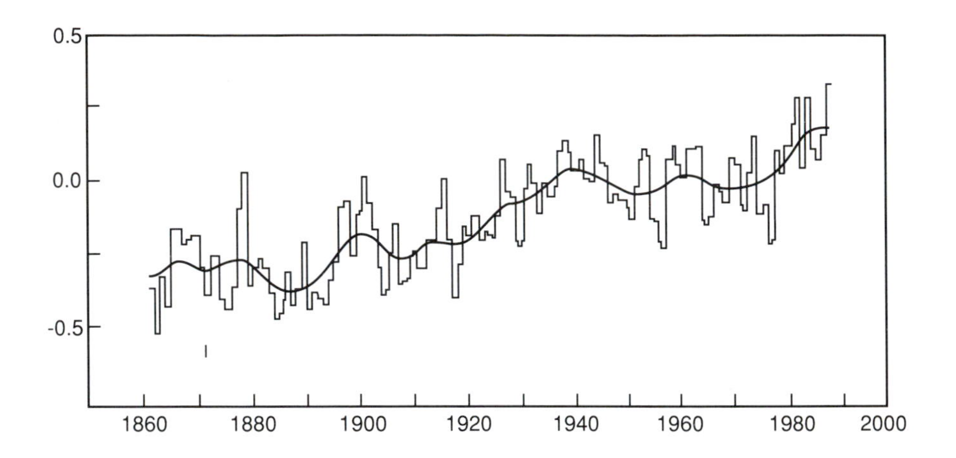

Fig. 8.1 Global Warming 1860-1990 Measured as deviations from the average between 1940-1960. A rising trend can be discerned despite major short term fluctuations but it should be noted that the total change of about 0.5 degrees Celsius is small and the possibility of error, particularly in old measurements, is large. (from CRU, East Anglia)

a correspondingly larger response by governments and industrialists and there is no doubt that even if some changes are inevitable their impact might be lessened if they come slowly rather than abruptly. As for the uncertainties themselves, they are manifold. The climate of the U.K., for example, is dictated more by ocean currents than by latitude. Winters in New York City are incomparably more severe than they are in London despite the fact that New York is much further south, at a latitude which the British would associate with winter sun in Spain. Should global warming deflect the course of the Gulf Stream away from the British Isles, the temperature of Britain could fall, despite overall global warming.

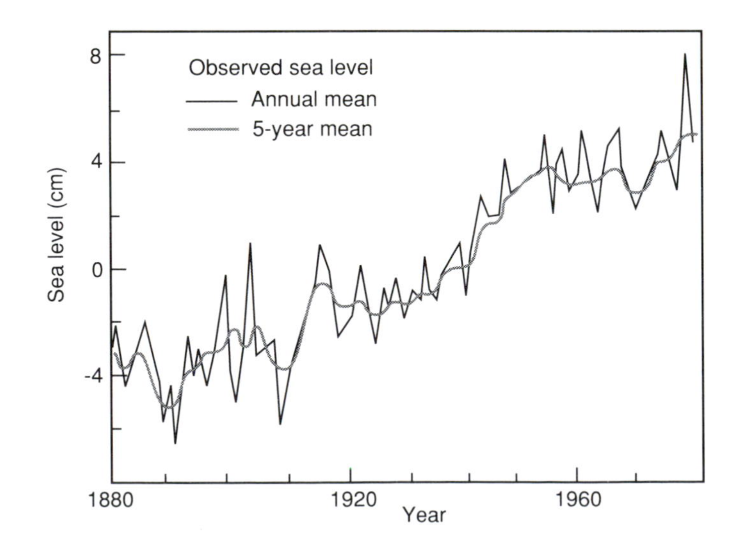

Fig. 8.2 Increases in sea level. The rising trend of global sea level appears to follow global warming (Fig. 8.1) but again the increase is small and the possibility of error is large. (from James Hansen/CISS)

Ice-out in Maine

SIR - The folks up in northern Maine have kept an accurate record of the time of ice-out on their largest lake, Moosehead Lake, for the past 142 years. In the spring, ice out comes suddenly, usually in one day. Moosehead Lake was the centre of the logging industry in the late nineteenth and early twentieth century, and ice-out in the spring meant fresh supplies by steam-boat to the isolated logging camps on the Northeast Carry. So ice-out, like Christmas, was celebrated, remembered and luckily recorded. More recently, ice-out contests have been sponsored by the Moosehead Lake Region Chamber of Commerce. Thus, an exact record of the date of this event each year since 1848 has been preserved. When the dates are plotted and the best fit line drawn, it appears that on average ice-out occurs about 10 days earlier now than in 1848. Perhaps these data will be of interest to those studying climate change, but up here in Maine we are going to take them seriously and start reducing greenhouse gases. Donald G. Comb

Frost Pond,
Maine, USA

NATURE - VOL 347 - 11 OCTOBER 1990

There are other uncertainties of much greater global importance. One is the arctic tundra. This vast wilderness is largely comprised of a few millimetres of vegetation overlying a metre or so of peat that has built up over millennia. Below the peat there is ice, in places 5,000 m deep, depressing the earths crust by its massive weight. This fascinating ecosystem is immense and extremely fragile. Drive across it, in a tracked vehicle, and the tracks do not mend as they would on grass or moorland or in forest. Instead, they persist and deepen. In twenty years time the tracks will have become a chain of lakes. If global warming were to tip this volatile system, from its present state into instability, the consequences could be incalculable.

8.2 Through a Glass Darkly

Inevitably, what we can learn from the past is limited. Some contemporary politicians are quick to claim that peace in the world (by which they mean lack of armed conflict between the "major powers") has been maintained for the last forty years by the threat of nuclear holocaust. Scientists, on the other hand, might question such a conclusion in the absence of the control experiment. Paleo-ecology, like history, is not open to experiment. We must distance ourselves from the concept that too much can be gleaned from climatic history by way of using warmer climates in the past as analogues for a future warmer world. Such "paleo-analogues" are *"beset with apparently insuperable problems"* and these problems are made immediately obvious if we are reminded, for example, that during the late Tertiary period (Table 8.1), some three to four million years ago, the Himalayas were about 3 km lower, the Atlantic and Pacific were connected at low latitude, and substantial evolution of organisms was still to occur. Accordingly, even though the late Tertiary was about as warm as the world might be expected to become in the next century, it hardly constitutes a credible analogue. Similarly, although the last interglacial period (125,000 years ago) and the mid post-glacial period (6,000 years ago) were also warmer, (at higher latitudes in the northern hemisphere) than today, they were periods of greater net solar radiation, distributed in an uneven, seasonal and latitudinal manner quite unlike the evenly distributed effects of greenhouse gases.

> *"However attractive the idea, paleoclimates cannot offer acceptable analogues for warmer climates of the near future".*

On the other hand, the most recent paleoclimates can provide valuable indications of the sensitivity of global climate to changes (other than changes in greenhouse gases) and these can contribute both to the validation of global circulation models and the mechanisms of climate change. In this context it is arresting to learn, or to be reminded, that the most rapid change in climate comparable to that predicted for the next century, which occurred 10 to 15 thousand years ago, was no more than 2.5°C warming *per millennia*. In other words, warming then was at least ten times *slower* than that envisaged for the next century. This brings us to an important conclusion. The paleo-ecological record shows that (within the last 12,000 years) forest trees, beetles, vertebrates and aquatic plants have migrated across the continents in response to deglaciation and subsequent rises in temperature during the post-glacial period. The most rapid rates of migration have occurred, during the last two millennia, at rates of hundreds of metres per year. In weighing this evidence it seems most unlikely that migration could occur at an order of magnitude faster than this. Moreover, any migration which could occur would be handicapped by barriers constituted by agricultural landscapes. Given the existence of these barriers it is difficult to envisage migration of woodlands at a rate of 20-30 m a month. Certainly migration of trees across Western Europe or N. America

GEOLOGICAL TIME SCALE

	Era	Period	Years before Present	Biological Features
PHANEROZOIC	**Cenozoic** Present to 63 million years ago	Quaternary	10,000-2 million	First true humans; *Homo erectus*
		Tertiary	2-10 million	First hominids; herbaceous plants more abundant
			10-25 million	
			40-60 million	
			60-63 million	First primates: prosimians
	Mesozoic 63 - 230 million years ago	Cretaceous	63-130 million	Angiosperms appear and become abundant; modern conifers arise Last dinosaurs
		Jurrassic	130-180 million	Gymnosperms, especially cycads, and ferns dominate First birds Giant reptiles reign
		Triassic	180-230 million	Forests of gymnosperms and ferns spread Primative mammals and early dinosaurs appear
	Paleozoic 230 - 600 million years ago	Permian	230-270 million	Conifers, cycads, and ginkgos appear; earlier types of forests wane
		Carboniferous	270-345 million	First gymnosperms Reptiles and insects appear
		Devonian	345-400 million	First large vascular plants Modern groups of algae and fungi appear First amphibians and bony fish
		Silurian	400-440 million	First vascular land plants First fish with jaws
		Ordovician	440-500 million	Plants begin to invade land First vertebrates: armoured fish without jaws
		Cambrian	500-600 million	First marine plants First shell-bearing animals
PRECAMBIAN	**Proterozoic** 600 million - 2.5 billion years ago		1000 million	First living things: blue green algae and bacteria; eukaryotic cells and multicellular oganisims appear
			1500 million	
			2000 million	
			2500 million	
	Archean 2.5 - 4.5 + billion years ago		3000 million	
			3500 million	Oldest dated rocks
			4000 million	
			4500 million	Formation of the Earth

at 200 or 300 metres a month is unthinkable. The implications are immense and they can be spelled out in some detail. Many ecosystems are likely to disappear and be replaced by quite different ones. Some species will face almost certain extinction. Conservation by Man will face entirely new challenges including artificial transport of species unable to migrate with sufficient rapidity to ensure survival. At the extremes there is the prospect of autocatalytic effects, the triggering of methane release:-

> *"initially from the peatlands of the tundras and by melting of the permafrost, and subsequently by decomposition of the vast deposits of methane hydrates".*

Such feed-back, if it occurred, would indeed spell global catastrophe.

8.3 Back in the Dark Days

> *"And God said "let there be light" and there was light"* ...Genesis

Throughout this book we have been pursuing a constant theme. By and large, Man is totally dependent on plants. Contemporary photosynthesis provides us with food and firewood; past photosynthesis with fossil fuels. Utilisation of fossil fuels on a large scale has changed the face of the world out of all recognition in a remarkably short time. Look at the New York skyline from the sea and contemplate what changes have been wrought in 300 years. Do the same from North Point (Sydney), shrinking the elapsed time from 300 to 200 years. Go on to Hong Kong and Singapore, marvelling and wondering all the while; marvelling at what Man can create out of nothing in next to no time, wondering where it will all end. Clearly, it cannot continue at this pace for long. The fossil fuels that have made it all possible will soon be gone, other resources are fast diminishing, a finite space for agriculture cannot support an infinitely large population. Pollution not only threatens habitats it already shortens our lives. We can't win, we can't break even and we can't stay out of the game.

Table 8.2

Chemical composition of the atmosphere

Gas	Volume concentration (%)	Molecular mass
Nitrogen	78.08	28.0
Oxygen	20.95	32.0
Argon	0.93	39.9
Carbon dioxide	0.03	44.0

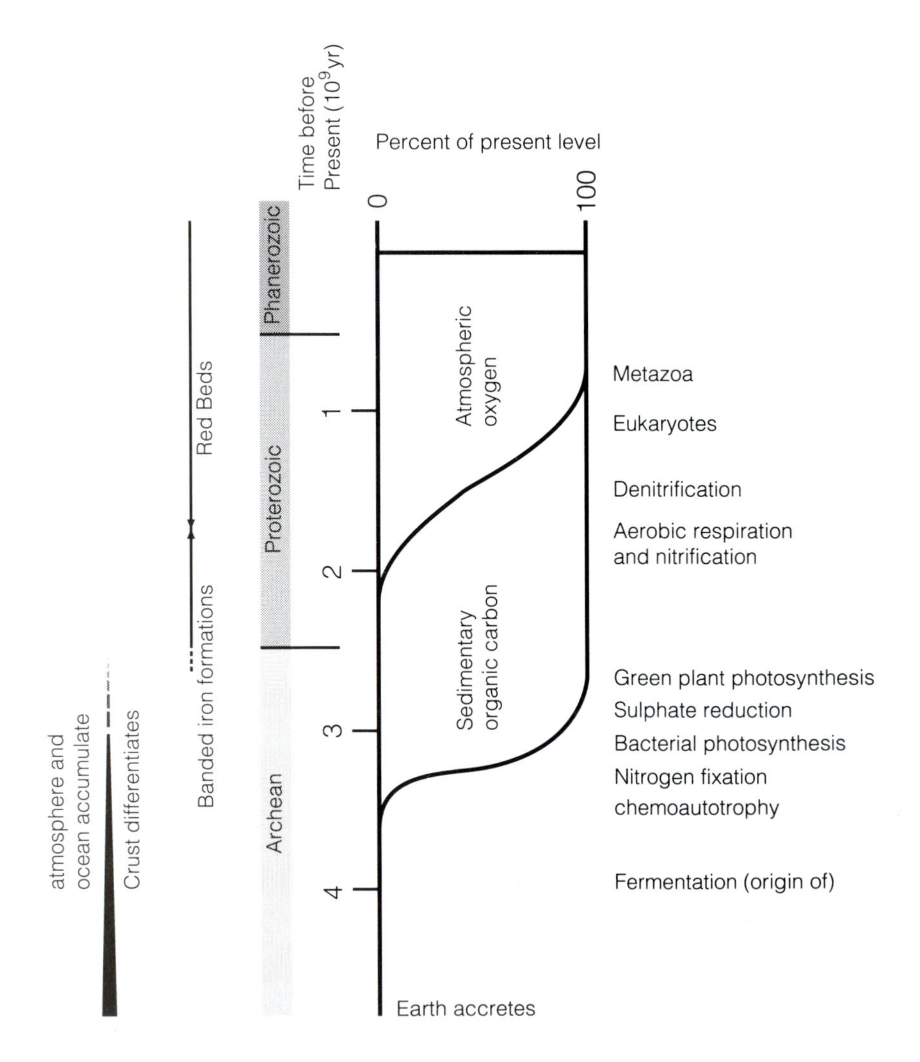

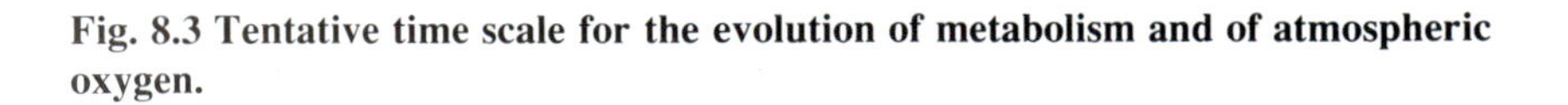
Fig. 8.3 Tentative time scale for the evolution of metabolism and of atmospheric oxygen.

ESTABLISHING A TIME SCALE

To put some of these things in proportion and in context we might just take a brief peep at the past to see where it all began. Present estimates put the formation of the Earth's crust at about 4,000 million years ago. Microbial life is very hard to date but it must have appeared well before the Cambrian Era, 570 million years ago, when fossil shells started to appear in the geological record (Table 8.1). The earliest "micro-fossils", which look very much like modern bacteria and blue green algae, are now thought to have originated about 3,500 million years ago. By the Carboniferous Period, which began about 350 million years ago, large trees (like *Lepidodendron,* long since extinct) were piling up the coal deposits which have

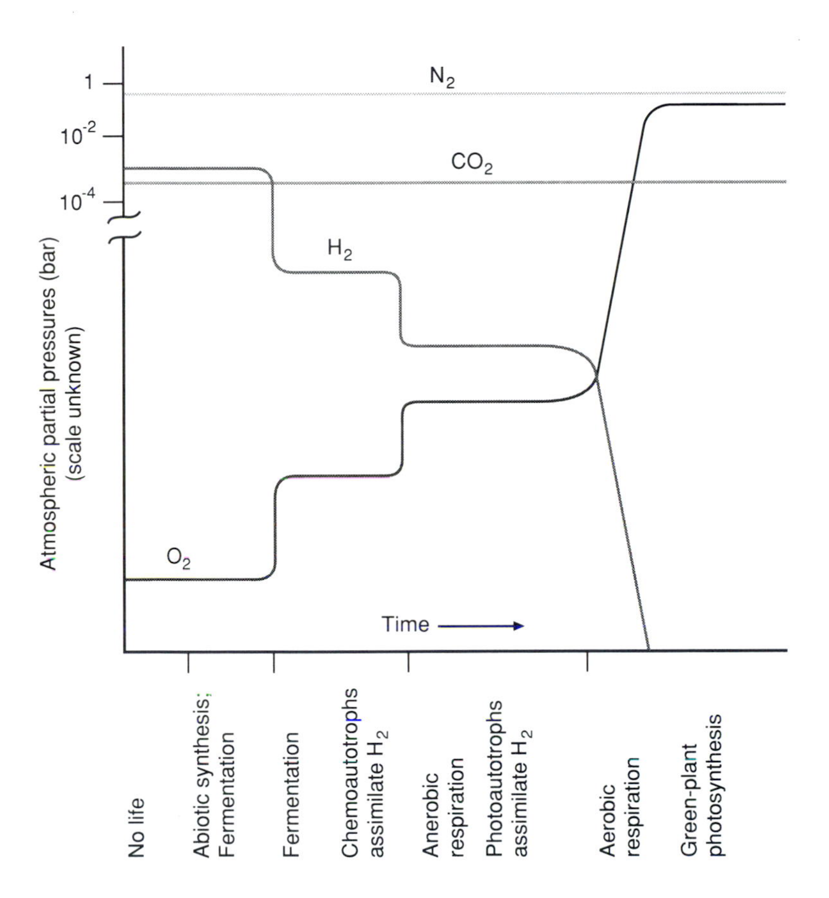

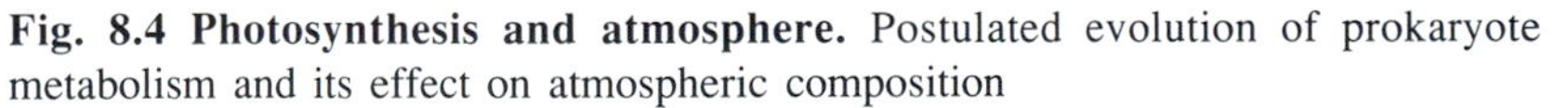

Fig. 8.4 Photosynthesis and atmosphere. Postulated evolution of prokaryote metabolism and its effect on atmospheric composition

played such a key role in Man's recent history. Only a handful of present day plants like the horse tails (species of *Equisetum*) and the maiden hair tree (*Ginkgo biloba*) have survived to the present day. Reptiles and insects abounded but mammals and dinosaurs had yet to appear (180-200 million years ago). Flowering plants and modern conifers did not arise until the dinosaurs were on their way out (63-130 million years ago). The first primate appeared 60-63 million years ago and the first true humans only in the last 2 million years or so. Millions of years are hard to envisage for organisms with the classic life expectancy of three score years and ten, particularly when there are many occasions when an hour can seem like an eternity, so here is another way of establishing a time scale.

> *"Imagine that all of time, from the moment when the earth's crust first formed right down to this moment, is compressed into one calendar year. Here is what geologic time looks like on this calendar. The earth's crust begins to form on January 1; the oldest rocks known date from about the middle of*

March. In May the first living things appear in the sea. By late November land plants and animals have appeared, and in early December, the vast swamps that formed the Pennsylvanian coal deposits have appeared, flourished, and disappeared after four days. By mid-December dinosaurs have begun to roam the earth; by December 26, just as the Rocky Mountains begin to rise up in the American West, they become extinct. In the evening of December 31, the first humanlike creatures have begun to appear. One minute and fifteen seconds before midnight on December 31, the glaciers of the last great Ice Age begin to retreat in America and Northern Europe. The Roman Empire rises, flourishes for five seconds, and disappears in the waning seconds of our year. Three seconds before midnight, Columbus discovers America." ...Laetsch

What is the relevance of all of this to our main theme? First of all there is the obvious point which should give us all pause for thought. On a geological time scale, Man is a very recent event. We have been around, if you like, for only a few hours on December the 31st. Even then it is only in the last second that there has been the explosion in number which has taken population from a few million to 3,000 million now and probably over 4,000 million by the end of the century. That, and all that goes with it, has only been made possible by events which took about 4-5 days early in December (i.e., 75 millions years or so in the Carboniferous Period).

Secondly there is the impact of photosynthesis on the atmosphere. Opinion about how much carbon dioxide and oxygen there was in the atmosphere when life started is divided and based on inference. The atmosphere itself is believed to have

"Free living chloroplasts"

resulted form "degassing" (a release of gases from the earth's interior) a process which continues today from volcanoes which release a great deal of water and carbon dioxide into the atmosphere. Most of this may have occurred within 100 million years of the earth's formation. It has been suggested, by various authors, that:-

(a) *"the water vapour and carbon dioxide contents of the primitive atmosphere were about equal to their present values after an initial accumulation period. Once water had started to condense and carbonate rocks to precipitate from the oceans, continued degassing led to an increase in the nitrogen content of the atmosphere, but caused no change in the carbon dioxide or water vapour contents."*

(b) *"that the carbon dioxide content of the atmosphere has never differed markedly from present-day conditions"*

(c) *"that the ocean carbon dioxide content has probably been close to constant for the last 3 billion years."*

but if impressions of constancy can be inferred from such statements they are not universally agreed (Section 8.4). There is no suggestion of course, that oxygen has remained constant since the "beginning". Some oxygen would have been formed by photo-dissociation of water but is not known whether this was as little as 0.001% of present atmospheric levels (PAL) or as high as 0.25 PAL. What does seem clear is that major changes commenced some 2,000 million years ago with the evolution of oxygenic photosynthesis (i.e. oxygen producing photosynthesis) by organisms capable of bringing about the photolysis of water. The organisms primarily responsible were blue-green algae (the cyanobacteria) and it was almost certainly the blue greens and other marine algae which increased atmospheric oxygen to something like its present levels.

[Cyanobacteria are, in a sense, like free-living chloroplasts. They contain chlorophyll within internal membranes just as the photosynthetic pigments in

"Free living chloroplasts"

chloroplasts are restricted to the thylakoid membranes. It is even conceivable that green plant chloroplasts evolved from these organisms. Currently, one group of workers finds that the closest relative of a chloroplast ancestor could be the protist *Cyanophora* whereas another puts their money on the *prochlorophytes* such as *Prochlorothrix*.]

The increase in oxygen during the evolution of the atmosphere may have been complete before evolving life colonised the land masses from the oceans. Oxygen, of course, is required for respiration but the crucial aspect of increasing partial pressures of oxygen in the atmosphere was a concomitant accumulation of ozone in the stratosphere. Ozone is formed in the stratosphere, and above, by the photo-dissociation of oxygen

$$O_2 + h\nu \rightarrow O + O \qquad \text{Eqn. 8.1}$$

followed by recombination of the atomic oxygen (O), so produced, with molecular oxygen (O_2)

$$O_2 + O \rightarrow O_3 \qquad \text{Eqn. 8.2}$$

Ozone (O_3) absorbs ultraviolet radiation in the 200-300 nm range. DNA (deoxyribosenucleic acid, the physical bases of heredity) is destroyed by UV radiation (Fig. 8.5) at about 250 nm and

> *" contemporary micro-organisms would be killed in a matter of seconds if exposed to the full intensity of solar radiation in this wavelength region".*

Because of the susceptibility of DNA to degradation by ultra-violet light evolution must have gone hand in hand with oxygen evolution and ozone formation. The oceanic cyanobacteria would have been afforded a measure of protection by water (and perhaps by a degree of shielding by organic molecules and DNA repair mechanisms) but there would seem to have been little possibility of invasion of the land until the ozone layer was in place.

Today, it is commonplace to visualise the green plants that we see around us as the source of our oxygen and it is true that oxygen consumption by animals, *plants* and micro-organisms is in balance with oxygen produced by photosynthesis. These, however, were not the plants which created the oxygen in our atmosphere, nor were those ancient species, now extinct, which formed our coal deposits in the Carboniferous period. All, or most, of this was done before they existed by the cyanobacteria and other forms of marine phytoplankton. In all of this, the role of carbon dioxide, released by degassing is uncertain. Some authorities (Section 8.4) think that carbon dioxide may once have been much higher than today (perhaps 100 times higher) and that this would have contributed to surface warming during the early period of Earth's history when solar luminosity was believed to have been 30% lower than at present. Because Earth is nearer to the sun than Mars, and further from the sun than Venus, it avoided both the lack of accumulation of water vapour in the atmosphere on Mars (which has left that planet frozen) or the "runaway greenhouse effect" (which has kept virtually all of the water vapour in the atmosphere on Venus and resulted in surface temperatures much higher than Earth). It should also be remembered that whether or not ancient atmospheres once contained 3% carbon dioxide or about 0.03% (as they do now) carbon dioxide must have been released from reservoirs, such as the oceans to permit this conversion of carbon dioxide to oxygen on a molecule for molecule basis

$$CO_2 + H_2O \rightarrow CH_2O + O_2 \qquad \text{Eqn. 8.3}$$

The total organic carbon content of the biosphere is about twenty times greater than that in the atmosphere (approximately 3% in living organisms, 25% in dead organic matter and 70% as coal and oil). This is still about 30 times less than would have been needed to account for 21% in the atmosphere. Moreover, oxygen is consumed in the oxidation of reducing materials when sedimentary rocks become exposed to erosion and weathering. Some of the "missing" organic carbon is dispersed in sedimentary rocks and it is also possible that the Earth's crust is not isolated from the underlying mantle and that significant quantities of material have been exchanged between the two throughout geological time. Comparisons of the isotopic content of the atmosphere and sedimentary rocks suggests that most of the

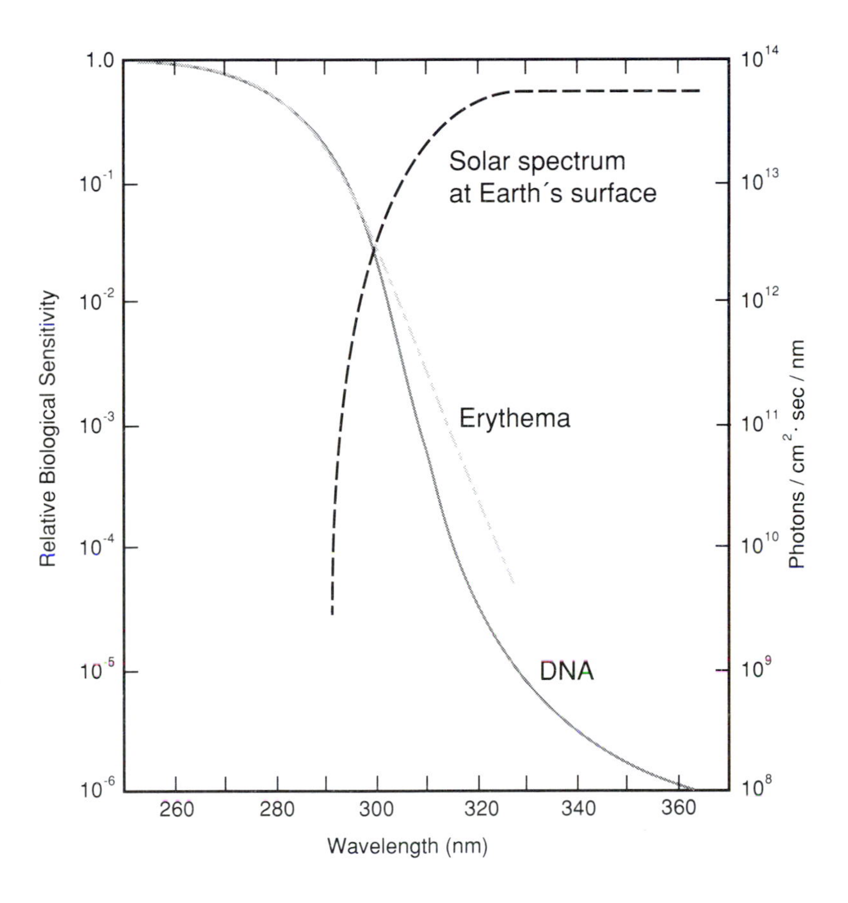

Fig. 8.5 Effects of ultraviolet light on DNA. The solid curve is the biological sensitivity spectrum for DNA, and the dashed curve is a typical Earth-surface solar UV spectrum. The response of human skin to sunlight ("sunburn") is shown as the curve labelled erythema.

organic carbon of biogenic origin in the latter had already accumulated there some 3000 million years ago. This is in accord with the fossil record because it would have given the cyanobacteria (etc.) about 500 million years to bring this about. Even so the oxygen content of the atmosphere did not arise to something like its present level until about 2,000 million years ago, presumably because photosynthetically produced oxygen was first used to oxidise iron, etc., in sea water.

[It is of interest, that paleoclimatogolists and atmosphere physicists have usually and remarkably tended to put more emphasis on oxygen than on carbon dioxide. Thus *"We shall concentrate on oxygen as the most interesting of the atmospheric gases, at least as far as life is concerned"*.]

Of course, life as we normally think of it, depends on the presence of oxygen and we are all aware of the physiological impact of thinning oxygen at high altitudes, the need to pressurise high-flying aircraft and possibly even the advisability of planning a honeymoon somewhere other than in the high Andes. Nevertheless, when we come to consider the sort of fluctuation which may have occurred during the Phanerozoic (changes in oxygen from about 0.25 to 1.5 present levels and tenfold changes in carbon dioxide) we can scarcely discount the importance of the changes in carbon dioxide which would, in otherwise stable environments, have profound effects on present day organisms, quite apart from the indirect effects of carbon dioxide on global temperature.

8.4 The Yo-Yo Concept

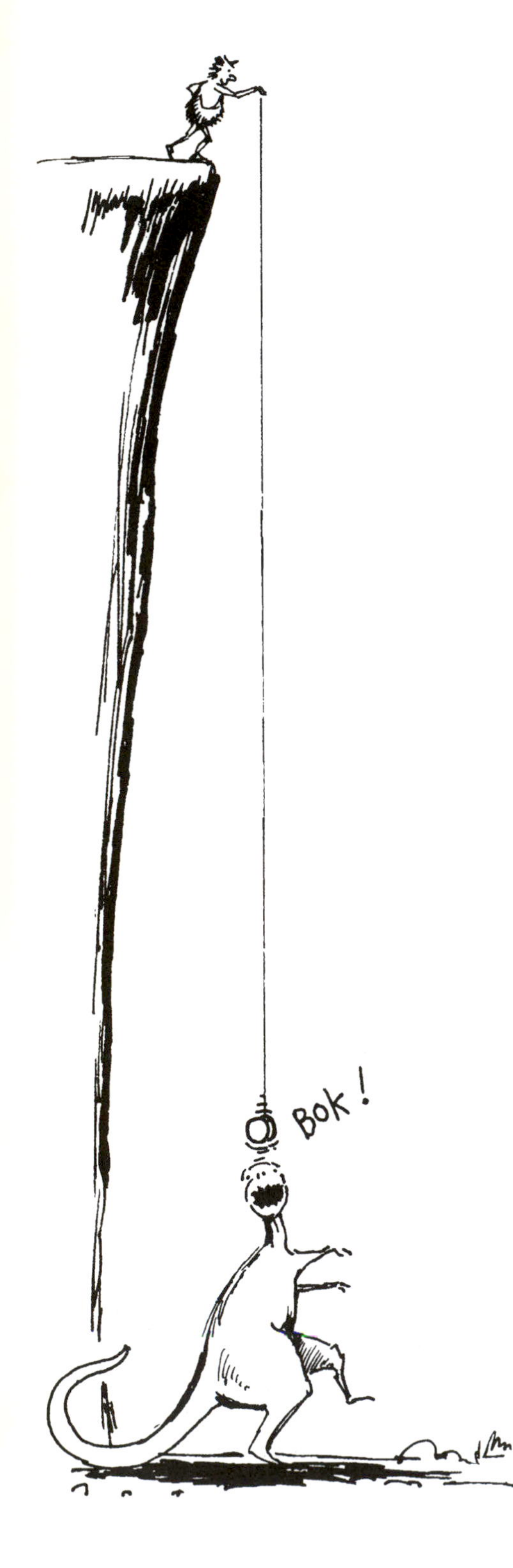

> " ***yo-yo,*** *n. Toy (introduced in 1932) consisting of pair of discs with deep groove between them in which string is attached and wound, rewinding itself and returning to the hand when thrown down.* ....Concise Oxford Dictionary.

What it does not say in the dictionary is that a yo-yo is supposed to go up and down most of the time. Reading Section 8.3, or some texts about the atmosphere, you might be left with the feeling that, after the first 100 million years or so of its existence, the Earth's atmosphere remained largely unchanged until the onset of photosynthesis and thereafter the oxygen rose smartly from next to nothing (to something like 21%) and stayed there, unchanged, until the present time. Some authors and research workers take quite a different view however, and Fig. 8.6 and 8.7 are distinctly yo-yo like in appearance, showing major oscillations in oxygen, methane and carbon dioxide. Sadly, *this* author lacks the competence to tell his readers which climatologists to believe but the possibility that the atmosphere has never been stable is interesting and certainly one which is relevant to contemporary change.

Figure 8.6 comes from a book by the Russian scientists, Budyko, Ronov and Yanshin. It shows massive oscillations in oxygen and carbon dioxide during a period in which there were correspondingly large changes in living organisms. These authors consider the possibility of a cause and effect relationship and pay homage to the French scientist, Geoffrey St Hilaire, who wrote (in 1883)

> *"Let us suppose that, in the course of a slow and gradual advancement of time, the proportions of different components of the atmosphere changed and it was an absolutely indisputable result that the animal world was affected by these changes"*

Similar thinking is implicit in proposals by McLean that the dinosaurs became extinct because of an enhanced carbon dioxide greenhouse effect (Sections 7.2 & 7.3) causing a sudden increase in global temperature, melting of the ice caps etc. Precisely *how* this might have affected the dinosaurs or the other genera which became extinct at this time (possibly as many as 75% of all of those in the Upper Cretaceous which became extinct - Table 8.1) is highly uncertain but there is little doubt that, either an abrupt warming, or an injection of carbon dioxide, would have had a profound effect on ecological stability. Because the dinosaurs were so spectacular, it is hardly surprising that their demise has attracted so much speculation but there were changes to plants, round about this time, which were no less dramatic. These were to have a much greater impact on Man who, while he was never to come face to face with a living dinosaur (except in Conan Doyle's "Lost World"), has always been dependent on the flowering plants. These flowering plants came in (give or take a few tens of millions of years) as the dinosaurs went out. Moreover, amongst these plants, C4 species may well have arisen as a response to *falling* carbon dioxide concentrations. As discussed in Section 5.10/5.11, C4 species have a built-in, carbon dioxide enriched greenhouse. The C4 syndrome (Section 5.11) may well have evolved more than once since the Cretaceous but the conventional wisdom is to perceive it as a response to *low* carbon dioxide concentrations. According to this view RUBISCO (ribulose bisphosphate carboxylase, Section 5.2) is an enzyme which evolved when carbon dioxide was more plentiful and oxygen less plentiful than at present. In such circumstances it did a better job than now, untroubled (in its work of carboxylation) by competing oxygenation at present levels. As the carbon

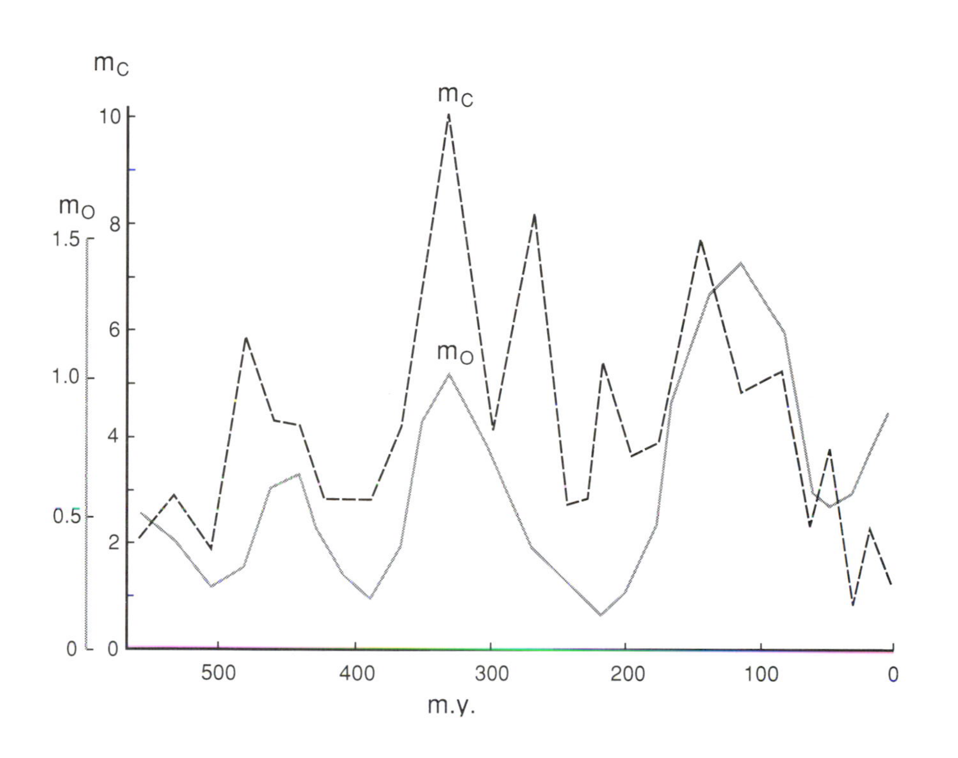

Fig. 8.6 The Yo-Yo in action. Changes in relative mass of carbon dioxide (mc) and oxygen (mo) during the Phanerozoic (After Budyko, Ronov and Yanshin).

dioxide fell and as oxygen rose, the task of RUBISCO became harder and harder. The C4 syndrome is seen as an evolutionary response to this situation as internal carbon dioxide-enrichment re-established, in some respects, a working atmosphere nearer to that enjoyed by the first photosynthesising organisms. Such propositions are not, of course, open to verification but reference to Fig. 8.6 (itself no more certain) gives credence to this concept in the sense that, at about the right time, it looks as though there could have been relatively abrupt decreases in carbon dioxide from concentrations an order of magnitude higher than those which exist now.

At the present time, mankind is concerned about much smaller *increases* in carbon dioxide and the consequences that these might have on his environment. Though much smaller, the rate of change is nevertheless very fast - so fast, indeed, that there is no question of an evolutionary response and real doubt that even migration might be too slow (Section 8.2) to keep pace.

Although there are major differences in scale between present and past changes, both in regard to carbon dioxide concentrations and time, contemporary changes still suffice to remind us how much life depends on climate and how much climate depends on carbon dioxide. It was Arrhenius who, in 1908, came to the conclusion that the oceans would be unable to absorb all of the carbon dioxide emitted by industrial processes as fast as it was produced and that, in consequence, global temperatures would rise and photosynthesis would flourish.

[In an early paper, entitled *"On the influence of carbonic acid in the air on the temperature of the ground"* (Philos. Mag 41, 237-276, 1896) Arrhenius credits Fourier with the first recognition, in 1827, of the possibility that rising carbon dioxide might bring increasing temperatures in its wake.]

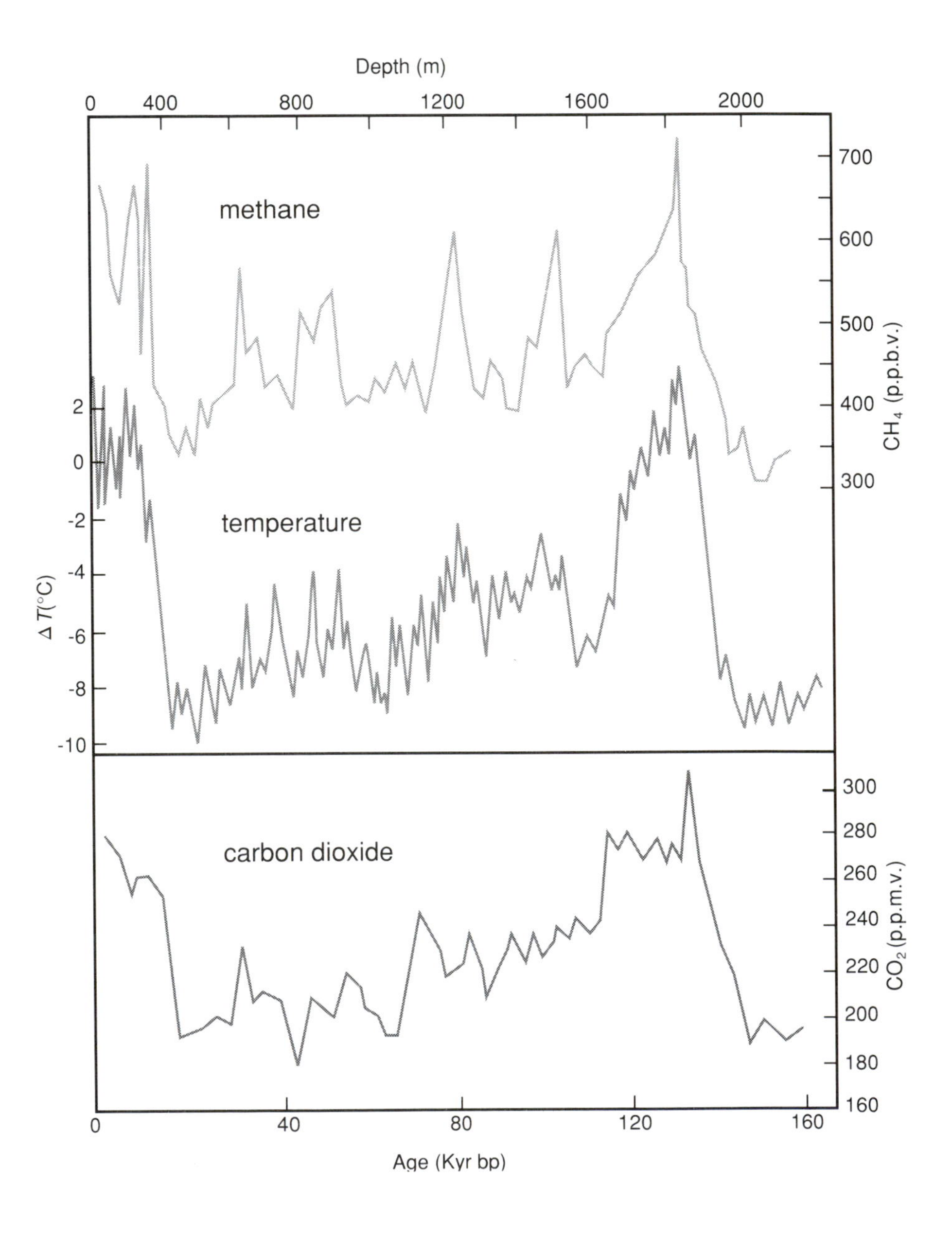

Fig. 8.7 Greenhouse gases and surface temperature. Changes in methane (top), temperature (middle) and carbon dioxide (bottom) over the last 160,000 years. Based on Vostock ice-core records. [After Chappellaz, Barnola, Raynaud, Korotkevich and Lorius (1990)]. An age of (e.g.) 80K yr BP means 80,000 years **before** the **present** time.

CLIMATE FORCING BY CO_2 AND METHANE

In 1938 Callender also warned of the possibility of climatic warming as a result of burning fossil fuels but, perhaps because of the imminence of world war, this warning also went unheeded and it is only in the last decade that the consequences of anthropogenic increases in carbon dioxide have been taken more seriously.

Figure 8.7 shows the Yo-Yo working like crazy in the relatively recent past (the last 160,000 years). It illustrates fluctuations in methane, carbon dioxide and temperature. The latter is based on deuterium content which is regarded as a good indicator of past temperatures in central Antarctica. It depends on the principle that heavy molecules of water (i.e. those containing deuterium) form snow more readily than those containing hydrogen alone. In this way the amount of "heavy hydrogen" (deuterium) in snow reflects the temperature at which it was formed.

Figure 8.7 is the work of Chappellaz *et al.* on the Vostock ice core (c.f. Fig. 8.1) and, although its important conclusions cannot be considered here at length, we may note immediately that two main glacial/interglacial transitions (at 150 - 133 Kiloyears before present and 18 - 9 Kyr BP) occurred during this period and, in both cases, the methane content rose from the lowest to the highest values recorded. Carbon dioxide taken from the same ice core behaved in the same way i.e. the highest and lowest values for both carbon dioxide and methane coincided with the warmest and coldest temperatures respectively, although there were important differences in profile which indicate that the carbon dioxide changes might have been driven by oceanic changes and the changes in methane by factors affecting continental sources. So what are we to make of this, particularly when we are aware that contemporary methane is already much higher, and increasing much faster (and exponentially), than before? Chappellaz *et al.* conclude that the methane (in the absence of an anthropogenic contribution) rose when the ice melted and it was released from wet-lands, methane hydrates etc. Thus the methane *followed* the temperature changes rather than causing them. On the other hand they point out the probability of positive feedback mechanisms, including the formation of *stratospheric* water-vapour from oxidation of methane (Eqn. 7.4).

> *"It has been estimated that a doubling of stratospheric H_2O from 3 to 6 ppm (by mass) results in a warming of 0.6°C at the Earth's surface".*

Together, the direct radiative effect of methane and stratospheric H_2O formation could have led to a warming of 0.15°C (about one third of the direct carbon dioxide effect).

> *"The combined climate forcing of carbon dioxide and methane (about 2.3°C) can therefore explain ~50% (~10% for methane alone) of the glacial interglacial warming!"*

There is no implication here that methane, (or for that matter carbon dioxide), initiated climate change, simply that once initiated for whatever reason, greenhouse gases can (and have) reinforced it. It also behoves us to believe (or at least not to dismiss) the possibility that the atmosphere may have been no less stable in the more distant past than relatively recently. However, when we come back to the present and we start to think once again in tens, rather than in thousands, of years it may not matter, all that much, whether the chicken comes before the egg or *vice versa*. We are currently engaged in a vast experiment in which we are set to double atmospheric carbon dioxide within the life-time of our children. We have not embarked on this because we are driven by scientific curiosity but by simple greed and ignorance.

Ignorance because, when the internal combustion engine and electricity generation from fossil fuels first came on the scene, there seemed to be no good reason for not using fossil fuels as fast as we could procure them (c.f. Section 9.2). Even twenty years ago, there was more concern about what we all might do when the oil had gone than for the consequences of burning it in so short a time. The consequences of greed and selfishness are self-evident. I am no more willing than the next to forgo the independence afforded by my car, to miss out on the pleasures of a full stomach, a room that is neither too hot nor too cold. I wish to be able to travel swiftly, and in comparative comfort, to Australia if it so suits me. I am very conscious, however, that I do these things at some cost to my environment. Yet there are those (Section 9.5) who write, persuasively, that we *can* have our cake and eat it. Perhaps they are wrong but one thing that is certain is that if the vast "cedar" forests of Australia had not been felled to the last tree by the early settlers, these forests would still be providing timber in large quantities. History is full of similar stories of plundered resources and species driven to near, or complete, extinction (including the genocide of Tasmanian man). Given a choice between conservation and terminal exploitation there should hardly be need for prolonged debate. Add the bare *possibility* that the consequences of exploitation could be incalculably nasty, even to politicians, and it is hard to see why we should hesitate for a moment to endorse conservation.

8.5 Will it Happen?

In the words of the late lamented Professor Joad, it all depends what you mean by "it". For the sake of discussion we can consider a number of possibilities which, although we can look at them separately are, in reality, inextricably linked. Firstly there is the likelihood of continuing increases in greenhouse gases, secondly the likelihood of global warming consequent upon the decreased emission of infra-red radiation from earth to space and finally the possibility that global warming could result in real, or near catastrophe.

"Life wasn't meant to be easy." ....English proverb

"Nichts ist schwerer zu ertragen als eine Reihe von guten Tagen" ...German proverb

The Mauna Loa record (Fig. 7.4) shows that carbon dioxide continues to rise inexorably and present extrapolation indicates doubling in the latter part of the next century. There is little doubt that carbon dioxide will continue to be emitted, at something like its present rate, in the immediate future. On the other hand, there is no reason why it *should* continue to be released indefinitely at present rates (Chapter 7) except for the fact that the history of human behaviour in this century gives us no encouragement to believe that there will be the international political accord, and determination, to put in train those measures which would slow carbon dioxide emissions to a manageable rate. Whether or not carbon dioxide will continue to *accumulate* in the atmosphere at present rates is quite another matter and one which will be influenced (indirectly as well as directly), by the rates of emission and the climatic consequences of emission.

CFCs (Section 7.7) are, in this context, quite remarkable because it is not as though refrigeration did not exist before their invention and will not continue to exist if the Montreal Protocol really succeeds in ending their use. For this reason it is not entirely improper, as is sometimes suggested, for the West to attempt to persuade "underdeveloped" countries not to use them. China is a nuclear power and has a successful commercial satellite industry. With a population of 1,000 million its brightest and best brains are unlikely to be inferior in quality or in number to those in Europe or America. At the same time it has recently elected to build 100 refrigerator factories (60% or more of Beijing households now have refrigerators) but based on such inadequate technology that their CFCs pose a problem (as does the generation of sufficient electricity to operate them). This is a problem of social mis-management and muddled thinking, rather than Western conservationists seeking to deprive the Third World of artefacts that they themselves enjoy. It should also be noted, in passing, that domestic refrigerators cause little problem while they actually function and would, in theory at least, cause none if ways could be found to safely decommission them. Conversely, a major contemporary concern relates to the production of frozen foods and inadequate technology which allows the more or less continuous release of CFCs, by leakage, during the operation of factories concerned in their production.

Methane will almost certainly continue to rise at its present rate of increase because much of it is so clearly anthropogenic in origin and mankind is unlikely to eat less rice or methane-producing ruminants or mine less coal (Section 7.6) in the immediate future. Moreover, methane is an increasingly popular fuel and some will continue to be lost on its way to fires and furnaces. When finally burnt it will also give rise to carbon dioxide.

8.6 Down to the Sea

"I must down to the seas again, to the lonely sea and the sky" ... John Masefield

Carbon dioxide dissolves readily in water and can, according to the pH of the solution and the presence of various other ions, enter into complex equilibria between carbonic acid, bicarbonate and carbonate. Some carbonates (e.g. calcium carbonate) are very insoluble, at least in waters of neutral and alkaline pH and are also used in skeletal structures by a variety of marine organisms. Carbon dioxide is also taken up, photosynthetically, by marine plants including microalgae which account for a major fraction of global carbon dioxide fixation. The entire system is immensely complex and still very poorly understood but some broad generalisations can be made. The equilibrium, between newly generated atmospheric carbon dioxide and oceanic surface carbon dioxide, takes less than a year to establish; that between the surface waters (the top 70m) and the deep sea, about 1,000 years. The distribution of carbon is thought to approximate to that in Table 8.3. This does not mean, however, that carbon dioxide released from oxidation of fossil fuels will distribute itself in these proportions. The equilibria between carbon dioxide, bicarbonate, etc., are such that only 9% of atmospheric carbon dioxide will enter the surface layers of the oceans, the remaining 91% will remain in the air.

> *"in the absence of transfer to the deep sea, the ocean does not constitute a significant sink for fossil fuel carbon dioxide".*

Transfer to deep waters does, of course, occur but the rate at which it occurs is still very uncertain. On the basis of necessarily uncertain assumption it is believed that about half of all of the man-made carbon dioxide now resides in the ocean. This largely neglects the role of sedimentary marine phytoplankton and other marine organisms which have directly or indirectly ingested photosynthetically fixed carbon dioxide. As a result of death and decay of these organisms, soluble organic carbon will be released, insoluble inorganic carbon will sediment. It should be borne in mind, however, that (on a much larger timescale) sediments also affect the complexities of $CO_2 \rightarrow$ bicarbonate $\rightarrow$ carbonate equilibria. Dissolved carbon dioxide penetrating deep waters should, within 100-200 years, decrease the pH and bring largely insoluble carbonates into solution. This, in turn, would *increase* carbon dioxide solubility. The rates at which this might occur are highly uncertain. Sedimentary carbonates, for example, are often *"surrounded by a matrix of silicate detritus"*. In all of these matters, and in others discussed in this chapter, we can begin to understand what infinitely complicated systems are involved and how large are the degrees of uncertainty about our influence upon our environment.

Recent ocean models from the Max Planck Institute in Hamburg, based on tritium measurements as indicators of the rapidity of mixing of surface and deep waters, suggest that, if future carbon dioxide emissions could be abated, so that they did not exceed present rates of output, uptake by the oceans could delay carbon dioxide doubling for two or three centuries. It can also be argued that if all fossil fuels reserves now known to be in existence were burned they would not double the carbon dioxide content of the atmosphere over this period assuming that, as now, only 50% of what is released would remain in the atmosphere. Of course, the definition of a feasible reserve (i.e. one which could be recovered with due economy and without unacceptable environmental impact) might easily change in a world increasingly desperate for fuel but these arguments do at least underline the fact that

Table 8.3

Distribution of Carbon dioxide between atmosphere, biosphere and ocean

Atmosphere	1.5%
Oceans	94.5%
Warm surface	1.5%
Deep Ocean	70.0%
Biosphere	4.0%

even current predictions about carbon dioxide doubling cannot go entirely unchallenged. It must not be forgotten that, regardless of what happens to carbon dioxide, CFCs will continue to pose a threat and that methane released by Man's activities has trebled in this century. Nevertheless even the apparently remorseless rise in carbon dioxide, which will remain a major concern, is still not so inevitable, nor its rate of accumulation so soundly based on fact, that there is no room for doubt or possibility of amelioration (Chapter 9).

8.7 The Missing Sink

"Error of opinion may be tolerated where reason is left free to contradict it" .....Thomas Jefferson

One of the many uncertainties about the greenhouse effect, perhaps the major uncertainty, concerns the missing carbon dioxide. According to different estimates, anything from 30-50% of the carbon dioxide released into the atmosphere simply gets lost or, in other words, cannot be accounted for. Since most of the surface of the globe is covered by oceans it is a good bet that this is where much of it must finish up but how it gets there is quite another matter.

Surprisingly, the biological factor in the uptake of carbon dioxide by the oceans is much more poorly understood than might be imagined. It is sometimes assumed that, since the oceans cover so much of the Earth's surface, their photosynthetic capacity (like their capacity to dissolve and release carbon dioxide), is commensurately large. In fact this is not so. In terms of accumulated dry matter per year the land is twice as productive as the sea (Table 8.4). Terrestrial photosynthesis per unit area is five fold greater than ocean photosynthesis. There are however, some mysteries to be solved. Photosynthesis per unit chlorophyll by marine phytoplankton is said to be six times greater than terrestrial photosynthesis although the chlorophyll content of both groups of plants, expressed as a percentage of dry weight, seem to be about the same. As Kelly points out there is no biochemical evidence in support of these latter claims, based on rather dubious techniques involving inappropriate use of radioactive carbon and possibly lack of appreciation of the role of pigments other than chlorophyll e.g. (phycobilipigments) in light-harvesting in marine phytoplankton (these could lead to an exaggeration of rates expressed on a chlorophyll only basis). Many marine algae also have "carbon dioxide-pumping" mechanisms (which do the same job as carbon dioxide enrichment mechanisms in C4 plants but in a different way). Conceivably these also might lead to the accumulation of tracer in inorganic carbon reserves. Conversely, accumulation of soluble organic carbon in surface waters recently reported by Japanese workers could, if confirmed, lead to upward re-evaluation of oceanic photosynthesis, expressed on an area basis.

Marine photosynthesis differs from terrestrial photosynthesis in several important ways. Most obviously the major limitation on the land is water. This is not difficult to see in temperate regions. In a comparison of deserts and rain forests it is self-evident. Marine phytoplankton are not, generally speaking, short of carbon dioxide (because of "carbon dioxide-pumping") nor often too cold to photosynthesise whereas coniferous forests, for example, are carbon dioxide limited in summer and mostly unable to fix much carbon dioxide in winter. Many land plants can be made more productive if given fertilisers. This, of course, is a central feature of western agriculture but, as famous Rothamsted experiments have shown (Section 1.5) crops manage quite well without. Some of the most massive accumulations of natural biomass (rain forests) stand on extremely infertile soils. Marine phytoplankton, by comparison, are normally desperately short of nitrogen, iron etc for the simple reason that these essential elements are present in such low concentrations in sea water. This accounts for the otherwise suprising productivity of cold waters which happen to be nutrient rich and of areas in which there is a continuous upsurge of nutrients from deep waters.

Table 8.4

Global net productivity of terrestrial and marine plants
(from Whittaker & Likens, 1975)

Environment	Production Giga tonnes dry matter.yr^{-1} (P)	
	Total	%
Marine		
Open ocean	42	24
Coastal	13	7
Terrestrial	118	69

[Marine phytoplankton fix significant quantities of carbon dioxide by carboxylation of phosphoenolpyruvate. This has led to ill-informed suggestions that PEP carboxylation (Section 5.14) could compensate for a perceived deficit in RuBP carboxylation (Section 5.12). More knowledgeable scientists and readers of this book will readily understand that this is a nonsense because *all* carbon dioxide fixation *ultimately* depends on RuBP. How else would a plant get its PEP?]

Land plants are often light limited and many grow in shade environments. On the other hand they do not, in general, suffer the constant attenuation of light, by water, experienced by marine algae.

8.8 What Rubbish!

"A good skip, a what you like, a dolly or a pepper ......" ..Traditional

Although I do not usually get into this sort of thing (except in a perfectly acceptable sort of way) I am told, on good authority, that Marks and Spencers now sell knickers on plastic hangers. This scarcely merits surprise. The other day, in pursuit of energy conservation, it took me a good ten minutes, a pair of pliers and more patience than I normally carry around with me, to force entry into the plastic container surrounding a time-switch. How things have changed. Screws and nails were once dispensed by iron-mongers in paper bags (following weighing in brass scoops which lasted for ever). Milk came in jugs (rather than fiendish micro-buckets of plastic, pre-programmed or predestined to launch their contents over your best suit or dress). Celery once came fresh from the earth. Am I the only one too shy to recruit the aid of an air-hostess to help me fight my way into a plastic meal wrapped in a plastic enclosure?

All of this is rubbish *par excellence;* refined trash that no one needs. This brings us face to face with the starving in Ethiopia. Perhaps you might feel that I exaggerate. But let me offer two examples from my own experience. Picture a scene in Port Jefferson, Long Island. A lunch served on plates and serviced by cutlery which would have won prizes for material and design in the United Kingdom of the 1950's and which would still be prized possessions in many a Third World home. What to do with these at the end of the meal? Commit them perhaps to some energy-devouring washing machine? Not at all; *"please put them in the garbage bag"*. Is this really less offensive than killing mink to wear as ornament? Of course not.

Picture the scene that meets the eye of the first-time visitor to India. What are those tiny enclosures, wrought in hardened mud, designed to accommodate? Not people, not people, *not people*? Go to the outskirts of Delhi and walk past the biggest, nastiest, rubbish pile that you might imagine. What covers its smoking and steaming surface? People? What are they doing there? They are living there, it is their home.

So there is rubbish and rubbish, garbage and garbage. Even the most modest and environmentally conscious person that ever existed generates rubbish. Even kings and presidents and priests must shit and eventually die. There are parts of what we buy or acquire as food which we would rather not eat. Friendly wood-fires produce carbon dioxide and a modest amount of ash. Clothes become rags. All of this is acceptable on a small scale. An equilibrium is established with the environment as organic materials are degraded by bacterial action, as iron rusts. Translate this into the context of "civilised" Western Society. We are back to plastic shrouds for everything from onions to hardware. We can drive past mountains of rusting cars. We are dimly conscious of comparable mountains of CO_2, of SO_2, and of NO_2. We curse the necessity of disposing of an immensity of Sunday newsprint, of beer cans and wine bottles. Some of us salve our consciences with trips to bottle banks and by the encouragement we give to Boy Scouts to salvage waste paper. How many of us take our bottles and cans back to the supermarket? How many of us complain about plastic wrapped vegetables, plastic wrapped pills, plastic wrapped contraceptives? Beaches thick with undegradable plastic flotsam and jetsam is regarded as an inevitability; how sad.

And yet there is a gleam on the horizon. Inevitably, since we are all human and equally culpable, it is not lack of concern or high principles but often necessity which causes us to make our contribution to the rubbish mountain. Given a chance we don't wish to share the fate of the garbage pile dwellers of Delhi. Self-interest beckons us in other directions. Slowly the penny drops. Loading garbage into skips and hawking it round the world suddenly seems a facile solution. Giving it to the British, a nation currently prepared (at a price) to take anything from dioxins to other people's used plutonium, suddenly seems just a shade immoral - one jot removed from selling it to Third-World countries, desperate to earn any sort of living.

So where *do* we begin? Turning to the ideas of Lovins (Section 9.1) we can begin to see that the real alternative to

"putting diapers on dirty power stations"

is not to generate rubbish in the first place, whether it is waste carbon dioxide or sewage. Immediately we must recognise degrees of realism. We cannot avoid producing rubbish or waste entirely but we can embrace the principle that the polluter must pay. It is a simple matter to make bottles, cans and used automobiles into objects of desire for those who would wrest a living from recycling rubbish (is it any surprise that waste vendors and recyclers are often amongst the most affluent members of society?). Smaller vehicles using lead-free petrol can be encouraged by governments, if they so wish. Collecting waste paper and reprocessing it is unlikely to be without environmental impact in its own right but it certainly has the edge on using trees for this purpose.

["Toilet" paper, made from unbleached recycled newsprint sometimes has an off-white colour which apparently causes no raised eyebrows in Continental Europe but is not yet acceptable to the fastidious British regardless of what they are just about to do with it. The time-honoured North American custom of using Sears Roebuck catalogues in rural privies presumably went out with the model T Ford. Perhaps the Indian custom of using a hand and water will soon come to acquire a Green image.]

Steps in this direction will have little real impact but they are helpful because they diminish the sense of real frustration that many of us feel when confronted with these problems. They also encourage thought and awareness. It is better to conserve energy than to build new electricity-generating plant or nuclear power stations. It is better to insulate buildings than to erode their fabric with acid rain. It is clearly better to avoid creating "unnecessary" garbage. We are left with what *is* "necessary" The only really effective way of dealing with *that* problem (Sections 8.10 & 8.11) is the ultimate Green solution of making fewer people.

The fundamentals are inescapable and the logic is inevitable. If we cannot dispose of our faeces in an acceptable manner we must produce less. Malthus is no longer a prophet without honour (Section 8.11). Of course we deny the evidence of our senses. People are said to be fat, not because they eat too much but because they are of sedentary habit or hormonely imbalanced. The fact that there were no fat people in concentration camps or amongst the starving in Ethiopia is somehow far removed from what comprises the most effective diet. In the same way we care not to think about the fact that there are too many of us, unless, of course, it is too many of *them*. In the end, burying our heads in the sand will no longer be a solution. We can start by banning or disapproving or taxing *unnecessary* garbage out of existence. All of it takes energy (which we cannot afford) to produce and thereby puts carbon dioxide, sulphur dioxide or whatever into our atmosphere. All of it takes space (which we cannot afford to squander), or pollutes our environment, or both. All of it is a literally disgusting waste of resources, both human and material. We can only hope that apples or nails in neat plastic containers will become as socially undesirable as thick fur coats or Semtex in radio/cassette recorders.

8.9 Will Temperatures Rise and Catastrophe Follow?

"Till a' the seas gang dry, my Dear
And the rocks melt wi' the sun" ....Robert Burns

Global warming is almost certainly inevitable. If it occurs it will not result from natural causes. It will be a consequence of Man's activities, as he rapidly restores to the atmosphere carbon dioxide removed by millennia of past photosynthesis. It will be because he augments this greenhouse effect with methane and synthetic gases such as CFCs. It should also be remembered that the world as we know it today has been largely fashioned by the use of fossil fuels and that, even if photochemical generation of electricity (Chapter 9) suddenly became so cheap and efficient that everyone could meet their total electrical requirements from inexpensive panels on their roofs, there would still be a growing need to find replacements for the coal and petrochemicals that we now use so extensively. There is no possibility, in the short term, of preventing the release of greenhouse gases. All that can be hoped for is that contemplation of the possible consequences brings such a world-wide and unified determination to diminish the present rates of release that Man will buy time to adjust to the changes which the greenhouse effect will impose.

8.10 Can Population be Maintained?

"I wish I loved the Human Race,
I wish I loved its silly face" ...Sir Walter Raleigh

Current predictions suggest that, by 2075, world population will have stabilised at between 8 and 10.5 billion. Looking beyond that could be regarded as

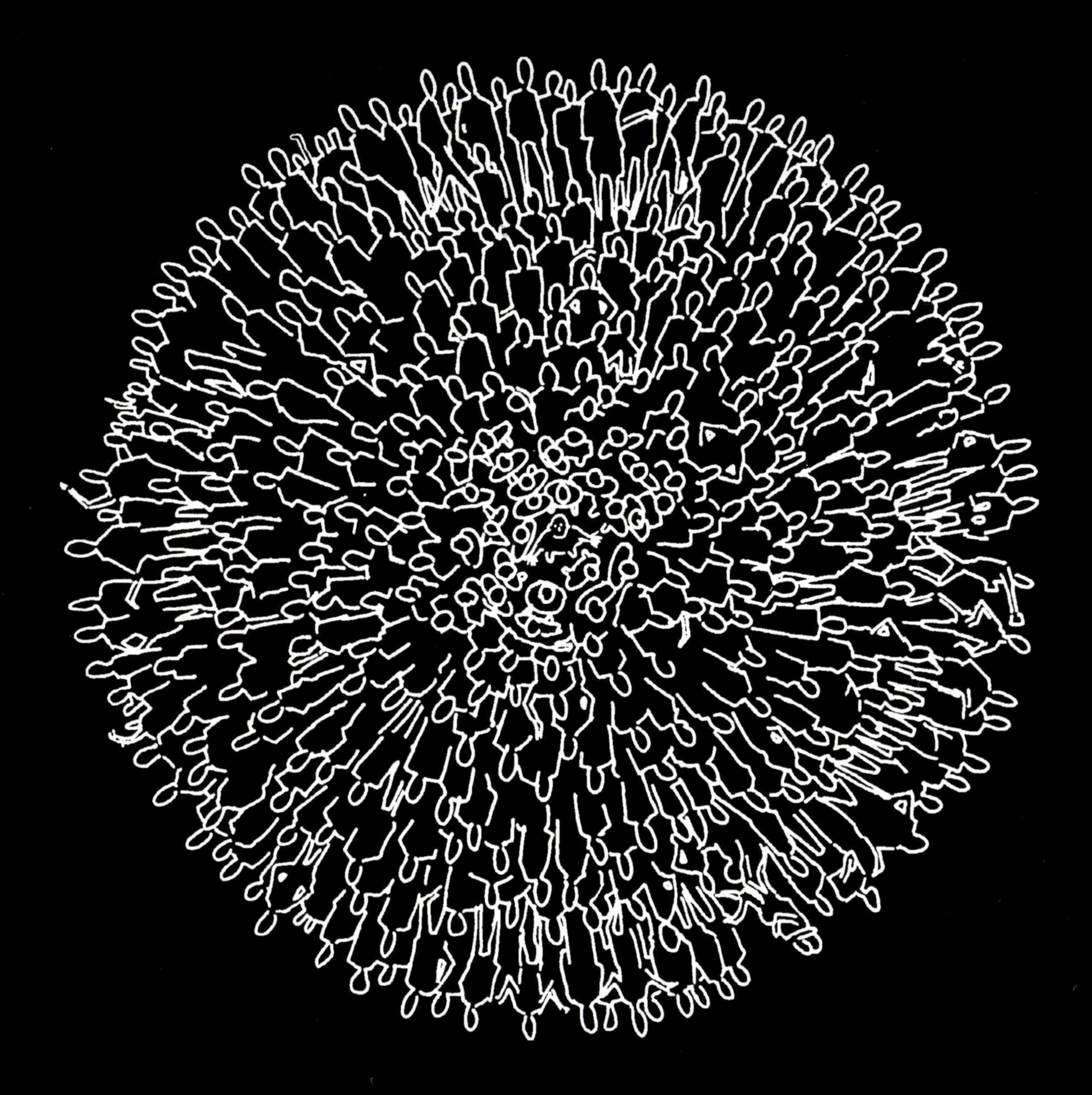

"The perpetual struggle for food and room"
.....**Malthus**

foolish or simply a waste of time, given the uncertainties involved. There are, however, related questions and facts which warrant consideration. First we might ask why there should be any doubt that things should not go on indefinitely as they do now. Common sense would tell us that continuing expansion is impossible. Everyone has heard of, or might be convinced by, the intrinsic truth of the story in which a pair of rabbits were put down on a tiny, uninhabited, but well vegetated tropical island. Some years later, passing sailors found the island, stripped of much of the vegetation and littered with the bones of hundreds of rabbits. Certainly it is self-evident that life of any form depends upon the consumption of resources and if resources are not renewable (or consumed faster than they can be renewed) there is no possibility of a continuously expanding population. Conversely, if we look at the population of the U.K. in the Middle Ages we find that although it was tiny by comparison with the present, it was almost entirely self-sufficient and in equilibrium with its environment. It had no need to import food, goods or materials from abroad. It did not need to consume fossil fuels, the native forests could provide wood or charcoal to meet every need. Pollution of the environment undoubtedly occurred but not at rates that were too large to be accommodated by natural processes of decay and erosion. That is not the present situation. We are now using irreplaceable reserves (most obviously fossil fuels but there are others) and putting them back into our environment, but changed in such a way that our environment itself is being changed in ways which are not consistent with life as we would prefer to see it. Ask a physicist or the man in the street what have been the most important scientific advances in the last half century. The answers might range from nuclear fission to television. Ask a biologist the same question and he might say antibiotics, selective herbicides and the contraceptive pill but he would also undoubtedly identify the work by Watson and Crick, which established the chemical basis of heredity, as the discovery most likely to change life as we now know it. All of these things would have been largely inconceivable to someone attempting to predict the future a century ago. Moreover, the rate of advance of modern science is such that it is almost inevitable that presently inconceivable discoveries will be made in the next century. The optimists take refuge in this, believing that new ways (the *"technical fix"*) will be found to cope with present problems. There is obviously some room for optimism but, this is not to say that we should do nothing now and simply trust that our unborn children will come up with hitherto undreamed of solutions. The following, based on facts and reasoning set out as far as possible in the preceding chapters of this book, should give us pause for thought.

8.11 Malthus Revisited

It is implicit in any consideration of the facts that the greenhouse effect results from Man's activities and that one certain way of bringing Man back into equilibrium with his environment would be to reduce populations to a very much lower level. A controlled decrease in population is not by any means the worst scenario that can be envisaged. Populations in the western world which threatened to double within the last half of the present century have already stabilised and there is little to choose between stability and gradual decline. Stability was not brought about by coercion or regulation but by the contraceptive pill, medical advances, economics and changing social attitudes which have encouraged young women to pursue careers rather than procreation. If every "married" couple produced only two children, populations would eventually fall for the simple reason that disease, disaster or famine would ensure that not every child would survive to procreate in turn. It is not difficult to imagine many societies in which a combination of economics, self-interest, and changed ethics could not combine to make families of more than two

children at least as socially unacceptable as smoking in aircraft or mugging old ladies.

[We might also be reminded of the science fiction story about the man who asked why an alien race was rapidly colonising the entire known universe. He was told that it was a matter of ever-increasing population combined with the political undesirability of controlling procreation. The man replied that if the aliens were unable to control their breeding, someone or something might have to do it for them. In this parable we could just as well envisage our environment as the instrument which might control our breeding. If we continue not only to use our non-renewable resources in a profligate manner but also to use them in a way in which they will affect and diminish our environment there is every prospect that our environment will turn on us and declare itself unable to sustain existing levels of population.]

8.12 What Will Happen If We Get It Wrong?

"This is the way the world ends,
Not with a bang, with a whimper" ....T. S. Eliot

No one knows what will happen if we get it wrong. Do not, dear reader, be persuaded otherwise. Not so long ago, Margaret Thatcher was Prime Minister of England. (I say England, advisedly, because the Scots took a somewhat different view even if they had to go along with the rest). She had ruled England for one fifth of my lifetime and, if there is one thing which can be said with present certainty, she would have liked to continue to do so for a few years more, just as I yearned to see what life would be like with someone else at the helm. When she came to power, inflation in the UK was a little over 10% and she consistently made the reduction of inflation her principle aim. At the end of her term of office, inflation was slightly greater that when she took over. Why do I say this? Well, of course, it is not that I feel a deep affection for Maggie but I do not doubt that she would have dearly liked to see us down to 1% inflation or that she had access to a great deal of well informed economic advice. It is quite simply that economics is an inexact science, entirely comparable to biology.

[In my youth I was obliged to spend two years of my life as a radar mechanic in the "Fleet Air Arm" (Naval Aviation). I was "called up" ("drafted") for the period of "Hostilities Only" although, in December 1946, hostilities had long ceased. Resenting this loss, of what might have been the best two years of my life, I consoled myself with the thought that if I had been marginally older it might have been six rather that two and I might never have returned to civilian life at all. Struggling with such thoughts I was amazed to discover that (in the Navy's view) biologists made better radar mechanics than physicists. The former viewed radar as they might regard living organisms, wayward and devoid of rationality. The physicists (raised on the behaviour of "ideal gases") were so disappointed by the failure of radar to obey some of the "rules" that they retreated into frustration and despair. Perhaps economists would make good biologists and *vice versa*.]

So *no-one*, knows what *will* happen. All we can do is to make some very poor guesses about what *might* happen. Maybe existing "evidence" is inadequate and suspect but, if we were in a poker game, we would have to settle for the probability of elevation in mean global temperatures by as much as 2 or 3 degrees in our

collective life times. On this basis, climates will become even less predictable (even given better technology and computers) than it is at present. The UK used to have "weather" whereas continents, like America (North & South), Europe and Australia had "climate". England, it seems, may soon export its weather, as well as its language to the world. Everything will become very unsettled.

Have you been to the Maldives? Flying in from Europe you see a representative sample of the 12,000 islands. Some of these are little more than sand spits, most of them uninhabited, few populated, none more than two metres above the sea. The first nation in the world to disappear if the waters rise. Maldives are about as near as we can get to the romantic concept of desert islands.

"with oval basins of coral-rock just lipping the surface of the sea, and each containing a lake of clear water" .....Charles Darwin

There are those, quite literally, which resemble the cartoon drawings that we all know so well. So they are very beautiful and great places for a pleasant holiday. It would be sad to see them go but, in global terms, would it matter all that much? Perhaps not (it all depends whether or not you take John Donne's view of life or if you would regret the casual destruction of one of the numerous paintings, by Monet, of the bridge over his pond). On the other hand, if you happen to be a native of Amsterdam, New York, London, Boston, San Francisco, Sydney, Tokyo, Singapore, Hong Kong or Shanghai you might feel some concern because, in the words of Tom Lehrer.

"the bomb that falls on you, will kill your friends and neighbours too"

In other words what is bad news for the Maldives is bad news for many of the principal ports and centres of population in the world. So if the waters do rise we can wave goodbye to the low lying areas of many major cities and to large areas of very productive agricultural land. But, will it happen? We are back to "we don't know but it *is* possible". If we return, for the moment, to politics we immediately find this weird ambivalence in thinking. The United Kingdom has spent, and is spending (despite the demise of a supposedly aggressive Soviet Union) vast amounts of money on "an independent nuclear deterrent". How much it has spent is hard to say because realistic estimates would have to take into account the disguised expenditure on establishments which were ostensibly concerned with the nuclear generation of electricity and, in reality, with the production of plutonium for bombs. But let us not get into arguments about whether or not this policy is laudable or ludicrous. Let us simply assume that it has worked. Why then are we not prepared to spend a fraction of this sum on energy conservation, knowing that what we spend in this way could slow global warming (however marginally) and would almost certainly make good economic sense and might even stave off global disaster?

One thing that we can be reasonably certain about, once we start to talk about spending money or cutting down on wasteful and damaging use of fossil fuels is that, if we are to do it at all, we had better start soon. London has recently spent a lot of money on a barrier across the Thames which was designed to hold back the waters when rare combinations of high-tides and on-shore winds combined to inundate large areas of the city. Flood Street, in fashionable Chelsea earned its name the hard way. It has all happened before.

For the moment the threat has receded but, as the sea levels rise, it will return. Since one in four of those living in England have their homes in within 25 miles of

Flood Street there is little doubt that money will be found but don't count on dry feet if you live near the Wash or in other low lying areas like Hull, New Holland and many parts of East Anglia.

8.13 How Will Rising Carbon Dioxide Affect Plants?

Of all the many intrinsic uncertainties of the greenhouse effect this is one of the largest, not least because it is equally to the point to ask how plants will influence the greenhouse effect. As we have seen (Section 8.6), about half of the carbon dioxide released to the atmosphere gets lost. If it does not remain in the atmosphere it can only go into the oceans or the land masses and there is no doubt that photosynthesis by phytoplankton and land plants makes the major contribution to carbon dioxide uptake.

[In figure 7.4, for example, the annual "zig-zags" in the Mauna Loa graph of atmospheric carbon dioxide are mostly attributable to diminished photosynthesis by land plants in northern hemisphere winters. The tropical forests of the southern hemisphere are essentially evergreen and temperatures in these forests are such that there is no comparable decrease in winter photosynthesis as there is in the coniferous forests of the north.]

It is by no means certain, however, that carbon dioxide uptake by land and ocean will remain unchanged as atmospheric carbon dioxide concentration rises. The biological component of uptake is unlikely to be indifferent to enhanced carbon dioxide or to increased temperature, nor is it inconceivable that entire ecosystems could change beyond recognition. Regrettably our understanding of the physiology of individual plants, let alone our knowledge of the physiology of ecosystems is so inadequate that we can do little more than speculate. Let us start, at least, from relatively firm ground and progress as cautiously as possible into the realms of science fiction.

8.14 Carbon Dioxide as a Fertiliser.

There is no doubt that some plants do better, in a lot of ways, in carbon dioxide enriched air. It might be supposed that using augmented carbon dioxide in greenhouses to promote plant growth came from an understanding of plant physiology. It did not and it is a matter for regret that the world at large knows little and cares less about the process of photosynthesis on which we all depend. So, who knows, there could have been someone who thought that, since plants require carbon dioxide to photosynthesise, it would have made good sense to offer them a little more. It is even possible to find references to suggestions of such commercial application in fairly venerable tomes like Rabinowitch's *"Photosynthesis and Related Processes"*. Sadly, it appears that contemporary carbon dioxide fertilisation came not from earnest perusal of the scientific literature but from Dutch tomato

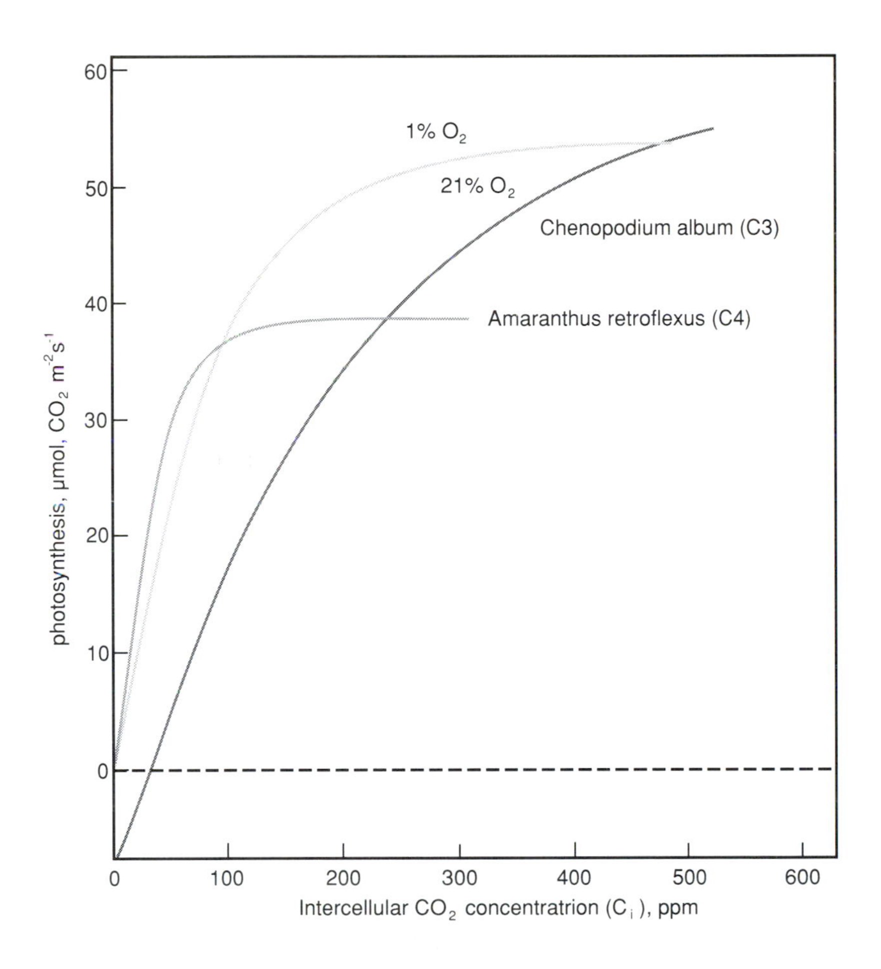

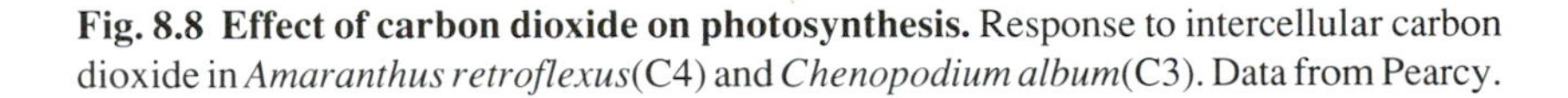

Fig. 8.8 Effect of carbon dioxide on photosynthesis. Response to intercellular carbon dioxide in *Amaranthus retroflexus*(C4) and *Chenopodium album*(C3). Data from Pearcy.

growers who burned propane in greenhouses (for heating) and found it led to better yields than other forms of heating which did not generate carbon dioxide in the same space as the growing plants.

Carbon dioxide fertilisation is an immensely complex business. Applied to tomatoes it can affect many aspects of plant growth (including the number of fruit per truss and their size and quality) but there is no doubt that it works in the sense that it changes the nature of the crop for the better. In modern commercial practice it can make the difference between profit and loss.

Similarly, there is no doubt that, if we were to compare the rate of photosynthesis by many plants in air with those in air containing twice as much carbon dioxide, we might expect to see about twice as much photosynthesis per unit time. Not all plants would show such a large response. Armed with our knowledge of C4 photosynthesis (Chapter 5) it would come as no surprise to discover that C4 plants, with their own built-in carbon dioxide concentrating mechanisms, are less responsive to augmented carbon dioxide than C3 species (which depend upon unaided diffusion of carbon dioxide from the external atmosphere to the site of fixation within the chloroplast).

[In this regard, C3 species provided with additional carbon dioxide will, in general, respond to changes in their immediate environments, such as increases in light intensity and temperature, much as though they were C4 plants (Chapter 5). In other words, if they are provided with sufficient carbon dioxide to suppress photorespiration and enhance photosynthesis, C3 species will do what C4 species accomplish unaided by virtue of a combination of leaf structure, biochemistry and internal transport.]

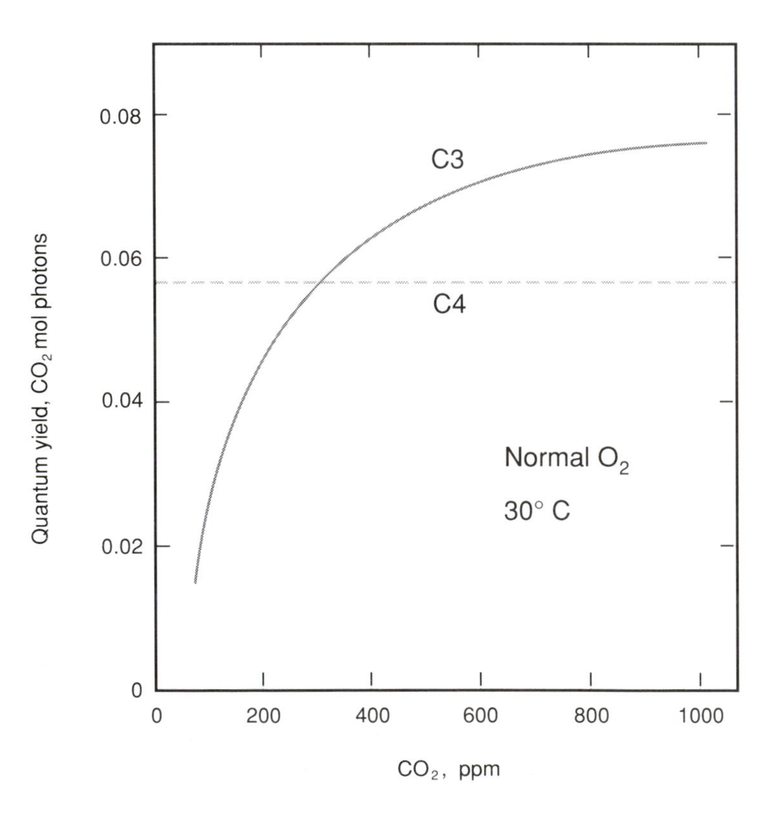

Fig. 8.9 Effect of carbon dioxide on quantum yield in C3 and C4 species. [After Osmond, Bjorkman and Anderson (1980).]

Much also depends on the experimental conditions under which measurements are made. If photosynthesis is limited by cold or by poor light, the response to increased carbon dioxide will be less readily detected than if all other external constraints have been removed. In a way this is self-evident but it has, in the past and in the absence of experience, led to the erroneous idea that (putting to one side debates about cost-effectiveness which have more to do with market forces than plant physiology) carbon dioxide enrichment only helps under otherwise favourable conditions. More importantly in the present context, it has added to uncertainties about the consequences (for photosynthesis) of growing plants in high carbon dioxide as compared with briefly exposing (to high carbon dioxide) plants

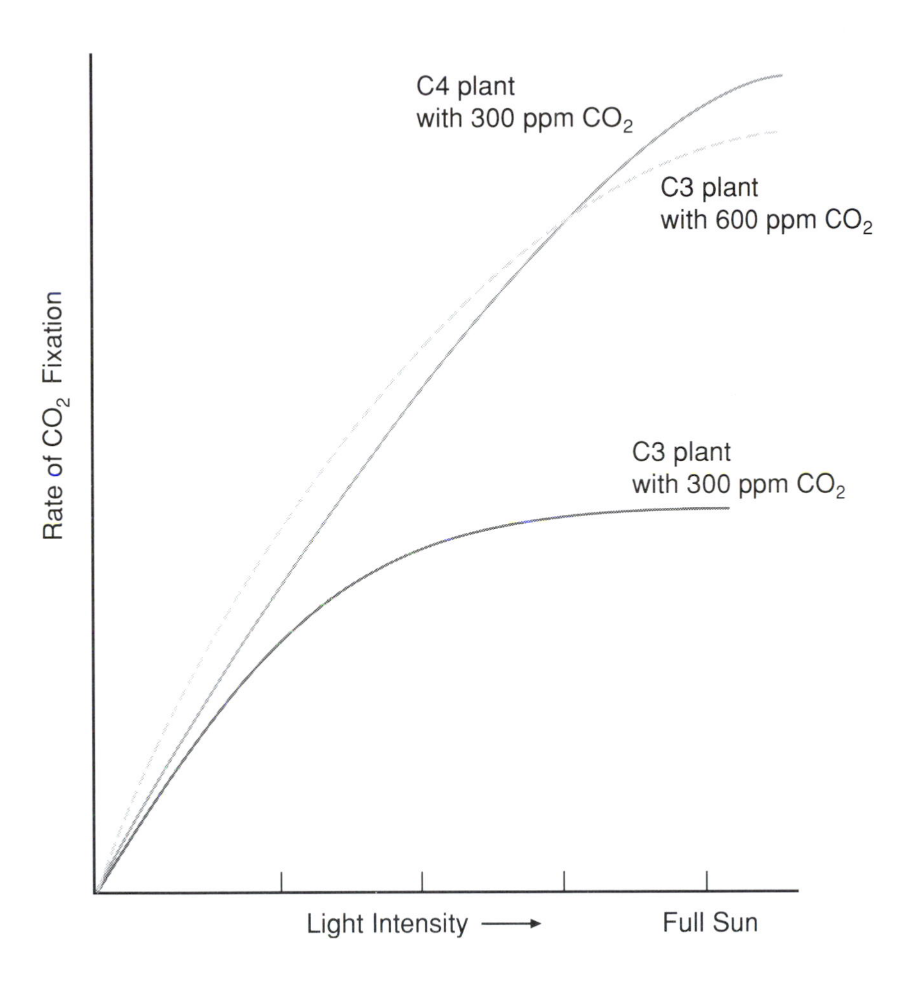

Fig. 8.10 Rate as a function of PFD: the effect of CO_2 concentration. A C3 plant behaves more like a C4 plant if the carbon dioxide concentration is doubled.

grown in normal atmospheres. It has been suggested that while tomatoes and cucumbers undoubtedly do better when grown in carbon dioxide-enriched atmospheres the same may not apply to other species which nevertheless show increased rates of photosynthesis if grown in air and assayed in high carbon dioxide. There may be some truth in the proposition. Plants are notoriously conservative. Spinach grown in water culture is more susceptible to wilting at high temperatures and in high light than if grown in pots, exposed to periodic drying, in an otherwise identical environment. It is as though a plant with its roots permanently in water gets the message not to make more roots than it needs. This, of course, can take us into the dangerous realms of teleology and gardener's folklore but that is not to say that such aspects of growth and form are no less real because they are poorly understood. It is not inconceivable, and there is ample evidence to support this supposition, that a plant which finds itself in high carbon dioxide will not put so much of its resources into manufacturing the machinery for carbon dioxide assimilation as it would in normal air. At the same time, like Robert Boyle, we must remain properly sceptical.

In the Robert Hill Institute in Sheffield, where we have long experience and a vested interest in growing good plants for experiments, we grow the best spinach in the U.K. Our spinach photosynthesises better than our maize. Conversely, our maize, grown at moderate temperatures and often inadequate light, compares poorly with maize grown in the field at the Maize Research Institute near Belgrade. When plants are grown in high carbon dioxide for experimental purposes it is therefore, important to

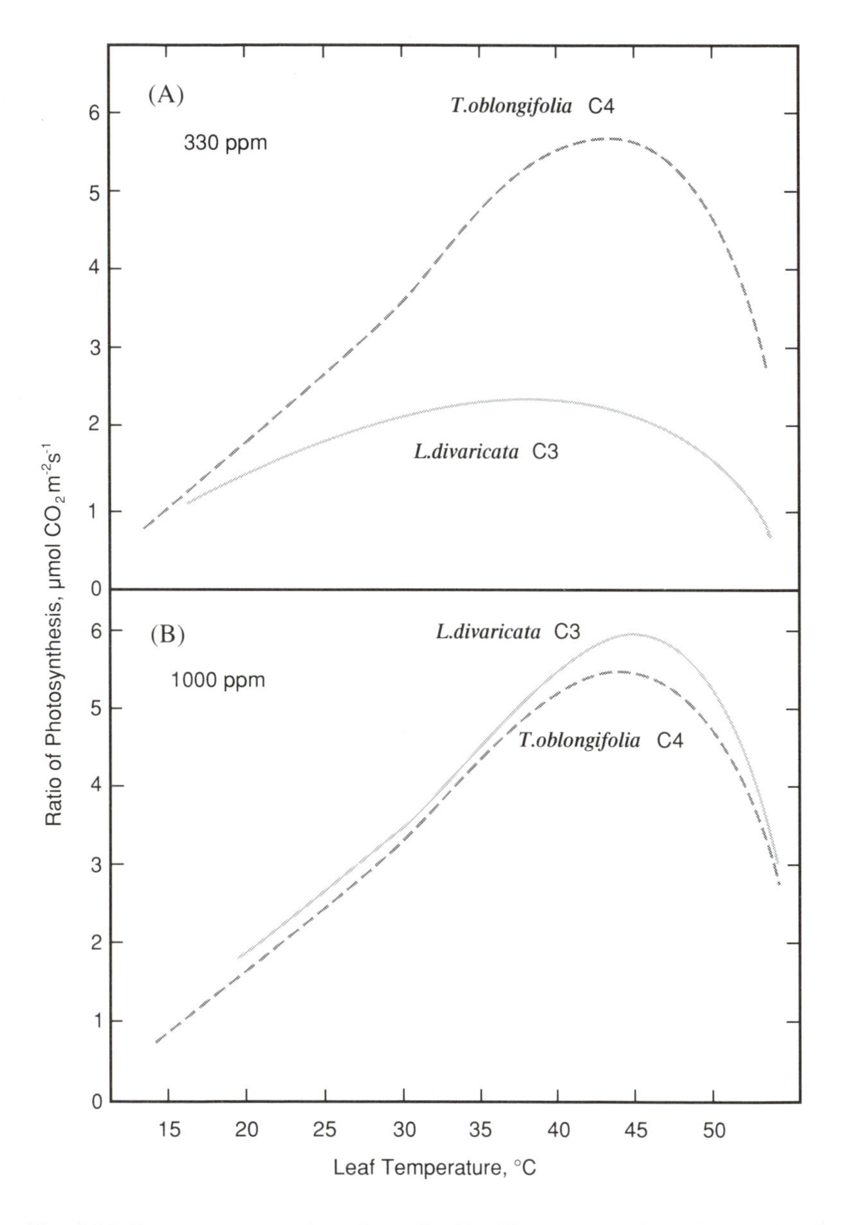

Fig. 8.11 Temperature and carbon dioxide. Temperature dependence of carbon dioxide uptake at (A) 330 ppm carbon dioxide and (B) 1000 ppm for *Larrea divaricata* (C3) and *Tidestromia oblongifolia* (C4). [After Osmond, Bjorkman and Anderson, 1980.]

ask if they have been inadvertently subjected to constraints which might have been less important in air-levels of carbon dioxide. Nitrogen supply, for example, is immensely important in this regard and can have major impacts on the amounts of RUBISCO in the leaf. All we can reasonably say at present is that we do not have enough facts to go on. It may be unwise to suppose that a plant grown and assayed in high carbon dioxide will show the same increased rate of photosynthesis as one grown in air and assayed in high carbon dioxide. It would be equally unwise to dismiss the vast experience of commercial growers of tomatoes and cucumbers or to suppose that only some species such as these would do better as atmospheric carbon dioxide rises. The likelihood, if not the certainty, is that photosynthesis in general will increase as atmospheric carbon dioxide increases and that the potential for increases will be amplified by higher temperatures and better water use efficiency (Section 8.15).

It should also be noted in this context that in arriving at conclusions from relevant experiments it is important to know what to measure. Working in a facility built in Maryland by the Carnegie Institute, Steve Long has recently shown that some species grown over a period of three years in augmented carbon dioxide (but otherwise in their natural environment) show the same increase in quantum efficiency when assayed in high carbon dioxide as do control plants grown in air. Not many leaves spend much of their time in saturating light. How they behave in low light may be more important.

Given two plants with identical rates of photosynthesis in saturating light but with different rates in low light (because of differences in quantum efficiency) it is self-evident that the more efficient leaves would accomplish most photosynthesis. Meaningful conclusions can only be drawn in such circumstances if really accurate measurements have been made of the low rates of photosynthesis which obtain in low light. Until relatively recently, the ability of most laboratories to make such measurements has been seriously curtailed by inadequate techniques. Once again the need for caution is apparent.

8.15 Increased Water Use Efficiency.

C4 plants use less molecules of water per molecule of carbon dioxide fixed than do C3 species (Chapter 5). It seems probable that, in high carbon dioxide, both C3 and C4 will be advantaged in regard to water loss.

What, you might ask, are the principles involved? C4 plants you might recall, often do well in semi-arid conditions because they first fix carbon dioxide in an outer compartment (the mesophyll) in a reaction catalysed by PEP carboxylase. This enzyme has a high affinity (Section 5.17) for carbon dioxide like RUBISCO which, in C4 species, does its work in its own carbon dioxide-enriched “greenhouse” (the inner, bundle-sheath compartment). Because PEP is not subject to oxygenation, PEP carboxylase works better than RUBISCO in very low concentrations of carbon dioxide and therefore, if the stomata (“pores”) in C4 leaves are barely open (the smaller the aperture the lower the diffusive loss of water vapour), PEP carboxylase can still function well and maintain a steep diffusion gradient (for carbon dioxide) between the mesophyll compartment and the external atmosphere. Obviously, C3 species do not have this advantage and are less well placed than C4 species to cope with the inescapable fact that stomata must be opened to permit carbon dioxide entry and that water is lost through open stomata. If external carbon dioxide is increased then, of course, any diffusive barrier imposed by stomata which are partly closed to prevent water loss) will be less advantageous. Here again, the situation is probably no way near as simple as indicated above. It used to be believed that photosynthesis was very dependent on stomatal movement in the sense that if stomata tended to close (in response to a water deficit or incipient wilting) the rate of photosynthesis would decrease in a commensurate fashion. This is still not questioned but, since the work of Cowan, Farquhar and others, it seems likely that, at least in some circumstances, it is not stomatal aperture which determines the rate of photosynthesis but rather rate of photosynthesis which determines stomatal aperture. All of this implies some very sophisticated signalling between chloroplast and stomata but it has been known for many years that, other things being equal, high carbon dioxide will tend to cause stomata to close and low carbon dioxide will tend to cause them to open.

[It should be noted that a number of other factors can be involved in stomatal opening such as the plant hormone, abscisic acid. Actively photosynthesising chloroplasts may also communicate with the cytosol and other cellular compartments and organelles both by demanding inorganic phosphate and affecting the release of inorganic phosphate from triose phosphate during cytosol sucrose synthesis. - Figs 4.10 and 4.11.]

"sophisticated signalling"

Whatever the mechanisms involved, there is good evidence that, in many circumstances, stomatal apertures tend to be controlled in such a way that stomata do not open more than necessary (in the sense that a more or less constant carbon dioxide concentration is maintained within the leaf). In other words, carbon assimilation and its associated machinery is in balance with carbon dioxide supply as though the leaf is smart enough to know that wider apertures will not help carbon dioxide assimilation already limited by other factors (such as the amount of RUBISCO, or the regeneration of ribulose bisphosphate) but it might increase water loss to a rate which would be damaging. It could be concluded that this is yet one more reason for supposing that photosynthesis would not increase with increasing carbon dioxide. The plant, however, appears to know when it is on to a good thing. No doubt it goes through the same fine balancing act (between distribution of resources and regulation of metabolism) at 600 ppm carbon dioxide as it does at 300 ppm carbon dioxide but, in the end, it appears to finish up with a faster rate.

8.16 Summary

Are we to be optimistic or pessimistic about the Greenhouse Effect and the prospect of global warming? In the final analysis it is anyone's guess. But that is not really the real issue which ought to concern us all. More to the point is that we ought not to be indifferent to polluting our planet and depleting it of so much that makes it a pleasant place, in which to live.

"dear to the heart of politicians"

Your author is a biologist, working somewhere in the area between plant physiology and plant biochemistry. He has laboured sufficiently hard and long in the field of photosynthesis to feel comfortable with that subject but unable to share the schoolboy's confidence that he really understands it. Most certainly he is not a climatologist or an economist but he has no doubt that all three disciplines (climatology, economics and photosynthesis) are vastly complex subjects, all well laced with physics and politics as well as with biology. Take economics, for example. Like the other two, it is a subject which impinges daily on our lives and is dear to the heart of politicians. Almost invariably they get it wrong. We may take for granted that many politicians delight in the prospect of ordering the lives of their fellow mortals, that they often represent sectional and vested interests and that they are no less greedy, avaricious, or dishonest, than the rest of us. However reluctantly, we must nevertheless accept that they do not get it wrong on purpose. In democracies, at least, getting it right would presumably give them a marginally better chance of staying in power even though we are all so prejudiced that politicians can seemingly make the most fearful mess of things before the electors have had enough.

So what has this to do with the contents of this chapter? Let me come clean. Having looked at the climatological evidence as a biologist (i.e. as a lay scientist in this regard). I simply cannot say with any sort of authority, that the world will grow warmer and that unpleasant consequences will follow. If I had spent my life working in climatology rather than photosynthesis, I suspect that I would feel equally uncertain. On balance, I think that there are very good reasons for supposing that warming *might* occur and that, if it does, we might wish that this had not happened. There my uncertainty ends because I can see no early end to Man's determination to liberate, from fossil fuels, all of the carbon dioxide which was once put there by primeval photosynthesis. To the question "does it matter" I am bound as, a biologist, to lament the irretrievable loss of any natural resource, (living or mineral) because the world, as we know it, exists best in some sort of slowly changing equilibrium. Of course it has never been stable but it is clear that biological organisms, including Man, can accommodate to almost unthinkable change provided that change is reasonably slow. Of all of the present uncertainties, the fate of the "missing" CO_2 is one of the most perplexing. If so much of what is being released gets "lost" there is obvious hope that given a lot more time we could lose a lot more - a sort of cosmic "kick it around until you lose it"

What CFCs have done to the ozone layer will very probably have dire biological consequences quite apart from any impact on global warming. Ultraviolet radiation does nasty things to eyes and to skin and most importantly to DNA, the very heart of the biological blue-print. Even so, growing political awareness and realism may soon preclude their production and much of the damage that they have done to the atmosphere will repair itself, given time. Fossil fuels, on the other hand will not replace themselves, (once burned) on any sort of timescale that it is imaginable to Man. This issue is inescapably linked to what has already been discussed in chapter 7 and will be returned to in chapter 9. Man is totally dependent on photosynthesis. Contemporary photosynthesis of the sort most important to Man (sophisticated agriculture) is totally dependent on fossil fuels, an inefficient way of converting oil into crops. There is a sad contradiction, not immediately evident to an observer admiring a field of wheat, that the food industry is responsible for releasing more CO_2 than it takes out of the atmosphere, and adds more than a little methane (via cattle and rice production) as a bonus. In my childhood the textbooks stated, with colonial certainty, that London was the largest city in the world, with a population of 8 million. Since then despite its own growth it has been overtaken by Tokyo and Mexico City, and who knows where, but it is still a large centre of population by any standards. At current rates the net increase in world population is sufficient to produce a new London every month. Once mature, the populations of these new London embark on procreation. The rate of increase accelerates, fossil fuels are used at an increasing rate, CO_2 builds up. So this much, at least, is very clear. This phase of world history cannot last. Moreover it is most unlikely to come to an end in a painless manner. It may be in Man's power to put the cork back into the CFC bottle but there is no possibility of governments stopping CO_2 release or population increase. If left unchecked these will sooner or later stop themselves. Fossil fuels will be spent and, even if alternative energy sources are exploited to the full, any biological population has an upper sustainable level. What does lie within Man's province, as a supposedly sentient animal, is his ability to recognise the problems as real and pressing and to do everything possible to make the last years of the great fossil fuel bonfire as painless as possible.

Chapter 9

Is There Another Way?

Looking for alternatives

Energy conservation

Renewable resources

Trees and people

9.1 An Excursion into Dangerous Ground

"....could thou and I conspire
to crush this sorry scheme of things entire
and remould it, nearer to the hearts desire"
.....Omar Khayam

The reader who has read this far will be one who has not wearied of a single recurring theme - i.e. where do we get the energy which we need for our metabolism and all additional aspects of "civilised" life, what role plants play (or have played) in this process, and what are the implications for the future. Chapters 8 and 9 carry the stark message that mankind cannot continue, into the foreseeable future, as it is at present. Small populations of mammals can live in reasonable equilibrium with their environment. Large populations create difficulties for themselves. Vast populations, increasing in an exponential fashion, cannot be sustained. Malthus had it right. People consume energy and create pollution. At present, most of our energy derives from past photosynthesis. It cannot even be argued that most of our metabolic energy comes from contemporary photosynthesis. Certainly contemporary photosynthesis provides virtually all of our food, whether we eat plants, or animals that eat plants, or both. But modern civilisation is mostly dependent on farming and modern farming is mostly dependent on past photosynthesis in the form of fossil fuels. Ergo, as fossil fuels are depleted, it will become increasingly necessary to look for alternatives. This is the subject of this chapter but it is evident at the outset that prevention is better than cure. "*What can't be cured must be endured*" but it is scarcely sensible to continue to add recklessly to mounting problems while simultaneously striving to solve them. Inevitably, if we wish to address these matters at all, we enter immediately into politics, economics, sociology and even into religion. Naturally, I have strong convictions of my own but my purpose in writing is not to proclaim my own values or remedies but rather to examine consequences. If there are those who believe that there are circumstances in which the use of nuclear weapons or germ warfare would be justified, that it is better to be "*dead than red*", that blasphemy should be punishable by death, that the clear and obvious rights of the unborn child and of animals transcend the rights of society as a whole, that there are political causes which justify the use of Semtex to blow apart people who only happen to be involved by virtue of birth, accident or religion, I know very well that no words of mine would ever give them even pause for thought. Nevertheless, there is now some awareness, once scarcely apparent in anything other than some of the most primitive societies, that living in harmony with the environment has potential advantages for us all. Those already wholly or partly convinced of this might find some profit in what follows because it is very evident that these areas of knowledge and understanding are strewn with all manner of misinformation and economies with the truth. Everyone knows, of course, that "truth" like "beauty" lies, at least to some extent, in the eyes of the beholder but if "a fact" is unpalatable it is often better to seek to challenge it, or refute it, than to be unaware of it. It is also easier to condemn perceived ills than to offer remedies.

I am no better placed than any other scientist to offer remedies. It is easy, for example, to say that we should all use trains and other forms of public transport rather than our own motor vehicles but I would not wish to do that myself, I am well aware that, despite its awful penalties, personal road transport is an aspect of civilisation which is extraordinarily liberating and one which is now virtually essential in Western rural communities. So there will always be a balance between what is desirable and what is feasible, what we can do voluntarily to make things less difficult for ourselves and our children and what is likely to happen if we allow matters to take their own course, indifferent to the consequences. All I seek to do, in this chapter, is to discuss some of the problems and some of the alternatives in the belief that it is better to debate them than to stick our heads firmly in the sand and hope that they will go away or that we can safely entrust all decisions to politicians and bequeath all consequences to our children. At the very moment of writing, George Bush has stated publicly and proudly, that he will not attend The Rio di Janeiro "Earth Summit' on the environment if, by so doing, there is any possibility that he will be obliged to "harm" the United States economy by limiting carbon dioxide emissions. In consequence the European Community Commissioner for the environment will also stay away, now believing that there is no real point in going, his Indian counterpart suggests that the Rio conference will be little more than a lovefest for ecologists, others that it will be "a monumental fiasco." Evidently we can not look to governments for a lead until disaster is almost upon us. In the present context all politicians have a limited life. Understandably, tomorrow's elections mean more to today's politicians than what is going to happen to the unborn. Urban deprivation is all around us now but here again the tendency is for the politicians to fiddle as Los Angeles burns and for the rest of us, if we are able, to distance ourselves as far as possible from the ghettos and to build bigger and better stockades.

9.2 Procreation and Population

"The biggest threat we face is our own success as a species" ....C. Tickell

If we go back to the apocryphal island stripped of vegetation and littered with the skeletons of rabbits (Section 8.10) we can see a possible but unlikely future for the human race. Perhaps, more realistically, we could look to the garbage mountain dwellers of Delhi (Section 8.8), eking out an unenviable existence in an entirely man-made environment of fearful squalor. Common sense tells us that, as sentient beings, we would never let things become that bad. But where then do we find comfort? Certainly not in the Holocaust, the genocide in Tasmania, the murder of three million Chinese by the Japanese, the more recent events in the Middle East, Jugoslavia and the Soviet Union or in the combination of evil and profligate exploitation of the environment which has created starvation in Ethiopia and is still extending the world's deserts. Can we find hope in the rampaging destruction of the rain-forests, the past decimation of bison and the present slaughtering of dolphins and whales, the extinction of more animal and plant species than many of us have ever seen? Suddenly the prospect of a future race of garbage dwellers becomes no less realistic than the possibility of nuclear winter. We can already find parallels between the garbage dwellers of Delhi and the conditions in which some people live in London and New York. The more affluent in these cities can literally live above the mess and enjoy a life-style which might have been envied by ancient kings and emperors. At the same time, we have seen that affluence depends on energy and that energy in the form of fossil fuels is not a renewable resource. Our renewable resources will only remain renewable if we can somehow continue to avoid Man's pillage of our environment.

Why then do we choose to continue to procreate like the rabbits on the island? Sex, of course, is only marginally less of a driving force than food. Prolonged starvation of the sort experienced by the inmates of concentration camps or castaways in lifeboats eventually eliminates strong sexual desire and may even inhibit reproductive function as it does in anorexic women. On the other hand, whole races of people who are either chronically under-nourished or who experience intermittent starvation (or both) are often amongst the most prolific. Perhaps it is all down to the "selfish gene" and there is some deep biological tendency in people, as in plants, to flower when stressed or when facing elimination but, in many societies, there is the simple but painful fact that children hold the only promise of survival for parents; particularly ageing parents. In the West, life may be very difficult for those who must rely entirely on what the State provides but in some Third World countries only the charity of relations, friends or neighbours lies between the aged (or incapacitated) and death by starvation. In primitive rural communities, what wealth there is often resides in the number of hands available to work the land. None of this accounts, of course, for the fact that when Victorian Britain was at the height of its wealth and power it was not unknown for women to give birth to more than twenty children and large families were as common amongst the affluent as amongst the poor. Nor does it account for the prevalence of families of four children in wealthy middle class families in the United States in the 1950's. So perhaps we all tend to seek our own immortality in our children or are moved by deep seated emotions which are not met by sex without issue. Religion cannot be dismissed as entirely irrelevant in this context. Where it sets its face entirely against contraception or abortion and finds no favour in sex which is not intended to induce pregnancy, it is not going to help to solve problems of over-population. This, no doubt, is why the Archbishop of Canterbury recently said that the Roman Catholic Church should reconsider its attitude to contraception. Some would argue that Catholic countries are no more prolific than non-Catholic countries, that the forces that encourage large families are many and varied and there are some rules that are more honoured in the spirit than in the observance. Nevertheless, there is little doubt that, if the Catholic Church were to set its stamp of approval on contraception, it would encourage population control measures in those countries in which it has a major influence on government.

That there is a need to limit population is self-evident. It is simplistic to suppose that there will soon be standing-room only, or that Earth could not accommodate many more than it does at present. The awful fact that the *net gain* in population is currently running at about 250 per minute, (and rising), speaks for itself. In the short term this will continue to happen without most of us being even aware of it. This is not to say that the majority of these newcomers will have an easy life but we are talking about net gains. When we talk about 250 per minute we are not talking about births but the extent to which numbers of births exceeds numbers of deaths. Despite the enormity of this problem we are not going to be overwhelmed by it, this year or next. The world could, and will, find space for the an extra New York next year, the year after, and for many years to come. Moreover, given modern technology, and despite dwindling oil reserves, it will be able to feed most of these extra mouths well into the next century. That much seems certain but it is equally certain that such a path cannot be followed indefinitely. The smallest constraint is space in which to live. Not all of Earth's surface is very hospitable, indeed much of it is extremely hostile but there is still lots of it. Race and religion are much bigger factors. Despite the demise of what was the Soviet Union there are many Russian Jews who are prepared to exchange the vast expanses of their native country for the tinyness and dangers of Israel. All over the world there are populations which would apparently prefer to engage in mayhem than to exist in peaceful coexistence with others. The motto "*what we have, we hold*" could be aptly applied to 99% of the worlds population. Already, the European Community reacts fearfully to the prospect of increased migration from the east and the south. Space for effective agriculture is also not limitless. Arithmetic in Chapter 6 led to the conclusion that 300 to 400 square meters were needed to supply each person with a modest sufficiency of potatoes. Meat eaters would need a great deal more. Multiply this space requirement by 13 million per annum and it points, once again, to an eventual (and perhaps not too distant) limitation. Pollution may seem trivial by comparison. It is not. True we all produce some excrement and many of us a great deal of garbage. This can do all manner of nasty things to our environment and is not to be lightly brushed aside but it pales into insignificance by its potential threat to our atmosphere. The impact of CFCs on the ozone layer (Sections 7.7 to 7.9) is no longer dismissed, even by the chemical companies which produce them. Rising carbon dioxide is very real even though there are immense uncertainties about consequences. At worst, this sort of pollution, by affecting agriculture, climate and sea levels could drastically affect living space and food supplies.

> "*In 1971, Georgescu-Roegen published a book entitled 'The Entropy Law and the Economic Process' which puts forth the proposition that waste, which may be considered to be degradation* (or an increase in probability or entropy in a thermodynamic sense) *of the environment, is the inevitable consequence of economic activity and that the production of "bigger and better" washing machines, automobiles, and superjets must lead to "bigger and better" pollution. On thermodynamic grounds therefore it is not possible to get rid of pollution simply by using nonpolluting industrial techniques. Degradation of the environment is a corollary to economic activity*" ....Geoffrey P. Glasby

All of these possibilities must be taken seriously but, in the end, it is energy which is the greatest constraint of all. Contemporary agriculture is a relatively inefficient way of converting oil into potatoes. Whether or not the line is drawn at the farm gate, or at the supermarket door, there is no way in which people can be fed without the use of large amounts of fossil fuels (Table 6.2). Moreover intensive agriculture brings related problems in its wake. A decade ago much of the agricultural research in the Netherlands was focused on agricultural pollution rather than production. The Dutch are clean and tidy people living in a small densely populated country. Nevertheless they earn a lot of money from the tomatoes, cheeses etc., that they sell to their neighbours. No self-respecting Dutchman likes to live down-wind from a pig-farm. Spreading pig shit on fields may be economically and even ecologically sound but, like so many other things, we would prefer it not to be done in our back yard. There are arguments about how great is the impact of manure and fertilisers on our drinking water but there is no doubt that the provision of clean and acceptable drinking water is a major problem world wide and that all manner of things, including pesticides, can find there way into areas where they are neither desirable or welcome. There are innumerable lakes, once teeming with fish, now green and stinking because of algal blooms made possible by phosphates and nitrates draining into them as a result of human activities. Methane released by grazing herbivores (page 169) is not entirely a laughing matter and even something as seemingly innocuous as growing rice can do substantial harm once it is on a large enough scale. The CFCs released from factories producing frozen foods are much greater in volume than those released from domestic refrigeration although these can also be justifiably regarded as part of the whole process that brings food to the table. In the end, however, it is CO_2 (by its effect on the atmosphere) which looms largest in regard to agricultural pollution just as it does in regard to industrial pollution. Geologically speaking we are actively engaged in putting CO_2, fixed into plant products by millennia of primeval photosynthesis, back into the atmosphere in a very short time. Clearly we cannot continue to do this indefinitely. Sooner or later there will be no more such CO_2 to return. Nor, by implication, will there be any more coal or oil; at least none that we can afford to burn.

> [Hubbert has calculated that Man will utilise 80% of the entire fossil fuels accumulated by 600 million years of photosynthesis in a mere 300 years.]

At this point, the agriculture, presumably needed to feed much larger populations than at present, will need an alternative source of energy. True, we could go back to horses and peasant labour. We could leave the combined harvesters rusting in the fields and use animal manure rather than artificial fertilisers. We could abandon selective herbicides and pesticides. Perhaps, in the end, the world would be a happier place. It was once written that the English country gentleman of the eighteenth and

early nineteenth centuries led the best life ever so far experienced by mankind. The poorer workers on his estates might not have have thought themselves so well blessed but a modern and more equitable rural society could still have much to recommend it, particularly if it proved feasible to retain some aspects of contemporary technology and medicine. What is quite certain, however, is that such a population would not be large. Living in equilibrium with the environment is as feasible now as it was before oil was thought of but only on the basis of the small population (an order of magnitude smaller than present population) that contemporary photosynthesis supported in the past. It is difficult, of course, to imagine a world in which oil is not consumed in vast quantities in internal combustion engines but edifying to look at how this has changed in a relatively short time. When oil was *re-discovered* in the UK. in the early part of the last century it was actually exported to the United States for use in cooking and lighting.

"About the year 1847 E. W. Binney of Manchester, England, called attention to the petroleum discovered at Riddings, near Alfreton in Derbyshire, and a few years later he, together with James Young and others, commenced the manufacture of illuminating and other oils from it. The supply of crude material from this source soon became inadequate, and they then commenced distilling the Boghead mineral that had been found near Bathgate in Scotland. The success attending this enterprise soon attracted attention in the United States of America and a number of establishments were in operation in the course of a few years, some of them being licensed under Young's patents."

The author of the above entry in the Encyclopaedia Britannica of 1885, not knowing about the forthcoming impact of motor vehicles and aircraft on life in the Twentieth century, struggled with prediction:-

"Looking towards the past, it may be said that petroleum has attained universal diffusion as a lighting agent; it is fast displacing animal and vegetable oils as a lubricator in all classes of bearings, from railroad-axles to mule-spindles, and also where other oils are liable to spontaneous combustion; it is very largely used as fuel for stoves, both for heating and cooking; it is very successfully used for steam purposes when other fuel is scarce and petroleum plentiful; it is likely to be used for production of pure iron for special purposes; and it has become a necessity to the apothecary as petroleum ointment. Looking towards the future, what assurance have we that these varied wants, the creation of a quarter of a century, will be satisfied? While it is not probable that the deposits of petroleum in the crust of the earth are being practically increased at the present time, there is reason to believe that the supply is ample for an indefinite period. Yet the fact is worthy of serious consideration that the production of petroleum as at present conducted is everywhere wasteful in the extreme."

He was right about the wastefulness but if he had looked into a better crystal ball he might not have wished to use the term "indefinite" except in its sense of "uncertain". Of course, prediction is a fools game. Who in the heyday of the English country gentleman could have foreseen radio, television, telephones, cars, word-processors, commonplace travel by air, let alone genetic engineering and all that is implied by that? So there may be a future "technical fix", quite unimaginable now, which would not (like nuclear fission and nuclear fusion - Section 9.4) create more problems than it would solve. All that can be said is that if there is to be such a technical fix it is not on the horizon now and that all existing alternatives, however desirable, would require some drastic downturn in the number of people that the planet could support. At the end of the day we are back to the laws of thermodynamics. These cannot be repealed or circumvented. There is a universal tendency for entropy (i.e. disorder) to increase. Living creatures are the only things which can reverse this tendency for a while. They can only reverse it, building order out of chaos, by utilising energy. Putting aside a few specialised bacteria, the source of that energy is (and always has been) nuclear fusion of protons in the sun. The process of energy transduction has been photosynthetic. Life, as we know it, is based as much on past photosynthesis as on present photosynthesis. The more of us there are, the quicker we will exhaust the products of past photosynthesis and the more strain we put on contemporary photosynthesis.

"Whats to do about it? Lets put out the lights and go to sleep"

In the circumstances this might not be the best remedy but restriction of family by decree, in the manner tried by the Chinese, would not be widely welcomed however well intentioned. Similarly, the Indian attempt to reward vasectomy with transistor radios, while understandable, is hardly laudable. We can really only hope for changes in society, such as provision in old age, which will make more than two children per "family" seem socially, morally and financially less desirable than the alternatives. In the West it ought not to be beyond the wit of politicians to arrive at taxing policies which help those on low incomes but do not, at the same time, offer financial rewards to those on high incomes who choose to have large families. Much of this could be less painful and less intolerable to some religious convictions than might be imagined. In the U.K. and Germany, populations are more or less stable, certainly not expanding in the way they were thirty years ago. Easier contraception and the availability of abortion on demand may have played some part in this but the situation has not been materially different in Ireland where such steps are frowned upon or even not permitted by the nation's constitution. It is economic and social considerations which have been more important. Western women now frequently wish to work outside their homes as well as having children. Usually they are encouraged to do so by their male partners. Many more, who do not work from choice, do so because it is the only way in which they can afford the things that they would like to buy for themselves and their families. For reasons such as these, families are postponed and the biological constraints on the age of child-bearing diminishes numbers. In countries like the U.K. large families are no longer regarded as "normal" or desirable. In such societies, having only two children is not regarded as a disaster. Some couples do not wish to have children or are not able to have two or even one. This is a prescription for stable or declining populations. It seems preferable to the converse in which an uncomfortably large population would become limited by factors quite outside its control.

9.3 An Inversion of Ends and Means

Until relatively recently, nations and communities have tended to evaluate their future energy needs almost entirely on the basis of existing *per capita* consumption and predictions about future growth in population. Little or no thought was given to the continuing availability of fossil fuels, to the impact of alternative technologies, to effects on the environment or on social implications. To quote Lovins:-

> *"this approach is unworthy of any organism with a central nervous system, much less a cerebral cortex. For those of us who also have souls it is almost incomprehensible in its inversion of ends and means"*

However incomprehensible, this is precisely the sort of argument used by the proponents of nuclear power who imply that there is no real alternative. Even those who are concerned about their environment and yet regard themselves as realists, are often genuinely puzzled about what might be done to pave the way to an acceptable future. In this chapter some of the alternatives will be touched upon but the most important first step is implicit in the above quotation. If we are running into debt, there are really only two things that we can do about it - acquire more money or spend less. There is always the possibility that we can improve our financial resources by earning more, by robbing a bank, by successful gambling or by stumbling on hidden treasure. Mankind as a whole has often been well-served by those who have produced great works of literature, or whatever, because they needed the money. As a general rule, however, the course of action which most recommends itself in the short term is to spend less on those things which offer no future return or amelioration of our debts. There seems to be a natural law which ensures that if you put your last pound, dollar, rouble or deutschemark on to a horse it will almost inevitably fall at the first fence. Robbing banks or old ladies has little to recommend it and chancing on nuggets of gold is a fairly infrequent occurrence. These facts are self-evident but past energy policies have rarely been better based. The first response to an energy shortage has been to seek more energy rather than to use less. The acquisition of more has often had consequences much worse than languishing in jail for theft. The concept that someone, somewhere, will come up with some totally new technological fix in the future is like counting on winning a lottery; it happens but don't hold your breath.

Imagine an immense stone wall in the North East of England, an area not noted for the clemency of its climate. Along the wall, having obviously once grown in profusion, the remains of peach trees. How could that possibly have been achieved? The answer lies on the other side of the wall which turns out to be hollow and constructed out of fire brick. Every few metres along the wall, a blackened aperture; running alongside the wall, the railway line that brought the coal wagons to feed the fires. The owner of the wall was a very wealthy man, the owner of the nearby coal mine. His answer to British winters was to heat his whole environment. This is not a course of action which is open to us all and it would, of course, have made its own minor contribution to the social deprivation of nineteenth century miners, to increasing carbon dioxide and to the pollution which was already changing the British landscape.

Today, of course, no one would contemplate anything so outrageously expensive, or would they? Sadly little has changed. Putting aside the Chernobyls, the pollution of the Irish Sea by the Sellafield nuclear reprocessing plant and the devastation caused by burning "brown" coal in Eastern Europe and we can still arrive at energy policies which are as indifferent to environmental considerations and as far removed from rational thought as the hole-in-the-wall solution. Currently, for example, an electricity generating company in the U.K. is trying hard to displace relatively clean coal with an imported product so rich in "nasties" that it has environmentalists apoplectic with outrage. British Gas offers a substantial discount to those who use most of what it produces - a natural and logical extension of discounting for bulk or wholesale purchasing which has been a common aspect of business procedures the world over for many years. Inevitably, of course, this is a real disincentive to energy conservation. A local education authority, bent on saving, succeeded in using 20% less gas for heating. It was rewarded by a 20% higher bill as a consequence of having dropped out of the discounted price range. It is reported that the response of the gas-producing utility was to tell the education authority to burn off the gas which would otherwise have been saved in order to get back into the lower price bracket. All of this could be dismissed as sound commercial practice, bringing benefits to industry but it has to be asked if it should be allowed. Burning more gas is not always bad. It is better, for example, (in terms of carbon dioxide released per Kjoule) than burning coal. It may make more sense, in the long term, than burning oil but it is still, basically, something to be discouraged. Those who live in the environment in which gas is burned have as much right to be protected from the consequences of burning it as they have to be protected against passive smoking; an expectation that their governments will not stand idly by and permit some to profit by inflicting unreasonable penalties on others. Even the ethics of using vast quantities of electricity to make artificial snow (for skiing) is questionable. For sure there is general acceptance of the fact that everyone should be allowed maximal freedom to spend their money as they wish but, equally, no one cares to have someone else's rubbish dumped in their backyard. If we accept the principle that "the polluter pays" it ought surely to follow that using energy should be discouraged and that saving energy should be rewarded.

The alternatives to present means of energy production are numerous. Some will be considered below. There is no doubt, however, that the first, practical and imperative alternative is to ask if we can manage with less. There was a time in Man's past when he lived in equilibrium with his environment. Some few in the depths of fast vanishing rain-forests still do. No doubt, in many ways, they have a hard life. Whether it is harder or more enviable than that of some "inner-city" dwellers, in otherwise affluent societies, is debatable. However, it does not follow that the only escape from the deprivations of "civilisation" is a return to an ancestral life style. We cannot in any case go back to what prevailed before the Industrial Revolution. The question that should be asked, however, is whether or not it is really unthinkable to have the best of all worlds. Asked what he thought of American civilisation, Gandhi apparently replied *"I think that it would be a good idea"*. He

had a point even if his own solutions were perhaps not eminently practical at that time. Today, in many respects, the "need" for urbanisation has gone. We no longer have to think of cottage industries only in terms of Harris Tweed. Human ingenuity is such that we can probably continue to squander our resources and pollute our planet for decades, or even centuries, to come but we cannot do these things indefinitely. If Man (with anything like the attributes which we currently admire in ourselves, or envy in others) is to continue into the future he will have to cry a halt to endless expansion, despoliation and pollution.

The twentieth century has seen all manner of political, religious and social orthodoxies challenged. On the other hand there has been scarcely any erosion of the very similar factors in these categories which divide people on the basis of culture, colour, language and religion. These, in turn, are exacerbated by pressures of population, opportunity and economics. Those members of the Muslim community of the United Kingdom who currently call for the death of Salman Rushdie cannot comprehend why a supposedly tolerant British population is so slow to understand why the "Satanic Verses" causes devout Muslims so much offence, why the British are seemingly so indifferent to blasphemy (in either a Muslim or a Christian context) and, at the same time, so deeply dismayed by the Muslim reaction. Nor would better understanding or education help. One person's tastes in food, the climate which he or she enjoys, or the values which he or she esteems, may be anathema to another. Even within families there are aspects of behaviour which no amount of love, tolerance, or understanding can accommodate. Such problems are mostly solved by a degree of separation. Offence only follows when people of different tastes, cultures or patterns of behaviour are thrown together, not from choice but because of economics. What the Muslims of Pakistan do in Pakistan is of little concern to the British in Britain, or *vice versa.* Aggravation only follows economically propelled association. Energy, economics and population levels are clearly inter-related. The world is large enough to support many hundreds of millions of people living in equilibrium with their environment and almost any number of groups with widely disparate beliefs and requirements. It is equally clear that its resources are not large enough to maintain indefinitely the present rate at which energy is being used. It is threatened by effects, on the environment, of energy use. In turn, the rate at which energy is being used is directly related to global population and to the competition between different populations for available energy and energy-related sources. To propose much smaller communities, capable of freely using much larger amounts of *renewable* energy and sufficiently isolated from one another to cause little real offence might seem a naive, simplistic and hopelessly optimistic solution for existing strife. The cynical and pessimistic alternative (continuing population growth, feckless squandering of resources and ever increasing pollution) would seem to offer little prospect of a world worth living in. Perhaps the real future will be somewhere between these extremes. The following sections consider some of the options that still remain open to us.

9.4 Going Critical

The present government of the U.K. is wedded to nuclear power just as it is wedded to its "independent" nuclear deterrent - in part, a cause and effect relationship.

> *"For the first decade of British nuclear activities, uranium was not a fuel and reactors were not for generating electricity. That was all to come later. In the 1940s and early 1950s uranium was bomb-material, and reactors were a way to turn uranium into a better bomb-material"* ....Walter C Patterson.

Against a background of increasing public concern about the environment, the fact that nuclear electricity generating plants, once built, would not add to the Greenhouse Effect came as a godsend to politicians desperate to bolster a nuclear image profoundly diminished by Chernobyl. Particularly if the enormity of radioactive waste disposal could be conveniently disregarded, nuclear power could be proclaimed as "green". Beneath the evident sophistry there is a real point. Nuclear electrical generation does not normally discharge carbon dioxide to the atmosphere (although there have been some inadvertent releases of radioactive carbon dioxide). It is, therefore, reasonable to ask how much carbon dioxide emission could be spared. Fortunately, the arithmetic is simple. Nuclear plants currently provide about 7% of the U.K.'s primary energy.

[Primary energy is the sum of all fuel inputs used in producing the energy purchased by final users. It is normally based on a fossil fuel equivalent. Thus, in 1991, U.K. primary energy was equivalent to 203 million tonnes of oil and nuclear power is put at 7% of this on the basis that British nuclear electrical output would have consumed about 14 million tonnes of oil if it had been produced by conventional power stations.]

By the turn of the century, given the real probability of more of the older, ageing, nuclear installations being closed, on grounds of safety and economy, it could be as little as 3%. It is just conceivable that, despite public opposition, despite an infirm economy and a diminished manufacturing capacity, the U.K. might find it possible to increase its nuclear capacity again by the year 2010. At present, about one third of U.K. carbon dioxide emission is directly attributable to generation of electricity (as it is in the United States) so, at the most, nuclear generation could save an output of carbon dioxide equivalent to about 6%. Doubling present nuclear generation would, therefore, save further carbon dioxide release by the same order of magnitude. A saving of 5 or 6% (say 25 million tons of carbon dioxide *per annum* is not inconsiderable but neither is it a lot (global carbon dioxide emission is about 5000 million tons per annum) The same sort of arithmetic can be applied *pro rata* to other countries, some like France and Taiwan with relatively large investment in nuclear power, others like the United States where development is currently stagnant or like Sweden where it is to be phased out. In total it seems unlikely that, in the foreseeable future, switching from fossil fuels to nuclear power would have anything other than a relatively modest impact on atmospheric carbon dioxide concentration. It can also be argued that, since nuclear energy is so much more costly than energy conservation, it would, for given expenditure, displace much less coal, gas, or oil-fired electricity generation plant and thereby *actually contribute to carbon dioxide emissions and global warming rather than the converse.*

At first sight, nuclear *fusion* would solve everything. There is absolutely no doubt that it is an extremely effective source of energy. All of our metabolic energy and all of our fossil fuels come, originally, from nuclear fusion of protons in the sun. This conversion of hydrogen to helium involves a slight decrease in mass (4 hydrogen atoms weigh 0.7% more than the helium atom that is produced by fusion). Remembering that energy and mass are interchangeable according to Einstein's famous equation:-

$$E = mc^2$$

(where E is the energy, m is the mass and c is the velocity of light) we can calculate that 0.7% of the mass of 4 protons is a very small number indeed it still represents an awful lot of energy, relatively speaking. Thus, if we do our arithmetic on the basis of *one gram* of hydrogen (i.e. Avogadro's number of hydrogen atoms) we get

$$E = (1 \times 0.07) \times (2.99 \times 10^{10})^2 = 6.39 \times 10^{18} \text{ ergs}$$
$$= 6.39 \times 10^{11} \text{ joules}$$

If, on the other hand, we wished to produce that amount of energy by oxidation we would need to burn 4,580 *Kg* of hydrogen or 15,270 *Kg* of fuel oil. Such figures suggest that we would be on to a good thing with fusion but the practical problems are immense. First of all we have to learn how it is done. Despite occasional bouts of optimism we have not yet discovered how to create conditions on Earth which would allow this process to proceed and give rise to more energy than it consumes. As a pure scientist, I applaud fundamental research of any sort. As a tax payer and a biologist, seeking support for research, I must always question costs. It may be parochial to suggest that the money spent on Concorde would have benefited the travelling public more if it had been spent, instead, on the London underground system but it could scarcely be denied that if the billions which have gone into the research and development of nuclear (fission) electricity had been invested in renewable energy technology the world would be a better place. We need to be as financially circumspect in regard to fusion as we should have been about fission even if we take the view that fission always had more to do with power politics than the generation of electricity. Then there is the question of pollution. If there were no other feasible alternatives we might have to regard fission as something worthy of consideration despite all of its associated risks and the remorseless accumulation of toxic products which cannot be made safe, or stored in a cheap and safe manner. It is tempting to think that nuclear fusion would not bring comparable problems in its train but *fusion also produces radioactive materials* because of the need to contain neutrons (page 4). On a site 90 million miles from Earth, fusion is fine. Nearer home it must, in the light of the fission story, be viewed with considerable suspicion.

CHAPTER 9 - IS THERE ANOTHER WAY?

It will have already become clear, from what I have written elsewhere in this book, that I am not wildly enthusiastic about nuclear power. To be blunt, I regard it as dangerous and expensive lunacy and I take some comfort in the fact that this view is widely endorsed by many others in the scientific community, even though there are those who still see no practical alternatives. Lest my view of the nuclear industry immediately casts me in the role of a muddled thinker, with all of the characteristics which that industry loves to attribute to its critics, may I restrict myself to a mere two quotations in my support.

Sir Brian Flowers chairing a Royal Commission on "Nuclear Power and the Environment" said:-

> *"We believe that nobody should rely for something as basic as energy on a process that produces in quantity a product as dangerous as plutonium ... We believe that security arrangements adequate for a fully-developed international plutonium economy would have implications for our society which have not so far been taken into account by the government in deciding whether or not to adopt that form of economy"* ....Sir Brian Flowers.

and in 1976, Gerald Ford then President of the United States said:-

> *"I have concluded that the reprocessing and recycling of plutonium should not proceed unless there is sound reason to conclude that the world community can effectively overcome the associated risks of proliferation ... I have decided that the United States should no longer regard reprocessing of used nuclear fuel to produce plutonium as a necessary and inevitable step in the nuclear fuel cycle"* ....Gerald Ford

But it was left to Tony Benn, himself extremely honest about his own basic understanding of the underlying science, to put the politicians' dilemma in a way which carries a message for us all.

> *"There is another set of factors to which reference has been made in public debate: I would describe them as domestic political factors arising out of two considerations. One is the problem of security and the risk of terrorism and the second arises from what happens when you have policies so complex that the democratic process finds it hard to come to terms with the choices that have to be made. Certainly as a Minister with these responsibilities... ...I have always found nuclear policy the most difficult because Ministers are not experts, they are not scientists, they are not engineers, they are not qualified to assess in any way the technical decisions that had to be made., And, yet whether you look at it in terms of the environment or safety or energy policy, it is essential that nuclear policy be preserved within the democratic framework of control and not subcontracted off to those whose only claim to reaching decisions might rest upon their technical qualifications. I think it would be very frightening indeed if we were to say that our fuel policy required us to adopt a technique of production like nuclear power which in its turn required the decisions to be taken from the process of government answerable to Parliament and the public, and put into the hands of those whose special qualifications for deciding them would rest upon their technical knowledge".* ....Anthony Wedgewood Benn

An economic footnote might be added. Margaret Thatcher and her successor, John Major, are great believers in "market forces". This is why British Coal can only look to a future if it can produce coal at the same price as imported coal. Why, it is asked, should the rest of British industry be asked to pay more for electricity, to stop British Coal going into terminal decline? As a political philosophy based on "market forces" this argument is hard to fault. Where market forces go out of the window is the fact that all British consumers pay a higher price than they need for their electricity because the distributors are obliged, by law, to buy electricity (from nuclear generators) which is twice as expensive, per unit (5.5 to 7.5/pKWh), as anything derived from coal. (3-4p/KWh for new coal-fired plant with flue gas desulphurisation).

In another respect "market forces" have redeemed themselves. Certainly in the U.K., Government consistently closed its ears to the cries of anguish from "Greenpeace", "Friends of the Earth", and the greater part of the scientific community all of which found no merit in nuclear power. Even Chernobyl, which contaminated British uplands for years to come, was dismissed as "*something that couldn't happen here*" despite the appalling example of Windscale (now Sellafield). But then Government sought to "privatise" the industry (i.e. to sell the plant owned by the State to private investors). Wise investors declined to buy because of the high running costs, the costs of decommissioning etc., and Government's refusal to add a further £2.5 billion (in the way of risk guarantees), over and above those already included in the privatisation deal. In short, we are left with nuclear power, not because it is simply too awful to contemplate but because no one in their right mind was prepared to buy it. Sadly, British nuclear power has been comprehensively protected since then by guarantees that its output will be bought irrespective of cost. These guarantees expire in 1998. If not extended (and there is immense pressure from the privatised electrical industry and consumer groups to deny extension) nuclear power will be simply unable to sell its output. Thereupon we might all breathe a sigh of relief if it were not for the legacy of the "Magnox" plants. Whatever happens it is most unlikely that these will be allowed by safety considerations, to operate beyond the year 2000 and then decommissioning will become a significant fraction of national expenditure. As it is, "safe" long term storage of existing radioactive materials has already been postponed by 50 years, a margin which might be perceived by the uncharitable as long enough to ensure that all those currently engaged in policy making will be safely buried themselves.

To authorities like Professor John Surrey (of the Science Policy Research Unit of the University of Sussex) a nuclear revival seems remote, only to be contemplated if circumstances altered such that nuclear power became much cheaper than the alternatives.

> *"Given the public disquiet expressed over problems at the back end of the fuel cycle, chiefly over reprocessing at Sellafield and the long-term storage of radioactive wastes, there are certain other preconditions if nuclear ordering is to be revived and accepted by the public. The first is clearly the absence of a further major nuclear accident as at Three Mile Island and Chernobyl, which might well lead to a moratorium on existing as well as new nuclear plants.*
>
> *The second precondition for re-launching nuclear power on a substantial scale is the development of a reactor design with inherently safer features (e.g. giving longer response times under fault conditions) so as virtually to eliminate the chance of a catastrophic accident. This would need large scale funding for R & D and prototype experience over about 20 years and it scarcely seems a realistic option at a time when no ordering is taking place in the UK and only a little overseas. Changing reactors' systems would again*

create upheaval and strain further the resources of the nuclear plant industry which has no orders after Sizewell.

The third precondition is the emergence of a broad social consensus on the issue of the long-term handling of radioactive wastes. This is essential as long as the UK continues to reprocess spent fuel. Were reprocessing to be abandoned, it would be necessary to devise publicly acceptable policies for the long-term storage and ultimate disposal of spent fuel and to write-off the new Thermal Oxide Reprocessing Plant at Sellafield, which has cost £2.7 billion to build, and forego its lucrative export earnings. Stopping reprocessing would also in the longer term prevent the UK stockpile of plutonium, which has no civilian use in the absence of fast reactors, from getting even bigger."

Only good fortune prevented Sellafield going the way of Three Mile Island and Chernobyl. It persists as a major threat to the environment. Nuclear energy is not the answer to our problems. There are , of course, still those who take a different view. Dr. Robert Hawley is one. According to "The Observer" of May 17, 1992, he states:-

"I think if the public were better informed people would be less misguided. It is really a question of getting people to listen to the facts. The simple, plain fact is that if you go back to the electricity side and took all the environmentalists' objections on board, we would be sitting here in the dark"

Who, might we ask is Dr. Robert Hawley? There are no prizes for correctly guessing that, at least in May 1992, he was the newly appointed chief executive of "Nuclear Electric".

SATISFYING FUTURE ENERGY NEEDS

9.5 The Virtues of Thrift

If the draconian solution of switching to nuclear energy as a means of delaying global warming is unlikely to work to advantage in the time available and may even bring economic as well as other disadvantages in its train is it conceivable that merely saving energy could do as well or even better? Perhaps surprisingly even the most conservative estimates put the potential contribution of conservation (to decreases in carbon dioxide emissions) much higher than the potential sparing effect of nuclear power. Indeed its most optimistic proponents see it as the complete answer to carbon dioxide increase in global warming. According to Lovins, for example, energy-saving (in the period 1984-86) expanded U.S energy availability by nearly thirteen times as much as did nuclear power despite the fact that nuclear energy was 80 times more heavily subsidised than efficiency improvements. Moreover, he contends that:-

"full practical use, in existing buildings and equipment, of the best electricity-saving technologies already on the market would save about three-fourths of all electricity now used, at an average cost certainly below 1c/kW.h^{-1} and probably around 0.6c/kW.h^{-1} much less than the cost of just running a coal or nuclear power station, even if building it cost nothing; and " -

"full practical use of the best demonstrated oil and gas saving

technologies (many already on the market and the rest ready for production within 4-5 years) would save about three-fourths of all oil now used, at an average cost well below $10/bbl and probably nearer $5-6/barrel less than the typical cost of just finding new domestic oil."

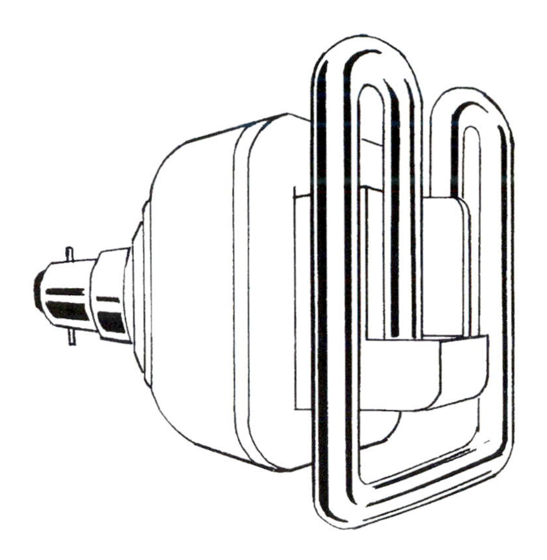

He also makes the point that (305)

"most of the best electricity saving technologies on the U.S. market today were not available a year ago and the same was true a year ago" - i.e. that the pace of technical innovation in energy conservation is manifestly rapid".

In terms of what can be readily understood he cites the following example.

"An 18 watt compact fluorescence lamp, producing the same light as a 75-watt incandescent lamp for ~ 13 times as long, will over its lifetime save about one-ten-millionth as much electricity as a huge (1,000-megawatt) power station generates in a year. A single such lamp will thus

avoid the emission from a typical U.S. coal plant of

a ton of CO_2, which adds to global warming, and
~ 8 kg of SO_2. which contributes to acid rain, plus NO_x, heavy metals, and other pollutants; or

avoid the production by a typical nuclear plant of

half a curie of strontium-90 and cesium-137 (two major components of high-level waste) and

~25 milligrams of plutonium - about equivalent in explosive power to 385 kg. of TNT, or in radiotoxicity, if uniformly distributed into human lungs, to at least ~ 2,000 cancer-causing doses.

Yet far from costing extra, the lamp will save

the cost of a dozen ordinary lamps and their installation labor (typically totalling ~$20), plus

the cost of generating 570 kW-h of electricity (typically ~ $20-30 worth of fuel),

and, during its lifetime, will defer

approximately $200-300 worth of generating capacity, reserve capacity, grid and fuel-cycle equipment.

Since the lamp's typical retail price is only $15-18, and its real resource cost is about a half or third of that, its use generates for society, at least tens of dollars of net wealth, not counting its avoided environmental costs."

This message has already been taken to heart by a number of electricity utilities in the United States. For example:-

"In a $215-million plant that is the first of its kind, Boston Edison, one of New Englands largest utility companies, will at least temporarily forestall the need for new power plants by paying instead for an array of energy

conservation measures. The conservation programme, negotiated by representatives from the utility, state government and local environmental groups, is hailed by all three groups as a "revolutionary shift" and "a model of enormous importance for the rest of the nation".

The plan is expected to reduce the demand for electricity in the Boston area by a thousand million kilowatt hours over the next five years, roughly equivalent to the amount of power generated by a small 120-MW power plant. It is particularly notable for its timing, announced shortly after the Seabrook nuclear power plant in neighbouring New Hampshire was cleared finally for its full-power licence after a gruelling and costly decade-long ordeal (see Nature 344, 96; 8 March 1990).

Boston Edison will invest immediately in measures such as making new buildings more energy-efficient than present standards require, retrofitting existing residential and commercial buildings and supplying energy-efficient light bulbs free or at low cost.

During the next five years, the utility expects to save two dollars' worth of electricity for every dollar invested. In an innovative economic incentive system, the utility will be allowed to retain a portion of this expected $400 million in energy savings as profit in addition to recouping its costs. The rest of the savings will be passed on to customers. With such a market incentive, conservation becomes equally if not more profitable for the utility as the construction of new generating capacity.

Although conservation plans are not new, observers say that the Boston Edison announcement marks a trend towards a distinctly new movement on the part of electric utilities towards conservation. The intense opposition to the Seabrook nuclear power plant may be an extreme case, but is indicative of the difficulty of bringing any new power plants on line."

....Seth Shulman ("Nature", 282, 344, 1990).

Similar economic considerations apply to the more complex issues of abating nuclear proliferation, and reducing dependence on Middle Eastern oil. *"Achieving these benefits by improved end-use efficiency lets one make money on the deal."* This is, in a variant on Marv Goldberger's memorable phrase, *"spherically sensible"*. Energy efficiency makes sense no matter which way you look at it. It should be done to save money, even if we're not concerned about such issues as global warming.

9.6 Would Energy Conservation Help?

The implication of what is set out in the preceding section is that not only is energy conservation a feasible and laudable strategy but that it might, unaided, solve the greenhouse problem entirely. There are, of course, as many uncertainties in this approach as in any other, not least the fact that

*"people and nations are unlikely to behave wisely (*until*) they have exhausted all other alternatives".*

Even so, it is both possible and informative to make educated guesses. These put possible future increases in saving efficiency based on existing technology at somewhere between 4 and 10 times those achieved in 1973. The more optimistic of these give considerable food for thought. One German study, for example, envisages a world of 8 billion people in 2080 "*uniformly industrialised to the level of the Federal Republic of Germany in 1973*" (at this time the FDR was one of the most heavily industrialised and energy-efficient countries in the world). Such uniform industrialisation would imply a 5-fold global increase in economic activity and a 10-fold increase in under-developed countries (i.e. the scenario is not one which is unrealistic about the natural aspirations of the third world). Nevertheless, if such a world had fully embraced the use of energy saving technologies already available in 1980 it would, in 2080, need only about one third of the energy actually used by the real world a century before, in 1980. Moreover, most areas of the world would be virtually able to meet all of their energy requirements from renewable sources. In Lovin's words there would be "*no energy problem - even on economic growth assumption that most would consider implausibly high*". Such a world would use 4 or 5 times less energy (than conventional projectives would allow), cost much less, stretch oil and gas for centuries, dispense with reliance on both the Middle East and the atom and by 2030 have "attained" a carbon dioxide level barely above today's and rising by 5ppm every three decades or so, making this part (~40%) of future global warming virtually vanish. Many other problems would also disappear such as those associated with the million bombs worth of plutonium it is proposed to extract *annually* and put into global commerce.

RENEWABLE RESOURCES

9.7 Looking at Alternatives

There are also other ways:-

Table 9.1
Renewable Resources

Solar	**Geothermal**
Photovoltaics	Hydrothermal
Active Solar	Electricity
Passive Solar	Direct Heat
Solar Thermal	Binary Cycle
	Hot Dry Rock
Biomass	Geopressured
Direct Combustion	Magma
Biogas	
Ethanol	**Ocean**
Gasification	Tidal
Liquefaction	Wave
	OTEC (p. 237)

From Renewable Sources of Energy OECD

"The modern world demands energy that is concentrated and readily available. The renewables, by contrast, are invariably the opposite. They are diffuse - the waves off the Hebrides have an average power density of around 45 megawatts a kilometre but it would require a 300 kilometre stretch of wave energy devices just to satisfy a fifth of Britain's present electricity needs. They are variable - the wind does not blow consistently over the face of the land. And their supply does not always coincide with demand - solar radiation, for example, is at its most intense in Britain in the summer months when the need for heating is at its least. Moreover, while they are often thought of as being totally benign, deploying the renewable energy technologies could exact an environmental price. Wind farms, for example, may be considered visually intrusive or too noisy; an over emphasis on energy forestry might upset the ecology of the countryside; and tidal power schemes could have a detrimental effect on the habitat of wildlife around river estuaries. The exploitation of any form of energy always has its drawbacks" ...U.K Dept. of Energy

The Department of Energy, might have added that most of the above drawbacks would pale into insignificance compared with the drawbacks of conventional sources but it does make a valid point. Nevertheless, despite the very real disadvantages of some renewable resources (Table 9.1) *they are renewable* and, collectively, offer real hope for the future.

9.8 Ocean Energy

"Water, water everywhere nor any drop to drink"
....Samuel Taylor Coleridge (The Ancient Mariner)

At present the most important renewable resource is wood (Section 9.14) still very widely used for cooking, heating and construction, and as newsprint. But, if we are to look to the future, water has considerable potential. If, as a citizen of the U.K., you visit the capital of Australia it comes as a shock to discover how little the people of Canberra pay for their electricity; that it is still possible to generate large amounts of electricity by conventional methods (water running down-hill, or from the base of dams, through turbines) in drought conditions in a country in which it might be more realistic to talk of the number of acres per sheep rather than the number of sheep per acre. In any country, with reasonable rainfall over hills, electricity can be generated in this fashion. In addition many countries are blessed with waves and tides. Let us take one very pertinent example. There are many sites in Great Britain which would lend themselves, admirably, to the harnessing of tidal power. Paramount amongst these is the Severn Estuary. Arguably this could displace nuclear power entirely in the U.K. if this declined by the year 2000 to the 6.0 GW capacity suggested by John Surrey. The Severn Barrage Development Project proposes 216 turbo generators each rated at 40 MV giving a total installed capacity of 8640 MW (8.6 GW). The direct capital costs are approximately £1/Watt (i.e. about £8 billion at 1988 prices - substantially less than the cost of the British "independent" nuclear deterrent). This includes the initial cost of construction and the length of the construction period before power is generated. Once up and running, the operating costs would be low, no CO_2 would be released into the atmosphere, there would be no accumulation of radioactive or chemical waste. That is not to say that there would be no environmental costs although present studies suggest that the estuary would support more birds and fish, in total, than at present even though some species might be adversely affected. Some 200,000 man-years of work would be involved in the construction of the barrage and the turbines and, although the arithmetic is clearly simplistic, it is worth noting that the present cost to the U.K. taxpayer of 200,000 man years of "idleness" (i.e. unemployment benefit etc.) is approximately £1.8 billion. Thirteen additional sites in England have been identified as having potential for tidal barrage schemes. It is not difficult, on this basis, to conclude that the argument that there is no practical alternative to nuclear power cannot be sustained or that renewable resources could not eventually displace fossil fuels. Sadly, the decisions involved do not rest only on feasibility, cost and environmental "friendliness". Possibly the greatest obstacle in the U.K. is still the strength of the nuclear lobby. For example, Britain is exceptionally well endowed with ocean waves. There are sites off the Western Isles which compare favourably with any in the world and the total potential has been put at about 40% (about 80 million tonnes of coal equivalent a year) of the U.K.'s primary energy requirement. Despite this vast potential of inexhaustible energy

> *"the Department of Energy concluded that large scale offshore wave energy was unlikely to be economic in the U.K. for the foreseeable future"*

and therefore suspended its research and development programme in this field after spending less than £20 million (about six orders of magnitude less than that expended on nuclear research). There have even been suggestions that some of those concerned in decision making have been economical with the truth when weighing the relative merits of wave and nuclear power. The "Salter Duck" (a moored device with a row of beaks which bobbed up and down and used internal hydraulic devices to capture energy and generate electricity) was held to be a dead duck on the basis of cost but newer, cheaper, versions still continue to offer real promise. In Britain, all that remains is a number of tiny, individual devices for powering navigational aids (such as tethered buoys) and a 180 KW prototype of a "Shoreline Rock Gully System" on Islay (better known for its whisky). This involves a "Wells Turbine" in which the rise and fall of waves drives air through a device which revolves in the same direction regardless of the direction in which the air flows. A larger (2GW) off-shore machine was deemed to have poorer economic prospects than other alternative generating technologies. Evidently the engineering problems are formidable but, given the success of oil companies in exploiting North Sea reserves it is difficult to believe that a potentially larger, renewable, energy supply did not warrant further development. In the United States there is continuing interest in a pneumatic wave energy conversion (PWEC) system which works by alternately compressing and expanding air and passing it through a unidirectional rotor. A possibility that 90% of a wave's energy could be extracted by judiciously placed sub-surface vertical flaps is also being investigated.

Ocean Thermal Energy Conversion Technologies (OTEC) operate on the principle that energy can be extracted from heat sources and sinks at different temperatures. Warm surface water is pumped through an evaporator and thereby used to vaporise ammonia or freon. The vapour drive a generator and cold water condenses it for recycling. Alternatively, in an open system, the warm ocean water itself is used as the working fluid. In the most advanced systems water is flash-evaporated in an evacuated tube and the rising mist condensed by the cold water. The condensate runs down a tube under gravitational pull and, in so-doing, drives a conventional turbine. This system, if we forget the fact that all is ultimately driven by a difference between hot and cold water (like conventional power plants in which fossil fuels are used to heat the water), seems to come perilously close to getting something for nothing. Nevertheless it does work and has much to offer to island communities in the tropics. As a by-product it produces distilled water for drinking and nutrient-rich cold sea water for fish farming etc.

9.9 Gone with the Wind

Go past Hollywood and away from Los Angeles towards Riverside and beyond. In the deserts there are wind-farms, not very pretty but effective. By the end of 1985 more than one gigawatt (GW) of wind capacity had been installed in the United States, nearly 99% of it (some 13,000 machines) taking advantage of the strong summer winds in California which represent 2.5% of installed capacity and producing 1% of demand. In Denmark, second only to the United States but still way

Table 9.2

WIND ENERGY PRODUCED IN 1985 BY LOCATION

CALIFORNIA INVESTOR OWNED UTILITIES[1]	
- Pacific Gas & Electric Co.	50.0%
- Southern Californian Edison Co.	38.0%
DENMARK	5.3%
CALIFORNIA DEPT. OF WATER RESOURCE	1.6%
OTHER EUROPEAN COUNTRIES	1.3%
OTHER UNITED STATES LOCATIONS	1.3%
REST OF THE WORLD	2.8%

[1]. Interconnected to, but not owned by three utilities.
Source: American Wind Energy Association

behind, wind farms provided up to 0.5% of electricity in 1985. Wind farms of 5 megawatt (MW) capacity are proposed for the Netherlands and Belgium. Denmark has plans to install 100 MW and to provide 8% of Danish consumption by 2000. The U.K. Department of Energy believes that:-

> *"if predicted costs can be achieved in principle it is possible that by the year 2025 on shore window power could be supplying 10% of the countries electricity needs and perhaps as much as 20% in the longer term"*
>
> ....U.K. Department of Energy.

The cost of electricity generated by wind turbines is similar to that from coal-fired power stations (i.e. half that of nuclear power). While the impact of wind farms on the environment shrinks into insignificance when compared with conventional or nuclear powered plant and coal mines, there is no doubt that they are noisy (audible up to 2000 metres downwind) and visually intrusive. There is radio and television interference extending up to 3 km or more from the largest turbines.

9.10 Let the Sun Shine In

We have all heard of surfaces which become so hot in the sun that can be used to fry eggs. Dudley Moore, on film, has graphically illustrated the problems of applying tender feet to hot sand. Along these lines, solar energy is widely used in "passive" applications (particularly in countries with ample insulation) such as heating domestic hot water and swimming pools. The best commercial hot water system operates with an annual solar conversion efficiency of 40%. Typical solar water heaters in the United States comprise 5-6 m^2 of collector area, and cost the user from $3000-5000 (1987 prices) installed and meet up to 80% of annual energy needs for this purpose. Solar cooling can be based on conventional closed cycle refrigeration techniques in which the high temperature needed to drive the cycle, more usually provided by electricity or gas, is replaced by solar energy. There are also systems based on open cycle evaporative cooling (i.e. ones in which the latent heat of evaporation causes cooling and solar energy is then used to dry or regenerate the desiccant). Simple but very effective evaporative cooling for glasshouses is based on water trickling over large porous surfaces (bales of straw, bark from trees, plastics) through which air is drawn by extractor fans at the other side of the house.

9.11 Photovoltaics

Robert Hill may be regarded as the father of modern photosynthesis in that the Z-scheme (Section 4.3) is the basis of our present understanding of the photochemical events concerned in energy transduction. The Royal Society, in recognition of his major contribution, to this field awarded him its prestigious Copley Medal a distinction previously afforded to such familiar names as Joseph Priestly, Charles Darwin and Francis Crick. In 1791 the Copley medal was also awarded to Allessandro Volta (1745-1827) who is repeatedly remembered, like Ampere and Watt, in electrical terminology - as in the terms "volt" and photovoltaics". The "photovoltaic effect" itself was discovered by Edmund Becquerel in 1839, some twelve years after Volta's death, when he noticed that when light shone on one side of a simple battery cell the generated current could be increased. The first selenium photovoltaic cell followed in 1883 but it was left to Albert Einstein to describe how the interaction of photons with the electrons in orbit about an atomic nucleus (c.f. Section 1.1) could generate a flow of electrons (an electric current). As figure 9.1 shows this is essentially the same, in principle, as electron transport in thylakoid membranes initiated by light excitation of chlorophyll (Section 3.1).

In 1954, Bell Laboratory Scientists produced a solar (photovoltaic) cell from pure silicon with an efficiency of 4%. Since then the efficiencies have risen and production prices have fallen (from US $20/Wp to US $5-8/Wp in 1985 and a prospect of US $1/Wp in 1992).

["A peak watt (Wp) is the power generated on a clear day when the sun's rays strike perpendicular to the cell".]

In practical applications, groups of cells are mounted on a rigid plate and bonded together to form a module, typically about $1m^2$ in size with a generating capacity of 50 - 100 Wp and an efficiency of 9-12% (90-120 $W.m^{-2}$).

Present day photovoltaic cells were largely a by-product of space research and large arrays of solar cells are used to power satellites. On Earth they are increasingly used in circumstances in which there is a need (in the absence of mains electricity) to power telecommunications, lighting, refrigeration, water-pumping, water-treatment, navigational aids etc. but the rapidly advancing technology holds out real hope that they could become major sources of renewable energy. There are vast areas of desert in Africa, Australia and elsewhere which could accommodate massive arrays of photovoltaic modules in areas of high insolation presently making no contribution to Man's energy requirements. At a more modest level the *"Forschungshaus Remscheid"* in Germany, has the distinction of being the world's first residence with photovoltaic roof tiles, generating up to 7800 Kwh/year. If it ever becomes technically and economically feasible to tile most houses in this way the direct savings in fossil fuels would be amplified by the enormous savings inherent in transmission losses through national grids and the erection and maintenance of such systems. No doubt, there would be an increase in the tendency, already apparent, of major industrial electrical consumers to generate power, on site, by more conventional methods. This, in turn, would lead to the efficiencies associated with the utilisation of waste heat for space heating.

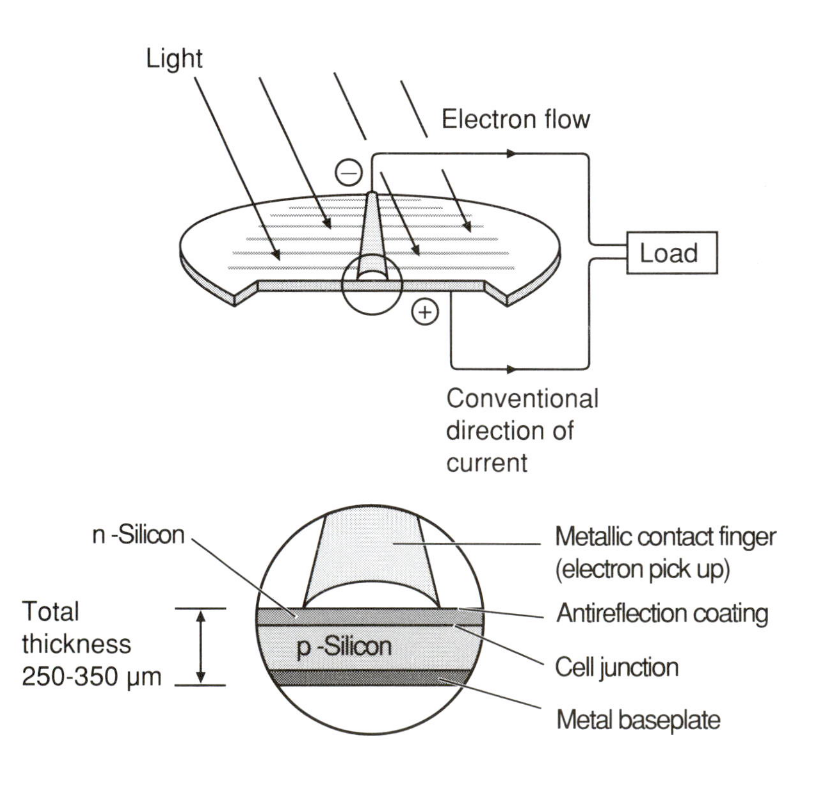

Fig. 9.1 A photovoltaic cell. (After McNelis, Derrick & Starr). The principle is similar to the generation of electron transport in photosynthesis (Section 2.8). Incident photons liberate (negatively charged) electrons thereby creating positively charged holes which constitute a sink. Thus an electric current is created, carrying electrons to fill the holes.

Despite its present usefulness and efficiency, photovoltaic generation of electricity on a really significant scale does not yet exist. To some, the prospect that it might seems fanciful. It would be difficult to argue, however, that the likelihood of appreciable generation by this mechanism is either remote or less probable than by fusion. Indeed it is a better bet, in almost every regard, than nuclear fusion.

"It is likely that a 15% efficiency for amorphous silicon (asi) thin-film cells can be reached fairly soon which would assure the viability of amorphous silicon for power applications in addition to its widespread use in consumer speciality products." ...OECD International Energy Agency.

9.12 Hot Rocks

Seek to order a Bourbon or a Scotch in a bar in the United States and you might well be asked if you want it "on the rocks" or "straight up" (i.e. with or without ice). But there are hot rocks as well.

> *"Geothermal energy derives from the heat flowing outwards from the interior of the earth, due to the cooling of the core and in some places the decay of long-lived radioactive isotopes of uranium, thorium and potassium. It can be tapped from aquifers - underground deposits of hot water in porous rocks - and from hot dry rocks "* ....U.K. Dept of Energy.

The Romans were into this with avidity, possibly learning from the Greeks, certainly combining (in Britain) their sybaritic pleasure in bathing with the availability of the warm aquifers which have kept the spa at Bath busy ever since. Go to Yosemite National Park in California and admire the geysers like "old faithful" which erupt periodically on time and throw jets of hot water many feet into the air. Go to Rotorua in New Zealand and see the spot where an erratic geyser terminally despatched a group of tourists some years ago. Marvel at jets of steam emerging from Maori graves or from lawns or flower beds. There, as in Iceland and elsewhere, this steam and hot water is put to use (Table 9.3) and indeed, in Rotorua, its unrestricted use is no longer allowed by the local inhabitants because such tapping has diminished and even threatened the continuing existence of the more lucrative tourist industry which it has attracted. All of these, like volcanoes, remind us that the centre of the earth is still extremely hot and that exploitation of such heat constitutes a source of power.

Table 9.3
Hydrothermal Electric Power Plants

Country	No. Units	MW
United States	56	2022
Philippines	21	894
Mexico	16	645
Italy	43	219
Japan	9	215
New Zealand	10	167
El Salvador	3	95
Kenya	3	45
Iceland	5	39
Nicaragua	1	35
Indonesia	3	32
Turkey	2	20
China	12	14
Soviet Union	1	11
France (Guadeloupe)	1	4
Portugal (Azores)	1	3
Greece (Milos)	1	2
Totals	188	4.7 GW

[Data for 1985 after Ronald DiPippo, *Geothermal Electric Power-Scale of the World*, International Symposium on Geothermal Energy.]

Future potential exploitation of geothermal energy may not be restricted to those few areas where hot water emerges from the ground or where there is volcanic activity. Nevertheless despite the long history of geothermal energy applications it was not until 1904 that natural steam was first harnessed for electricity generation by Prince Piero Ginori Conti at Larderello in Italy. Table 9.3 lists the number and capacity of such plants in 1985. Hot Dry Rock (HDR) technology seeks to heat surface water by pumping it into hot impermeable rocks through fissures enlarged by hydraulic fracture. One such project has been undertaken at Rosemanowes, in Cornwall, where the heat flow on the Carnmenellis granite batholith is twice the U.K. average. A three stage programme was based on examining the feasibility of a viable HDR reservoir. Bore holes were sunk to 300 metres and subsequently to 2500 metres with a view to eventually going to six metres. This latter would tap the 200°C heat required for efficient and economical electricity generation. Similar geothermal research is being undertaken in France, Germany, Sweden and Japan and at the Los Alamos National Laboratory (New Mexico) in the United States.

9.13 The Rain Forests

"Look guys, the cure for cancer could be out there in the forest and you're about to destroy it.' And they would say 'Dream on"

....Wendell Wood (Oregon National Resources Council)

The inclusion of a section on rain forests here, in a chapter which addresses the question of how we might satisfy future energy needs, carries no implication that they constitute a major energy resource. Rather it springs from the fact that the following section (9.14) summarises some of the ways in which contemporary biological materials, rather than fossilised biological materials can, and do, contribute to Man's energy requirements and that once we start to think about wood these days our next thoughts, if we have any concern at all for the environment, ought to turn to the rain-forests. It is clear that these are areas of immense beauty which deserve all the protection which we can afford them and that any exploitation should be carried out in the manner which would sustain them and guarantee their future existence. If there were ever a need for Man to exist in real equilibrium with his environment it is here. Nevertheless there are misconceptions. Thus it has sometimes been said that the destruction of tropical rain forests is to be deplored because these vast tracts of vegetation act as huge sinks for carbon dioxide. There are many compelling reasons for not destroying the rain forests but this is not one of them (see page 243). Trees temper the impact of heavy rain which otherwise denudes top-soil, and forests may absorb 20 times as much rain as bare earth. Transpiration returns much of this water to the atmosphere, so that it is effectively recycled, In these ways forests have a major effect on climate. But perhaps the most compelling reason for conserving the rain forests, even if we only view things from a totally selfish standpoint, is that they are the repository of a huge wealth of plant products and genetic information which, once lost, might never be recovered. Rubber is a perfect example. Once it was of no consequence and now, although still extremely important, is not something for which there are not satisfactory alternatives, at least in some uses. However, throughout the early development of electrical devices and road transport it was a key commodity in the development of Man's world as we know it. Manaus, formerly a place of no consequence on the Amazon, became as wealthy as an oil sheikdom because of rubber and its opera house was even able to pay Caruso to spend many weeks travelling between there and his native Italy. Eventually the perfidious British stole rubber plants and established plantations in what is now Malaysia, world wars promoted the search for alternatives and that aspect of rain forest economy is no longer what it was. Nevertheless, it is entirely feasible that there are, in addition to all of the plant products (like rubber) which are currently known to be of value to Man, others which could play a vital role in some not yet imaginable future technology - provided that, at this future time, they still exist.

Pacific yews, once felled and burned to near extinction simply to provide logging companies access to Giant Redwoods have been found to contain a chemical regarded as a most important agent in the treatment of breast and ovarian cancer. Drugs from plants are currently worth $40 million a year. Moreover, Man does not live by bread (or rubber) alone. Happily, there is now a large body of public opinion which believes that it is at least as important to cherish our natural environment and conserve it as it is to maintain artefacts of our own history and culture. Properly "managed" rain forest could not only give those who live there a better economic return than forest either totally untouched or totally destroyed but could make its own contribution to the sparing of the greenhouse effect. Fully mature forest does not do this. It is in near equilibrium with its environment, producing as much carbon dioxide and methane by respiration and decay as it fixes or withdraws from the atmosphere, in photosynthesis. Young, actively growing forest, on the other hand, consumes more carbon dioxide than it releases and this is equally true of northern, temperate and continental forests. Of course, if an existing forest is burnt, this releases carbon dioxide to the atmosphere in the same way as the burning of fossil fuels but forests, in themselves, are not the answer to the Greenhouse Effect. They fix enormous quantities of carbon dioxide, so much that northern forests are largely responsible for the annual "teeth" that are seen in the remorselessly rising carbon dioxide curve (Fig. 7.4) - i.e. the difference in the rate of photosynthesis between summer and winter, even in northern, evergreen, forests is sufficiently large to be detected in the annual records of carbon dioxide accumulation. On the other hand, mature forests *produce* equally large quantities of carbon dioxide.

It is clear then

(a) that burning forests adds to the carbon dioxide in the atmosphere

(b) that mature, stable forests are more or less in balance, consuming about as much carbon dioxide as they release, and

(c) that only actively growing forests (i.e. newly established forests or forests in which orderly felling of old wood permits active regeneration of new) act as substantial sinks for carbon dioxide.

Japan makes one of the world's largest contribution to the destruction of rain forests and uses much of the hardwood that it imports to make plywood cases to contain concrete poured into buildings during their construction. These plywood boxes are then stripped away and burned. Wood is preferred for this purpose only because it is cheap. Moreover, it is "cheaper" to destroy irreplaceable rain forests for this purpose than it is to use freely replaceable softwood. In the United States, which has a better record than most in some respects, President Bush calls upon scientists to provide more and better facts on global warming before seeking to make a real impact on his country's contribution (about one fifth of world totals) to carbon dioxide emissions. All of these and countless other arguments or evaluation of costs miss the point. We have to do what is *"spherically sensible"* (Section 9.5) because it is "globally sensible". It is worth doing because it will bring benefit in its own right. Burning rain forests is one of the largest contemporary contributions to global catastrophe but, irrespective of this, the rainforests should be preserved because they are home to an immense number of species including Man. As areas of great natural beauty, they are as worthy of preservation as the Taj Mahal.

"worthy of preservation as the Taj Mahal"

Even at a purely material level they could, if properly managed, continue to provide resources and a livelihood for their native human populations *ad infinitum*. Similarly, a major acceleration of energy conservation in the United States would, at least in the eyes of some economists (Section 9.5) make that nation even wealthier and more competitive than it is at present. A diminution in smog would be as welcome to the inhabitants of Los Angeles as would the prospect of relatively stable sea-levels to the inhabitants of the Maldives. There is nothing wrong with the accountants' approach, provided that we remember that accountants know the cost of everything but the value of nothing. We must all pay our bills but it is perfectly proper to insist that accountancy includes accountability. Saving money by energy conservation is laudable because everyone is likely to benefit in the long term. Saving money by fouling up our environment is unacceptable. Saving money in the short term, by exporting our CO_2, SO_2, plutonium, dioxins, into someone else's environment, is criminal.

9.14 Biofuels

> *"at current energy prices, biofuels could make a potential economic contribution to United Kingdom energy supplies of over four million tonnes of coal equivalent a year (4Mtce/yr) And given modest increases in energy prices, the size of the contribution could rise to fifteen Mtce/yr by 2025."*
>
> ....U.K. Department of Energy

BIOFUELS

"Biofuels" are derived from "biomass". "Biomass" is everything derived, directly or indirectly, from photosynthesis carbon assimilation. Globally, this comprises some 150-200 billion tons of dry mass annually whereas consumption is nearer 1 billion tons.

[Half of the world's population uses firewood, or recently produced organic matter, such as dung, for cooking and heating. The average daily consumption is about 1 Kg of dry biomass per person per day (i.e about 150 W on a continuous basis which is equivalent to about 300 GW, worldwide). Thus the poorest inhabitants of the Third World use half as much energy again, for cooking and heating, as for metabolic purposes, largely because of the inefficiency (5%) of open fires. It will be remembered that the corresponding figure in the wealthy West (for all non-metabolic purposes) is 5-10 KW.]

As a small boy it was my task to procure the family firewood. I scoured local shops for surplus wooden crates and chopped them into neat bundles of sticks. These were used with newspaper to pass the flames to the coal which was burned on open fires as the sole source of warmth and hot water. We all bathed once a week, not because we were basically unclean but because our hot water was strictly limited by rationing or hard economics. Had I known it, I was doing much the same task, at that time, as many in the Third World (where firewood is still immensely important for cooking and heating) continue to do on a daily basis.

Now, in partial retirement, the wheel has come full circle and, in a rural retreat, my wife and I rely on wood-burning stoves for warmth and hot water. A hectare of adjacent woodland will make us more or less self-sufficient in this regard. In the northern hemisphere, at least, wood warms four times (i.e. in felling, leading, chopping and burning) and happily it does nothing, whenever it is used, to adversely influence the level of CO_2 in the atmosphere, provided that any generation of people burns only that which it grows or allows to grow. Only actively growing forests remove more CO_2 from the atmosphere than they return. Only burning vast areas of forest increases atmospheric CO_2 and, in this regard, there is a clear comparison with burning fossil fuels. Both return, in short measure, CO_2 removed from the atmosphere over long periods. Burning the products of contemporary photosynthesis is simply a part of a re-cycling process. As David Hall and his colleagues have pointed out, a gigajoule of biomass substituted for coal would reduce CO_2 emissions by the carbon content of 1 GJ of coal (about 0.025 tonnes) and that, since biomass with a heating value of 20 GJ per tonne is 50% carbon, growing 1 GJ of biomass would sequester 0.025 tonnes of carbon. Moreover,

"biomass can play a larger role in global warming when used to displace fossil fuel than when used for sequestration because land can be used indefinitely in displacing fossil CO_2 whereas removal ceases at forest maturity in the "sequestration" strategy" ....D. O. Hall

[The "sequestration strategy" is the concept of growing trees, not to use, but as living repositories for CO_2. Actively growing trees abstract much more CO_2 from the atmosphere than they return. Mature trees do not. In the short term, therefore, *massive* planting of trees would tend to offset rising CO_2 to some extent. In the longer term a new equilibrium would be established. Using biomass, as suggested by Hall, is a better strategy.]

Even dung can play a role. The new visitor to India is surprised that sacred cows do not bring clouds of flies in their wake. Rather climate and poverty combine to ensure that cow "pats" are dried, often in neat and characteristic designs, on every available wall and surface and then used to heat cooking-pots. The women and children most exposed to the secondary products of this combustion in enclosed spaces do not benefit from this experience but otherwise it is an ecologically sound practice. Much research into more sophisticated systems is being undertaken elsewhere. Sweden does not squander its forest wastes and has looked seriously at the selection of fast growing willow varieties which could be coppiced for fuel and building. Even the feasibility of burning sawdust in a vehicle's engine has been contemplated. Yields of 25 dry tonnes per hectare have been achieved without nitrogen fertiliser, in Hawaii, by interplanting Eucalyptes with nitrogen fixing *Albizia* trees.

When bent on procuring firewood as a child I might well have chanted a little doggerel which included the immortal condemnation *"my mother doesn't like her, because she comes from Byker"* but Byker is now one of the jewels in the crown of the U.K. Department of Energy's scheme for developing biofuel technologies. A mile or two from the spot where a sad figure, contemplating execution, once wrote altogether better words (about taking the high road or the low road to Scotland), the Byker Reclamation Plant, produces densified RDF (refuse derived fuel) which can have a gross calorific content equal to about 60% of that of British coal.

Straw is also something to be conjured with. It suffers from the fact that it is very light compared with coal or even with wood, a drawback that it shares with many other forms of biomass which might otherwise be used more widely as fuel. The problem with lack of density has to do with difficulties in transport in the first place and problems with burning in the second. Oil, by comparison, does not contain water which must be transported willy nilly (thereby adding to costs) and feeding oil from a container into a furnace involves little more than a pipe and a valve. On the other hand straw contains a great deal less water than many forms of biomass and lends itself to use on the farms on which it is produced, thereby solving the additional problem of how else to dispose of it.

> *"Of the straw produced in the U.K. each year, some seven million tonnes, with an energy content of 3.6 Mtce, finds no saleable outlet and is largely burnt in the field. However straw could be put to cost-effective use as a fuel on the farm , in local industry and in rural institutions. Already some 170,000 tonnes a year are used on farms and this could rise to around 800,000 tonnes by the year 2000."* ...U.K. Department of Energy

We are all too aware of biogas (c.f. Section 7.6). Apart from the traces of extremely poisonous hydrogen sulphide which, when present, can give it such an unpleasant odour, biogas is usually a mixture of methane and CO_2 so that burning it is ecologically desirable in that methane, once burnt, becomes CO_2 which is less damaging than methane itself in terms of greenhouse effect. In the U.K. a great deal of garbage goes into holes in the ground ("landfill") where it generates methane by bacterial action. Demonstrations have established the credibility of landfill gas as a practical energy source equivalent to 120,000 tonnes of coal a year. This is expected to rise quickly to about 400,000 tce/yr.

Elsewhere sewage, in pits, provides methane for lighting. Intelligent use of biomass in the Third World includes, for example, burning sugar cane wastes (bagasse) in distillation of alcohol (produced by fermentation) as a fuel for vehicles. This is being increasingly imitated in the West. Sewage is a problem world-wide and in places where temperature permits, as in Israel and California, algal ponds are often used in sewage treatment. The algae photosynthesise and, at the same time, utilise the organic nitrogen and minerals from the sewage. Although the algal end-product has been used as animal feed and for other purposes (some too extraordinary to be discussed in polite society) the principal by-product is clean water. This, of course can be immensely valuable particularly in arid environments. So is the general principle of yet again making a virtue out of a necessity.

Ethanol it would seem at first sight, has even more going for it than straw, refuse or dung. Certainly Omar Khayam echoed a fairly common sentiment when he wrote

"Come fill me with the old familiar juice,
methinks I might recover by and by"

and

"I often wonder what the vintner buys one half
so precious as the goods he sells"

Moreover, fermentation techniques are almost as old as Man and there is no shortage of suitable biomass for this purpose. Brazil was producing around 8 billion litres/year by the mid 1980s and, by now, total output will have been well in excess of the equivalent of 300 million barrels of gasoline. Brazil uses sugar cane and some *Cassava* as feedstock. The much smaller United States output (1.6 billion litres in 1984) is predominantly from corn *(Zea mays).*

The density and calorific content of ethanol (24 GJm^{-3}) is less than that of petroleum (39 GJm^{-3}) because ethanol is more oxidised. It burns well and modified engines running on ethanol can develop 20% more power and less polluting exhaust gases. There, regrettably, the advantages stop because fermentation ceases at relatively low alcohol concentration as the yeasts succumb to drunken stupor and give up the ghost. Thereafter there is a need for distillation and this, involving (as it does) a change from liquid to gas-phase requires very considerable energy inputs. Certainly if arithmetic takes every aspect of the overall process into account it is clear that ethanol production cannot constitute net energy gain. In the special circumstances of places like Brazil this hardly matters. There are billions of acres on which sugar cane will grow happily, like a weed, and the cane residues left after sugar extraction make an excellent fuel which can be utilised in distillation plants. Conversely in a country like Scotland (in which there are not too many arable acres or warm days) barley, a C3 species, grows much less rapidly than does C4 sugar-cane in subtropical climates. The products of distillation are then best stored for a while in oak casks and finally savoured in front of a log fire.

9.15 Hydrogen Power

"I believe that water will one day be employed as fuel, that hydrogen and oxygen which constitute it, used singly or together, will furnish an inexhaustible source of heat and light, of an intensity of which coal is not capable. I believe then that when the deposits of coal are exhausted, we shall heat and warm ourselves with water. Water will be the coal of the future."

....Jules Verne

The fact that it takes energy to split hydrogen-oxygen (H-O) bonds and that energy is released when H-O bonds are re-formed, has also been a central theme of this book. Most of Man's energy is still derived (indirectly from nuclear fusion in the sun) by re-joining hydrogens and oxygens separated during primordial photosynthesis. In some regards gaseous hydrogen (H_2) is an ideal fuel. Its combustion:-

$$2H_2 + O_2 \rightarrow 2H_2O \qquad \text{Eqn. 9.1}$$

releases much energy and the combustion produce (H_2O) is good enough to drink (indeed it was drunk by cosmonauts). The trick, of course, is to produce hydrogen in the first place. If we suppose that, by whatever future, unthought of, technical fix, or by judicious use of renewable resources, we have succeeded in generating enough electrical energy to maintain large populations in the manner to which they have become accustomed, we would still have a need for a fuel as convenient as petrol (gasoline). It is conceivable that improvement to batteries will soon make electrically powered vehicles more of a practical proposition than they are at present but long distance road haulage, tractors, combine-harvesters, ships and aircraft will seemingly need some appropriate petroleum product, or equivalent, into the foreseeable future. In many respects hydrogen would be better than hydrocarbon fuels. Apart from the fact that it would not give rise to pollutants in the form of oxides of carbon, sulphur and nitrogen, it produces 2.8 times as much energy per gm as hydrocarbons and the propulsive efficiency of hydrogen-burning jet engine is significantly higher because of the lower molecular weight of the exhaust products (18 g for water as against 44 g for the CO_2 which is also produces by burning hydrocarbons). Against this there is the fact that liquid hydrogen occupies ten times as much space as conventional aviation fuels and, it is argued, can only be feasibly carried in refrigerated tanks.

> [Hydrogen can also be stored as a metal hydride. In principle, this is highly desirable because the reaction is freely reversible and therefore allows a hydride store to be replenished by hydrogen gas. Moreover the hydride can accommodate large quantities of hydrogen so that its energy content becomes 20 MJl^{-1} compared with 1.7 MJl^{-1} for hydrogen gas (at 150 atmospheres pressure) and diesel oil at 30 MJl^{-1}. The weight of the hydride, like the weight of cylinders of refrigerated tanks, becomes, a problem when portable hydride stores for use in vehicles is contemplated.]

Notwithstanding these very real drawbacks, feasibility studies have shown that it is possible to use hydrogen to power both motor vehicles and aircraft and it might be supposed that the relevant technology will improve as soon as the oil companies' predictions about petroleum prices suggest to them that they should make a major investment in this area. (As oil reserves dwindle, and population continues its inexorable rise, it is clear that, apart from short-term fluctuations, oil prices will rise to a level at which it will be too expensive to use oil as a fuel rather than as a source of petrochemicals). It is also conceivable that, if the worst aspects of the rising carbon dioxide syndrome become reality, the gratuitous emission of CO_2 will become as unwelcome as the use of CFCs or smoking in public places. Companies like British Petroleum already have a solar division which produces photovoltaic panels to power vaccine refrigerators etc., for use in areas without mains electricity. It is possible, in principle, to use photovoltaics to drive electrolysis of water and indeed a practical version of Fig. 1.2 has been used for ten years or more as a demonstration in an Environmental Biology course in the University of Sheffield. Producing hydrogen in this way, or by utilising other renewable sources of electricity for this purpose (in order to provide a substitute for petroleum in vehicles) may not be round the corner; neither is it unimaginable. It is much more difficult to contemplate unchanging trends in petroleum usage well into the next century as sources diminish (and become more difficult to process, e.g. petroleum from coal and oil-bearing shales) and as pressures on the environment become less and less acceptable. The writing is already on the wall and it is beginning to be heeded.

9.16 Epilogue

"When the Earth's last picture is painted and the tubes are twisted and dried"
....Rudyard Kipling

In 1939 it was still true to say that "*the sun never set on the British Empire*" and although, as a consequence of the first world war and its aftermath, Imperial Britain was not even then the country of vast wealth and industrial innovation that it had been in Queen Victoria's day, it was still a major power. After the (1939-45) war, Britain was devastated and impoverished but certainly not in worse state than Germany and Japan. Today, despite the supposed advantage of North Sea Oil, Britain's economy is still a matter for major concern whereas that of Germany is strong and that of Japan has come to dominate the world. Without detracting from Germany's famous flair for organisation and thoroughness, or the fabled Japanese work ethic, it is evident that there must have been other factors which led these two "vanquished" countries to prosper while Britain has merely muddled through to its present economically vulnerable situation. The reasons are complex and hardly capable of analysis (certainly not as an aside, in what is primarily a biological text) but they cannot be ignored entirely because they illustrate, and are central to, the contribution which politics might make to the environment in the next half century. (It is worth noting immediately in this context that most of our present world leaders will be dead in 50 years time and that, in Western democracies there are few politicians who can look beyond the next election let alone the next 5 or 10 years).

So, although it would undoubtedly be challenged and contested by those of some political persuasions, there is little doubt that the major factor which have contributed to the economic miracles of Germany and Japan has been investment - positive investment in new industrial plant and technology and negative investment in armaments. Faced with the destruction of most of their industrial base and the need to make reparation to the victorious allies, Germany and Japan had to start afresh on what was bound to be a new and better basis. In so doing, they were bolstered by massive aid programmes from the United States born of the perceived threat of world communism. Moreover, they were bound, by treaty, to strictly limit their expenditure on conventional arms and totally forbidden to make or acquire their own nuclear weapons. Conversely, Britain and France were encouraged, by reparations, to persist with old technologies and anxious, because of their reluctance to lose world-power status, and the widely accepted concept of Soviet Imperialism, to invest heavily in "defence". Most of all there was the concept of nuclear deterrence, still claimed by those who favour it (but in the obvious absence of the "control" experiment) to have kept the peace in Europe for the past 40 years. Whether nuclear missiles have kept the peace or have been a threat to peace is a matter for argument and future historians but there are few historians, even now, who can regard the "independent" nuclear weapons of Britain and France as anything other than attempts by these nations to maintain a major role in world affairs. The economic price of these status symbols has been, and continues to be immense. Germany, so long a symbol of economic stability and power, now struggles with the huge burden of integration but the price, as a proportion of gross national product, is certainly much less than that of the United Kingdom's ill-advised marriage to nuclear fission and fusion.

By now, the reason for this excursion into history, economics and politics will have become self-evident. It is intended simply to illustrate the fact that, within recent history, putting aside the influence of world figures like Hitler, it has been possible for governments to have a profound impact on economics (itself an issue central to energy and hence to the future of the environment). In addition, even though governments and elected representatives would be expected to be primarily driven by self-interest and material well-being, there is absolutely no guarantee that their interests will not be subordinated to what might be most charitably described as matters of great principle. It is such matters that led to the recent awful conflicts in the Middle East, the intervention by the United States in Vietnam and any number of relatively minor conflicts. It is against this background that we can not count upon our elected representatives to do what is eminently logical in the face of the vast consequence of carbon dioxide increase. It is one thing to sign the Montreal Protocol on CFCs because the political kudos is considerable and the economic cost relatively trivial. It is quite another to contemplate action which would involve the deflection of substantial funds from an area of perceived orthodoxy (such as defence, or nuclear power) into the maintenance of the environment. Those who support or advocate such change are immediately classified as ill-informed or unworldly. For example in the U.K. the whole tenor of advertisement by the pro-nuclear lobby has been to dismiss alternative approaches as unrealistic, ill-conceived, ill-informed or eccentric. Immensely expensive demonstrations have been mounted to show that trains carrying containers of nuclear waste could crash at high speed without spillage, that the excessive changes within reactors could be accommodated without the prospect of a second Chernobyl, and so on.

So latter day Noahs must expect questioning and virulent attack not only by those with vested interests but even by the innocent (who prefer not to think about the future in the way that most of us prefer not to dwell upon the likely time and manner of our departure from the land of the living). None of this means that so-called "Green" politics are absurd or unworldly. None of us need to have sleepless nights about the eventual failure of nuclear fusion in the sun. It is far too distant in time to concern us but the decline in fossil fuel reserves and associated rise in population and in atmospheric CO_2 and methane is here and now. If what is written here serves any useful purpose at all, it might convince some that these processes cannot and will not continue indefinitely. In fact, no scientific argument should be necessary. Expansion within a finite space is necessarily limited. Man cannot continue to exist as Man except in equilibrium with the environment. At present we are very far from equilibrium and curving away from it rather than towards it. Personally, I take the optimistic view that Man will survive. In view of what is already possible in the realms of genetic engineering, the likelihood that *Homo sapiens* will survive, as such, is quite another matter. When we considered the laws of thermodynamics, back in Section 1.9, we learned that we can't win. There is a universal tendency towards disorder (increased entropy) which is only fleetingly reserved by living organisms. As Glasby has also emphasised, both an increase in world population and a corresponding increase in economic activity inevitably means increased environmental degradation and pollution (i.e. increased disorder). However smart we are we can't get round the laws of thermodynamics. Not surprisingly this is a bitter pill to swallow. We cling desperately to the view that we can have our cake and eat it. Those who enjoy life in the relatively wealthy parts of the northern hemisphere and have any feeling at all for those in the poorer parts of the southern hemisphere are glad to accept the concept that the "under-developed" should develop. Those in the Third World demand this as of right. No doubt they should. No doubt the rich should help the poor if only out of self interest. The rich and the poor make uncomfortable neighbours and it is facile to suppose that the juxtaposition of wealth and poverty does not affect us all adversely. But it should be clearly recognised that if some genial entity waved a magic wand and embarked us on a master plan which would make us all as affluent as the average citizen of the United States by the end of the next century, we would sink beneath the weight of our own iniquities long before that happy day. Two to three hundred million people consuming fossil fuels and producing CO_2 at rates near to those that currently obtain in North America have already created problems which will be solved only if there is a major, rapid and concerted response by the international community. There is simply no possibility that 10,000 million people (a figure fast being approached) could use and pollute, at the same rate as these chosen few, without inviting catastrophe. Mankind can only survive if the peoples of the Earth contrive to live in equilibrium with their environment. At present the environment staggers under the impact of a relatively small number of major exploiters and polluters. A better life for all is not compatible with the sustained and universal growth which is central to most political thought. The gulf between North and South will continue to widen. The present trickle of economic migrants will become a flood. Starvation, pestilence, pandemics and war will surely follow. Most of us have been acutely conscious, during our lives, of the threat of nuclear holocaust. As that dark and dire possibility apparently recedes we must become aware of an equally bleak future if we fail to limit population growth. It is already too late to avoid calamity but we might still spare ourselves total catastrophe. We all know that we will die. However, we also like to think that, with the help of modern technology and a degree of good fortune, we might manage a little more than the proverbial three score years and ten. While we *are* here, most of us see some sense in making the best of what the world has to offer and would take comfort in the thought that, in so doing, our children might be assured of an environment that they could find joy in. If the worst is to be avoided, now is the time to strive towards equilibrium rather than to accelerate our movement away from it.

Finally, at the end of this tale, I wish to return, very briefly, to the question of obesity and diet (page 127 *et seq.*) because our attitudes to these encapsulate many of our responses to much greater problems. In the richer countries there are large numbers of people who would like to weigh less. Avoiding excess weight is a complicated matter. Volumes have been written about diet but we are no nearer solving this problem than solving unemployment, destitution, crime or preventing fearful acts of war. We still yearn for some technical fix and prefer not to believe that body fat only accumulates when calorific input exceeds calorific output. But, in the end, even though we may hate the thought we come up against the laws of thermodynamics. Energy and mass are one, there is no escape. Our attitude to our planet is similar to our attitude to our bodies. We abuse both, careless of the consequences. We prefer not to think about what we are doing at all or to heap scorn on those who have the temerity to urge moderation. The latter is as it should be. Scepticism is to be applauded. No one can look into the future and the way ahead is littered with the remains of discredited predictions. Nevertheless we must recognise and accept the bottom line drawn by thermodynamics. My position in this is simple yet I fear that a casual reader might conclude that my Malthusian stance implies that I see over-population as the cause of all evil. I do not. I need to look no further than my own country. During a relatively short period, biologically speaking, the population of the British Isles has not merely doubled it has increased thirty fold. Everyone knows that this increase was made possible by the Industrial Revolution (i.e. by technology). Putting aside the (not unrelated) fact that more have died in warfare during this period than the total original population it is self evident that there have been immense benefits. We live longer, we are well fed, we are spared a great deal of pain and suffering. We have leisure. While we have undoubtedly changed our environment out of all recognition we still live in a green and pleasant land. Moreover, for many years now, our population has been stable. So why do I conclude that the rest of the world cannot do the same? Why should it not continue to increase (as it most certainly will), stabilise sometime in the middle of the next century (at somewhere between 8 and 12 billion), and then live happily ever after. Perhaps it will. At this moment we could feed probably 20 billion if we put our minds to it. We will not run out minerals in the foreseeable future. There will still be oil in the ground when we have ceased to use it. But, there's the rub. The oil will be there because it has become to expensive to extract. There may well be vast reserves of oil and coal still to be discovered but *the future is infinite and reserves of fossil fuels are finite*. What is worse, as we return all of this fixed carbon to the atmosphere in a perilously short time we clearly risk major adverse effects on our environment. We must then ask "for what?" It can be successfully argued that there is no past correlation between increasing population and decreasing human welfare. On the contrary, better times have come with more people. That the two have gone hand in hand in the past, however, does not mean for one moment that they will continue to go hand in hand in the changed circumstances of the future. Nor can it be reasonably argued that the present exponential increase in population and its associated pollution will make it easier to solve the inevitable problems that are following in its wake. If earth science is right and fossil fuels are fast diminishing, despite the fact that we currently float in a pool of oil, it is the decline in fossil fuels which will first threaten increasing starvation. As we have seen, we eat the products of past photosynthesis as we eat the products of contemporary photosynthesis and, if that is to reflect the pattern of future agriculture, the two will decline together. O.K., so we may come up with alternatives to oil, coal and gas and everything will be fine again. All that CO_2 and methane that we have blown off in the meantime may not be so bad anyway and who cares if we never get to see a whale or a rain forest? Such questions scarcely deserve answers. Instead we might reasonably pose others. Much has been made in these pages of renewable, alternative energy sources. Given the inevitability of short term increases in population, can it be doubted that new and alternative technologies would not benefit from a longer period of development? As we go through the looking glass into the future will we not have to run to keep in the same place? Come what may, what is left of our flora (and its dependent fauna) will have to cope with high CO_2, and the distinct possibility of significant changes in climate. Can it be maintained that a longer period of adaptation to these changed circumstances would not be desirable? And, in the end, (why beat about the bush) would not life be altogether sweeter if we could once again walk through relatively safe and pleasant city streets rather than perilously and nervously riding the subway from one foul patch to another?

Chapter & Verse

CHAPTER AND VERSE

Rationale

In writing a scientific paper it is often necessary to give the authority for almost every important statement that is made. This can be carried to extremes and there is the story about the author who wrote "In the beginning..." and then cited Genesis as his authority. Conversely there are lazy authors who are inclined to skip a few references or cite reviews by someone other than the person who should be credited with the original statement or observation. Ambitious and unscrupulous scientists have also been known to deliberately neglect previous work in the hope, surprisingly often realised, that they will reap the rewards for seeds sown by some earlier research, thereby denying the original authors the modest enjoyment of being credited with a new idea or observation. Others may simply, or understandably, be unaware of publications that they should mention or, growing long in the tooth, quite genuinely forget that they had read the seminal work on a particular topic many years before. Everyone's papers and books, these days, gets a mention in "Citation Index" which lists those authors who have cited their work in previous years and decades and all manner of judgements about a scientist's worth are sometimes made on this basis. Naturally this can be very misleading because individual scientists inevitably tend to cite their own work on every conceivable occasion and someone who has invented a useful technique may be cited by all and sundry for evermore even if they might wish to be remembered, or acknowledged, for other things.

All of this is mentioned here for two reasons. One is the hope that this book may be read by prospective researchers who have yet to put pen to paper and should be warned of what fate awaits them if they do. The second is to explain to a reader, who has no intention of engaging in such matters, what policy has been adopted in this book. It will be immediately obvious that although many names are mentioned in the text there are very few specific citations such as:-

Hill, R., Nature Lond., 139, 881-882 (1937).

Citations of this sort have not been made frequently because, if this is to be done at all, it should be done properly and the work involved in checking every reference is so immense that it would have added at least one more year to the writing of the book and might even have tipped the author into an early grave. There is also the still unsolved problem of how to insert lots of reference into a text without making it unpleasant to read. Not every reader wants chapter and verse for every statement and to have almost every sentence peppered with unasked for references can be a pain. At the same time most readers might wish to read other related works and some, out of interest or proper scepticism, might wish to know where the author got his "facts". What I have done is to offer some sort of compromise. Naturally, like all compromises, what follows is far from perfect.

PHOTOSYNTHESIS

Since the book is much concerned with photosynthesis it aims to be self-sufficient for the present purpose but:-

"Photosynthesis" by D.O. Hall and K.K. Rao. Published by Edward Arnold, London (1987).
is an excellent, and comprehensive, introductory text and:-

"Light and Living Matter" by R.C. Clayton, McGraw-Hill Book Company, New York (1970)

illuminates the underlying physics in a very clear fashion.

Even if it is slightly dated and out of print (but still obtainable through any good library), it will not be too hard to guess why

"C3, C4, Mechanisms, and Cellular and Environmental Regulation of Photosynthesis" Edwards, G.E. and Walker, D.A., Blackwell Scientific Publications Ltd, Oxford. (1983).

is listed here. Much more recent, but rather more general, information can be found in:-

"Molecular Activities of Plant Cells" by John W. Anderson and John Beardall, Blackwell Scientific, Oxford (1991).

C_4 Photosynthesis: "An unlikely process full of surprises" by Hal Hatch, Plant & Cell Physiol **33,** 333-342 (1992) and Biochim Biophys Acta **895,** 81-106 (1987).

This is authoritative and well presented. For every last thing you might wish to know about photosynthesis, the key to the older literature is the monumental:-

"Photosynthesis and Related Processes" by Eugene Rabinowitch, Interscience Publishers Inc. New York in three parts; Vol I (1945), Vol II 1 (1951) and Vol II 2 (1956).

Similarly comprehensive multi-author reviews and lists of references to the more recent literature can be found in:-

"The Encyclopedia of Plant Physiology" (New Series), edited by A. Pirson and M.H. Zimmerman by Springer Verlag, Berlin, Heidelberg New York.

Of the many volumes of this, 5 (1977), 6 (1979), 12A (1981), 12B (1982) and 19 (1986) are the most immediately relevant.

PLANTS IN RELATION TO THEIR ENVIRONMENT

For those who wish to learn more about this subject:-

"Plants" by Watson M. Laetsch, published by Little, Brown and Company, Boston, (1979).

is a delight.

GLOBAL WARMING, ENERGY, ETC.

Such is contempory interest in matters "green", fostered by splendid organisations like "Greenpeace", "Friends of the Earth", and "The World Wide Fund for Nature" that newspapers and magazines are currently overflowing with information (not invariably accurate) about global warming, the demise of the rain forests (fast following the dinosaurs) and the culpable follies of politicians intent on re-election like George Bush. Here, in short, you are spoiled for choice at the popular level but, if you do wish a relatively recent overview:-

"Hothouse Earth -The Greenhouse Effect and Gaia"- by John Gribbin, Bantam Press, London, New York, Toronto, Sydney, Auckland. (1990)

is gripping and very readable (even if you might not care to embrace the Gaia bit). To get into the hard stuff, the most recent comprehensive source of information undoubtedly lies in the three volumes of:-

"Energy Technologies For Reducing Emissions of Greenhouse Gasses" Proceedings OECD "Experts Seminar", Paris (1989)

Like any multi-author work, some chapters are better than others but that by Lovins ("Energy, people and industrialisation" Vol 1, pp. 301-326), is one of the most electrifying, significant and philosphically amusing articles that I have ever read. For this compelling reason I have drawn from it extensively in Chapter 9.

If rising carbon dioxide happens to by your forte, then the font of all knowledge is:-

"Trends '9O" A Compendium of Data on Global Change from the Carbon Dioxide Information Analysis Centre of the Oak Ridge National Laboratory Tennessee Ed. by T. A. Boden, Paul Kanciruk and M. P. Farell.

This invaluable work is sponsored by the United States Department of Energy but be warned that it is what the title suggests and not light reading. Finally if we turn to renewable resources, the U.K. Department of Energy publishes excellent and well presented information (in colour) on this subject which is intended for the public at large. These come from the:-

Renewable Energy Enquiries Bureau, ETSU, Building 156, Harwell Labotatory, Oxfordshire OX11 ORA.

A wealth of more technical information can be found in:-

"Renewable Energy Resources" by J. Twidell and A.D. Weir, E & F. N. Spon Limited, London & New York (1986).

and

"Renewable Sources of Energy" International Energy Agency, Paris (1987).

OTHER READING

I also gladly acknowledge the use that I have made of the following books etc.:-

"History of the Earth's Atmosphere" by Budyko M.I., Ronov A.B., Yanshin A L, Springer Verlag - Berlin, Heidelberg, New York, London, Paris, Tokyo (1987).

"Options for Mitigating the Greenhouse Effect" by K. Currie, Seminar on Climate Change, the Department of the Environment, London (1989).

"Saving the Tropical Forests" by J. Gradwohl and R. Greenberg, Earthscan Publications Ltd, London (1988).

CHAPTER AND VERSE

"Vegetation History" by B. Huntley and T.W.Webb III, Kluwer Acad. Publishers Dordrecht, Boston, London (1988).

"Energy and Food Production" by G. Leach, IPC Science and Technology Press (1976).

"Planets and Their Atmospheres - Origin and Evolution" John S. Lewis and Ronald G. Prinn. Academic Press, Inc., Orlando, San Diego, San Francisco, New York, London, Toronto, Montreal, Sydney, Tokyo, San Paulo (1984).

"Solar-powered Electricity - A survey of photovoltaic power in developing countries" by Bernard McNelis, Anthony Derrick & Michael Starr, Intermediate Technology Publications, London (1988).

"Can Britain Feed Itself?" by K. Mellanby, Merlin Press, London (1975).

"IPCC (Intergovernmental panel on climate change)" Meteorological Office, London Road, Bracknell, RG12 2SZ (1990).

"An Assessment of the Effects of Climatic Change on Agriculture" by M. L. Parry and T. R. Carter, Climatic Change 15: 95-116, Kluwer Academic Publishers (1989).

"Going Critical" by W.C. Patterson, Paladin Grafton Books, London, Glasgow, Toronto, Sydney, Auckland (1985).

"The Energy Fix" A. Porter, M. Spence and RR. Thompson, Pluto Press, London, Sydney, Dover-New Hampshire (1986).

"Scales and Global Change" by T. Rosswall, R.G. Woodmansee and P.G.Risser, John Wiley and Sons, Chichester, New York, Brisbane, Toronto, Singapore (1988).

"Nuclear Power: Phoenix or Dodo?" by J. Surrey, British Gas Key Environmental Issues 10 (1991).

"Technological Responses to the Greenhouse Effect" edited by G. Thurlow. Twenty-sixth Consultative Conference, The Watt Committee on Energy, Savoy Hill, London (1990).

"Probing Plant Structure" by John Troughton and Leslie A. Donaldson, Chapman & Hall, London (1972).

"Evolution of the Atmosphere" by James C. G. Walker, Macmillan Publishing Co., Inc., New York (1982).

"Scientific Assessment of Climate Change and its Impacts" by T.M.L. Wigley, Seminar on Climate Change, the Department of the Environment, London (1989).

"Hydrogen power: An introduction to hydrogen energy and its applications" by L.O. Williams Pergamon Press, Oxford, U.K. (1980).

INDEX

"Begin at the beginning" the king said, gravely, *"and go on to the end; then stop"*.